Autodesk Revit 2023 Architecture Certified Professional Exam Study Guide

Elise Moss

SDC Publications
P.O. Box 1334
Mission, KS 66222
913-262-2664
www.SDCpublications.com
Publisher: Stephen Schroff

ISBN-13: 978-1-63057-508-3
ISBN-10: 1-63057-508-9

Printed and bound in the United States of America.

Table of Contents

Lesson Two
Families

Lesson Three
Documentation

Preface

This book is geared towards users who have been using Revit for at least six months and are ready to pursue their official Autodesk certification. You can locate the closest testing center on Autodesk's website. Check with your local reseller about special Exam Days when certification exams can be taken at a discount. Autodesk has restructured their certification exams so they can be taken in your home if you have the proper equipment (web cam and microphone). This edition of the textbook has been restructured as well to align with the new certification exams.

I wrote this book because I taught a certification preparation class at SFSU. I also teach an introductory Revit class using my Revit Basics text. I heard many complaints from students who had taken my Revit Basics class when they had to switch to the Autodesk AOTC courseware for the exam preparation class. Students preferred the step by step easily accessible instruction I use in my texts.

This textbook includes exercises which simulate the knowledge users should have in order to pass the certification exam. I advise my students to do each exercise two or three times to ensure that they understand the user interface and can perform the task with ease.

I have endeavored to make this text as easy to understand and as error-free as possible…however, errors may be present. Please feel free to email me if you have any problems with any of the exercises or questions about Revit in general.

I have posted videos for some of the lessons from this text on YouTube – search for Moss Designs and you will find any videos for my book on that channel.

Exercise files can be accessed and downloaded from the publisher's website at:
www.SDCpublications.com/downloads/978-1-63057-508-3

I have included a sample practice exam for the Professional certification exam. It may only be downloaded using a special access code printed on the inside cover of the textbook. The practice exam is free and does not require any special software, but it is meant to be used on a computer to simulate the same environment as taking the test at an Autodesk testing center.

Please feel free to email me if you have any questions or problems with accessing files or regarding any of the exercises in this text.

Acknowledgements

A special thanks to Rick Rundell, Gary Hercules, Christie Landry, and Steve Burri, as well as numerous Autodesk employees who are tasked with supporting and promoting Revit.

Additional thanks to Gerry Ramsey, Will Harris, Scott Davis, James Cowan, James Balding, Rob Starz, and all the other Revit users out there who provided me with valuable insights into the way they use Revit.

Thanks to Zach Werner for the cover artwork for this textbook along with his technical assistance. Thanks to Karla Werner for her help with editing and formatting to ensure this textbook looks proper for our readers.

My eternal gratitude to my life partner, Ari, my biggest cheerleader throughout our years together.

Elise Moss
Elise_moss@mossdesigns.com

Introduction

FAQs on Getting Certified in Revit

The first day of class students are understandably nervous and they have a lot of questions about getting certification. Throughout the class, I am peppered with the same or similar questions.

Certification is done through a website. You can learn more at https://home.pearsonvue.com/autodesk. The website currently has exams for Revit Architecture and Revit Structure. They are only offering the Professional exam at this time.

In the past, the exams required you to use Revit during the exam to walk through problems. The exam is now entirely – 100% – browser-based. All the questions are multiple choice, fill in the blank, true/false, or point and click. You do not have any other software open beside the browser. This means you have to be really familiar with Revit's dialog boxes and options as you are relying on your memory for most of the exam.

You now have the option to take the exam at a Pearson Vue testing center or in the comfort of your own home. You can go on-line to determine the location of the testing center closest to your location and schedule the exam.

If you opt to take the exam in your home, you still need to "schedule" the exam as a proctor will be monitoring you remotely during the exam. In order to take the exam at home, you need a laptop, workstation or tablet device connected to the internet equipped with a microphone and a webcam. Prior to the exam, you will be required to test your system to ensure the microphone and webcam are functioning properly. You also need to upload your identification and take a "selfie" so the proctor can verify that you are the one taking the exam.

When I took the exam at home, I had to clear off my desk, turn off and unplug all monitors except for the laptop I was using for the exam. The proctor made me pick up my laptop and rotate the laptop 360 degrees so she could see the area I was taking the exam and verify that I had no papers on my desk and nothing that could help me during the exam. You are not allowed to take notes or refer to any reference material during the exam. The only device you are allowed to use is the device you are using to take the exam. Your cell phone is to be stored away from your work area.

The Professional certification requires about 1,200 hours or two years of experience with the software.

The Autodesk Certified Professional Exam is 45 questions and 120 minutes. This means you have an average of two and a half minutes per question. If you have a slow internet connection, there may be a lag and this will slow you down. So, make sure you have a good internet connection prior to scheduling an exam outside a testing center.

In the past, you could be certified as a User or as a Professional. In order to be certified as a Professional, you needed to pass both the user and the professional exams. Then, you could opt to take one or the other exam. Now, you can only be certified as a Professional. The user exam is no longer being offered.

What are the exams like?

The first time you take the exam, you will create an account with a login. Be sure to write down your login name and password. You will need this regardless of whether you pass or fail.

If you fail and decide to retake the exam, you want to be able to log in to your account.

If you pass, you want to be able to log in to download your certificate and other data.

Autodesk certification tests use a "secure" browser, which means that you cannot cut and paste or copy from the browser. You cannot take screenshots of the browser.

If you go on Autodesk's site, there is a list of topics covered for each exam. Expect questions on the user interface, navigation, and zooming as well as how to place doors and windows. The topics include collaboration, creating Revit families, linking and monitoring files, as well as more complex wall families.

Exams are timed. This means you only have one to three minutes for each question. You have the ability to "mark" a question to go back if you are unsure. This is a good idea because a question that comes later on in the exam may give you a clue or an idea on how to answer a question you weren't sure about.

Exams pull from a question "bank" and no two exams are exactly alike. Two students sitting next to each other taking the same exam will have entirely different experiences and an entirely different set of questions. However, each exam covers specific topics. For example, you will get at least one question about family parameters. You will probably not get the same question as your neighbor.

At the end of the exam, the browser will display a screen listing the question numbers and indicate any marked or incomplete questions. Any questions where you forgot to select an answer will be marked incomplete. Questions that you marked will display as answered. You can click on those answers and the browser will link you back directly to those questions so you can review them and modify your answers.

I advise my students to mark any questions where they are struggling and move forward, then use any remaining time to review those questions. A student could easily spend ten to fifteen minutes pondering a single question and lose valuable time on the exam.

Once you have completed your review, you will receive a prompt to END the exam. Some students find this confusing as they think they are quitting the exam and not receiving a score. Once you end the exam, you may not change any of your answers. There will be a brief pause and then you will see a screen where you will be notified whether you passed or failed. You will also see your score in each section. Again, you will not see any of the actual questions. However, you will see the topic, so you might see that you scored poorly on Documentation, but you won't know which questions you missed.

Results	100	200	300	400	500	600	700	800	900	1000
Required Score										
Your Score										

Section Analysis			Final Score	
Modeling and Materials		82%	Required Score	700
Families		100%	Your Score	691
Documentation		70%		
Views		71%	Outcome	
Revit Project Management		82%	Fail	✗

If you failed the test, you do want to note in which categories you scored poorly. Review those topics both in this guide and in the software to help prepare you to retake the exam. You can sign into the Pearson Vue website and download the report of how you did on the exam and use that as a study guide.

How many times can I take the test?

You can take any test up to three times in a 12-month period. There is a 24 hour waiting period between re-takes, so if you fail the exam on Tuesday, you can come back on Wednesday and try again. Of course, this depends on the testing center where you take the exam. If you do not pass the exam on the second re-take, you must wait five days before retaking the exam again. There is no limit on the number of re-takes. Some testing centers may provide a free voucher for a re-take of the exam. You should check the policy of your testing center and ask if they will provide a free voucher for a re-take in case you fail.

Do a lot of students have to re-take the test or do they pass on the first try?

About half of my students pass the exam on the first try, but it definitely depends on the student. Some people are better at tests than others. My youngest son excels at "multiple guess" style exams. He can pretty much ace any multiple guess exam you give him regardless of the topic. Most students are not so fortunate. Some students find a timed test extremely stressful. For this reason, I have created a simulated version of the exam for my students simply so they can practice taking an online timed exam. This has the effect of "conditioning" their responses, so they are less stressed taking the actual exam.

Some students find the multiple-choice style extremely confusing, especially if they are non-native English speakers. There are exams available in many languages, so if you are not a native English speaker, check with the testing center about the availability of an exam in your native language.

Why take a certification exam?

The competition for jobs is steep and employers can afford to be picky. Being certified provides employers with a sense of security knowing that you passed a difficult exam that requires a basic skill set. It is important to note that the certification exam does not test your ability as a designer or drafter. The certification exam tests your knowledge of the Revit software. This is a fine distinction, but it is an important one.

If you pass the exam, you have the option of having a badge displayed on your LinkedIn profile verifying that you are certified in the Revit software. This may help you convince a recruiter or prospective employer to grant you an interview.

How long is the certification good for?

Your certification is good for three years. It used to be that certification was specific to a release. In other words, you would be certified in Revit 2023 and be required to take the exam using Revit 2023. Now, since the exam is entirely browser-based and doesn't require you to even use the software, it is no longer tied to a specific release.

Most employers want you to be certified within a couple of years of the most current release, so if you wish to maintain your "competitive edge" in the employment pool, expect that you will have to take the certification exam every three years. I recommend students take an "update" class from an Autodesk Authorized Training Center before they take the exam to improve their chances of passing. Autodesk is constantly tweaking and changing the exam format. Check with your testing center about the current requirements for certification.

How much does it cost?

It costs $200 USD to take the Professional exam.

Do I need to be able to use the software to pass the exam?

This sounds like a worse question than intended. Some of my students have taken the Revit classes but have not actually gotten a job using Revit yet. They are in that catch-22 situation where an employer requires experience or certification to hire them, but they can't pass the exam because they aren't using the software every day. For those students, I advise some self-discipline where they schedule at least six hours a week for a month where they use the software – even if it is on a "dummy" project – before they take the exam. That will boost the odds in their favor. However, you do not have to have Revit installed or operating on your computer in order to take the exam. You do need to be very familiar with the software.

What happens if I pass?

You want to be sure you wrote down your login information. That way, you can log in to the certification center website and download your certificate. You also can download a logo which shows you are a Certified Professional that you can post on your website or print on your business card. Autodesk now offers a badge which you can add to your LinkedIn profile.

How many times can I log into Autodesk's testing center?

You can log in as often as you like. The tests you have taken will be listed as well as whether you passed or failed.

Can everybody see that I failed the test?

Autodesk is kind enough to keep it a secret if you failed the exam. Nobody knows unless you tell them. If you passed the test, people only see that information if you selected the option to post that result. Regardless, only you and Pearson VUE will know whether you passed or failed unless *you* choose to share that information.

What if I need to go to the bathroom during the exam or take a break?

You are allowed to "pause" the test. This will stop the clock. You then alert the proctor that you need to leave the room for a break. When you return, you will need the proctor to approve you to re-enter the test and start the clock again. Even if you take the exam at home, you are monitored during the exam. Just open a chat window and tell the proctor you need to pause the test for a bathroom break.

How many questions can I miss?

A passing score is 700. Because the test is constantly changing, the number of questions can change and the amount of points each question is worth can also vary.

How much time do I have for the exam?

The Professional exam is 120 minutes.

Can I ask for more time?

You can make arrangements for more time if you are a non-English speaker or have problems with tests. Be sure to speak with the proctor about your concerns. Most proctors will provide more time if you truly need it. However, my experience has been that most students are able to complete the exam with time to spare. I have only had one or two students that felt they "ran out of time."

What happens if the computer crashes during the test?

Don't worry. Your answers will be saved, and the clock will be stopped. Simply reboot your system. Let the proctor know when you are ready to start the exam again, so you can re-enter the testing area in your browser.

Can I have access to the practice exams you set up for your students?

I am including a Professional practice exam in the exercise files with this text.

What sort of questions do you get in the exams?

Many students complain that the questions are all about Revit software and not about building design or the uniform building code. Keep in mind that this test is to determine your knowledge about Revit software. This is not an exam to see if you are a good architect or designer.

The exams have several question types:

One best answer – this is a multiple-choice style question. You can usually arrive at the best answer by figuring out which answers do NOT apply.

Select all that apply – this can be a confusing question for some users because unless they know how *many* possible correct answers there are, they aren't sure. The test tells you to select 2/3/4 correct answers out of 5 and will prompt you if you select too many or not enough.

Point and click – this has a java-style interface. You will be presented with a picture, and then asked to pick a location on the picture to simulate a user selection. When you pick, a mark will be left on the image to indicate your selection. Each time you pick in a different area, the mark will shift to the new location. You do not have to pick an exact point…a general target area is all that is required.

Matching format – you are probably familiar with this style of question from elementary school. You will be presented with two columns. One column may have assorted terms and the second column the definitions. You then are expected to drag the terms to the correct definition to match the items.

True/False – you will be provided a set of sentences and asked to determine which ones are true or false.

Any tips?

I suggest you read every question at least twice. Some of the wording on the questions is tricky.

Be well rested and be sure to eat before the exam. Most testing centers do not allow food, but they may allow water. Keep in mind that you can take a break if you need one. If you are taking the test at home, you can have water on your desk or work area, but that is it.

Relax. Maintain perspective. This is a test. It is not fatal. If you fail, you will not be the first person to have failed this exam. Failing does not mean you are a bad designer or architect or even a bad person. It just means you need to study the software more.

Remember to write down your login name and password for your account. The proctor will not be able to help you if you forget.

Practice Exams

I have created professional practice exams at the end of each lesson. Do not memorize the answers. The questions on the practice exams will not be the same as the questions on the actual exam, but they may be similar. Some users have complained that they were marked wrong when they gave the correct answer, but they still were able to pass the exam. Some of this has to do with the way questions are worded. Some of the wording of the exam questions is vague or misleading, so be sure to read each question carefully.

Autodesk also provides preparatory courseware, including videos, and practice exercises to help you tune up your skills for the exam. This is an excellent free resource.

Notes:

Modeling and Materials

This lesson addresses the following certification exam questions:

- Create and Modify Walls
- Create and modify floors, roofs, and ceilings
- Work with stairs, ramps, and railings
- Cut openings in building elements, such as walls and floors.
- Work with columns
- Understand model and detail groups
- Using topography and site elements
- Associate a material with an object or a style
- Create and edit a material
- Load a material library
- Create, edit and load selection sets
- Create and modify rooms
- Create and modify areas

Users should be able to understand the difference between a hosted and non-hosted component. A hosted component is a component that must be placed or constrained to another element. For example, a door or window is hosted by a wall. You should be able to identify what components can be hosted by which elements. Walls are non-hosted. Whether or not a component is hosted is defined by the template used for creating the component. A wall, floor, ceiling or face can be a host.

Some components are level-based, such as furniture, site components, plumbing fixtures, casework, roofs and walls. When you insert a level-based component, it is constrained to that level and can only be moved within that infinite plane.

Components must be loaded into a project before they can be placed. Users can pre-load components into a template, so that they are available in every project.

Users should be familiar with how to use Element and Type Properties of components in order to locate and modify information.

There are three kinds of families in Revit Architecture:
- system families
- loadable families
- in-place families

System families are walls, ceilings, stairs, floors, etc. These are families that can only be created by using an existing family, duplicating, and redefining. These families are loaded into a project using a project template. You can copy system families from one project to another using the Transfer Project Standards tool.

Loadable families are external files. These include doors, windows, furniture, and plants.

In-place families are components that are created inside of a project and are unique to that project.

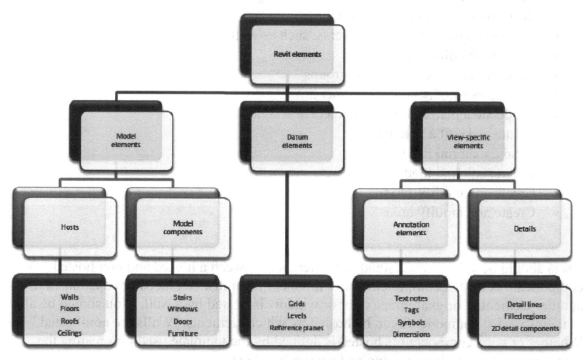

Revit elements are separated into three different types of elements: Model, Datum and View-specific. Users are expected to know if an element is model, datum or view specific.

Model elements are broken down into categories. A category might be a wall, window, door, or floor. If you look in the Project Browser, you will see a category called Families. If you expand the category, you will see the families for each category in the current project. Each family may contain multiple types.

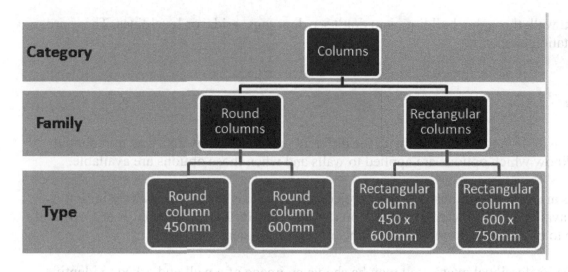

Every Revit file is considered a Project. A Revit project consists of the Project Environment, components, and views. The Project Environment is managed in the Project Browser.

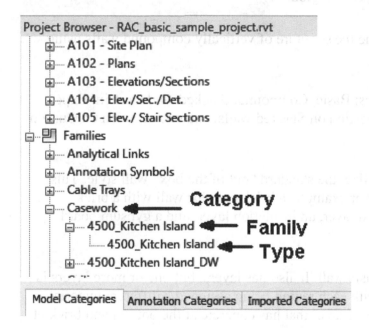

There may be one question on the exam asking you to describe the hierarchy of elements in a Revit project.

Elements are organized by Category, then Family, then Type, then Instance. This is easier to visualize if you look in the Project Browser.

If you look in the Visibility/Graphics Overrides dialog, elements are divided into Model Categories and Annotation Categories.

You should be able to identify whether an object is a model element or an annotation element.

One way to think of it is that model elements are actual physical items. If you walk through a building, you can see a door or column or floor. These are all model elements.

If you walk through a building, you don't see door tags, grids, or level lines. These are annotation elements.

Walls

Users will need to be familiar with the different parameters in walls. The user should also know which options are applied to walls and when those options are available.

Walls are system families. They are project-specific. This means the wall definition is only available in the active project. You can use Transfer Project Standards or Copy and Paste to copy a wall definition from one project to another.

On the Professional exam, you may be shown an image of a wall and asked to identify different wall properties.

Just as roofs, floors, and ceilings can consist of multiple horizontal layers, walls can consist of more than one vertical layer or region.

You can modify a wall type to define the structure of vertically compound walls using layers or regions.

Revit has several different wall types: Basic, Compound, Stacked, and Curtain. Expect one question on Basic walls, one question on Stacked walls, and one question on Curtain walls.

A Basic Wall is just what it sounds like, the standard "out of the box" wall style. This wall type may have several layers. For example, a brick exterior wall with a brick exterior layer, an air gap layer, a stud layer, an insulation layer, and a gypsum board layer.

A Compound wall is similar to a Basic wall. It also has layers, but one or more layers is divided into one or more regions, with each region being assigned a different material—for example, a wall that has an exterior layer that has concrete at the bottom and brick at the top.

A Stacked wall is two or more basic and/or compound walls that are stacked on top of each other. While Basic and Compound walls have a uniform thickness or width defined by the layers, a Stacked wall can have a variable thickness or width.

A Curtain wall is defined by a curtain grid. Mullions can be placed at the grid lines. Panels are placed in the spaces between the grid lines.

Exercise 1-1

Wall Options

Drawing Name: **i_firestation_basic_plan.rvt**
Estimated Time to Completion: 10 Minutes

Scope

Exploring the different wall options

Solution

1. 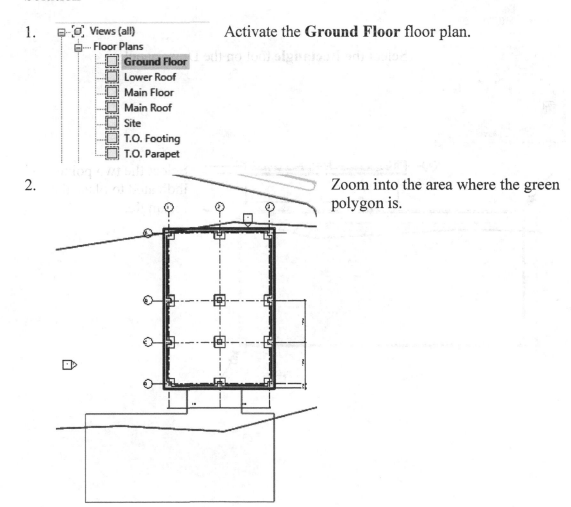 Activate the **Ground Floor** floor plan.

2. Zoom into the area where the green polygon is.

3.   Select **Wall** from the Architecture tab on the ribbon.

4.   Set the Wall Type to **Generic – 6″** in the Properties pane.

5.   Set the Location Line to **Core Face:Exterior**.

6.   Select the **Rectangle** tool on the Draw panel.

7. 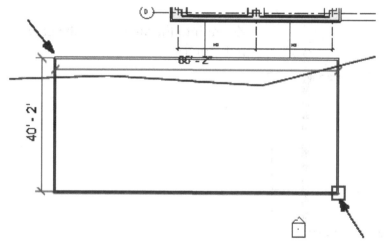 Select the two points indicated to place the rectangle.

8.
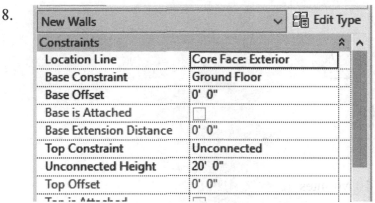

Notice that if the Top Constraint is set to Unconnected, you can define the Unconnected Height.

9.   Close the file without saving.

Attaching Walls

After placing a wall, you can override its initial top and base constraints by attaching its top or base to another element in the same vertical plane. By attaching a wall to another element, you avoid the need to manually edit the wall profile when the design changes.

The other element can be a floor, a roof, a ceiling, a reference plane, or another wall that is directly above or below. The height of the wall then increases or decreases as necessary to conform to the boundary represented by the attached element.

You can detach walls from elements as well. If you want to detach selected walls from all other elements at once, click Detach All on the Options Bar.

Exercise 1-2
Attaching Walls

Drawing Name: **i_Attach.rvt**
Estimated Time to Completion: 10 Minutes

Scope
Create a wall section view.
Attach a wall to a roof or floor.

Solution

1. Open *i_Attach.rvt*.

2. 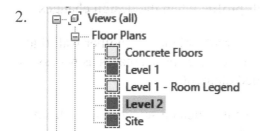 Activate Level 2 Floor Plan.

3. 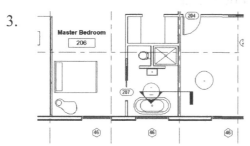 *Place a wall section as shown.*

    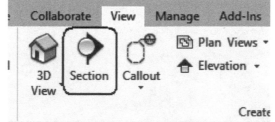

    Go to the **View** ribbon.
    Select the **Section** tool.

4. Set the view type to **Wall Section**.

5.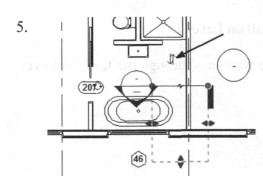

The first point selected will place the section head.
The second point selected will place the section tail.

Use the Flip controls if needed to orient the section head to face down/south.

Double left click on the section head to open the section view.

6.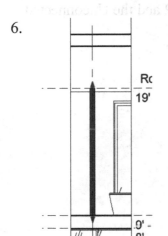

Select the wall on Level 2.

7.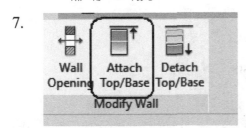

Select **Attach Top/Base** from the ribbon.

8.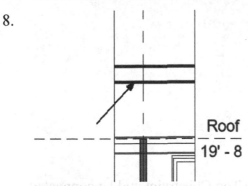

Select the roof.

9.

What is the volume of the wall after it is attached to the roof?

Select the wall and then go to the Properties panel to determine the correct volume.
It should be 88.83 CF.

10.

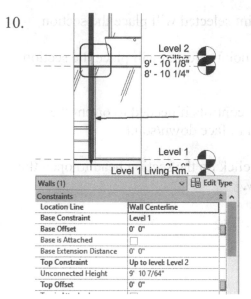

Select the wall on Level 1.

Note that the wall goes through the floor on Level 2.

Check the Properties palette and note that the Top Constraint is set to Level 2 and the Unconnected Height is grayed out.

11.

Select **Attach Top/Base** on the ribbon.

12.

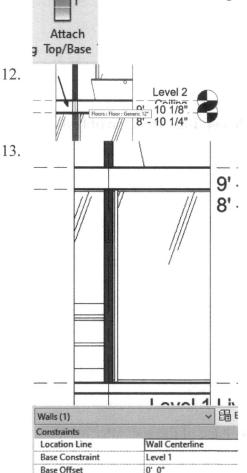

Select the Level 2 floor.

13.

The wall adjusts.

Note that the Top Constraint and Unconnected Height values did not change.

Exercise 1-3
Stacked Walls

Drawing Name: **i_Footing.rvt**
Estimated Time to Completion: 10 Minutes

Scope
Defining a stacked wall.

A stacked wall uses more than one wall type. Stacked walls may have varying widths.

Solution

1. 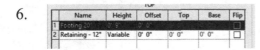 Open *i_Footing.rvt*.

2.  In the Project Browser, locate the Walls family category.
Expand the Stacked wall section.

Select the **Concrete with Footing** wall type.

3.  Right click and select **Type Properties**.

4. 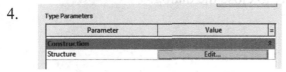 Select **Edit** next to Structure.

5. Insert   Select the **Insert** button.

6. Set the type to **Footing 20'** for Layer 1.
Set the Height to **9"**.

Note the Retaining – 12" Concrete wall is set to a Variable Height.

7.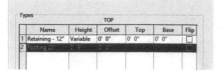

    Highlight Layer 1.

    Use the Down button to move the Footing 20'
    below the Retaining – 12" Concrete.

8.

    Select the Preview button to expand the dialog and see what the wall
    looks like.

    Click OK twice to close the dialog.

9.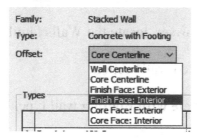

    Set the Offset to **Finish Face: Interior**.

10. *Notice how the wall adjusts so that the interior
    faces of the walls are now flush.*

    Click **OK** to close the dialog.

Exercise 1-4

Placing a Cut in a Wall

Drawing Name: **walls.rvt**
Estimated Time to Completion: 20 Minutes

Scope
Placing a horizontal or vertical cut in a wall using Reveal.
Creating a cut in a wall using Edit Profile.

Solution

1. 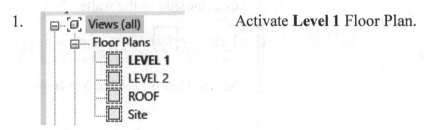 Activate **Level 1** Floor Plan.

2.  Select the **Wall** tool from the Architecture tab on the ribbon.

3.  Set the wall type to **Exterior - Brick on Mtl. Stud** using the Type Selector on the Properties pane.

4. Location Line: Finish Face: Ext ▾ ☑ Cha

   Wall Centerline
   Core Centerline
   Finish Face: Exterior
   Finish Face: Interior
   Core Face: Exterior
   Core Face: Interior

   Set the Location Line to **Finish Face: Exterior**.

5. 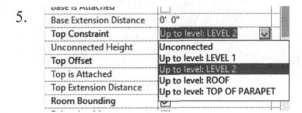 Set the Top Constraint to **TOP OF PARAPET**.

   Set the Top Offset to **0"**.

6. 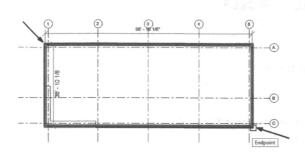 Select the **Rectangle** tool from the Draw panel.
Select the four green lines.

7. 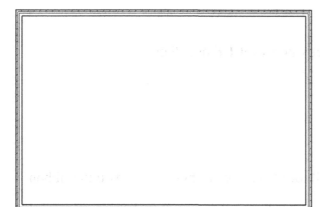 Select the corner by Grid 1 as the first point of the rectangle and the corner by Grid C as the second point of the rectangle.

Exit out of the Wall command.

8. The lines should be aligned to the exterior side of the walls.

Set the Detail Level to **Medium**.

9. Select a wall.

Note if the exterior side of the wall is oriented correctly. If it needs to be re-oriented, use the Flip Arrows to position it correctly.

10. Elevations (Building Elevation) — Switch to a South elevation.
- East
- North
- South
- West

11. Select the South wall.

Click **Edit Profile** on the ribbon.

12. 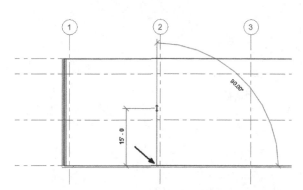 Select the **Line** tool from the Draw Panel.

13. Start the line using the endpoint located to the left of Grid 2.

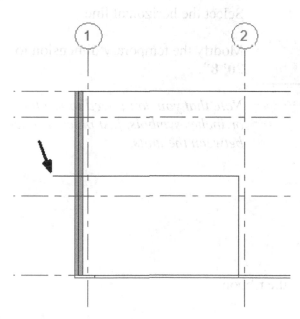

Draw the line vertically and select an endpoint above Level 2.

14. Drag the mouse to the left to place a horizontal line and click outside the building to select the endpoint for the horizontal line.

Exit the line command.

Your profile should look like this.

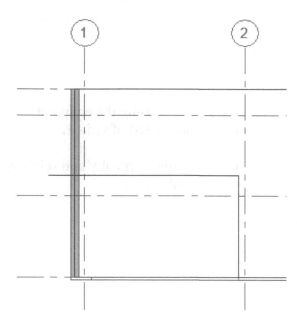

15.

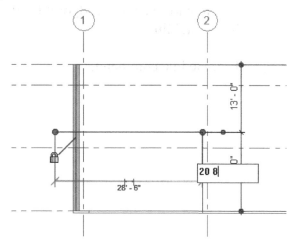

Select the horizontal line.

Modify the temporary dimension to **20' 8"**.

Note that you don't need to use the feet or inches symbols, just place a space between the units.

16.   Select the **TRIM** tool from the ribbon.

17.   Trim the two corners to create a single continuous wall boundary.

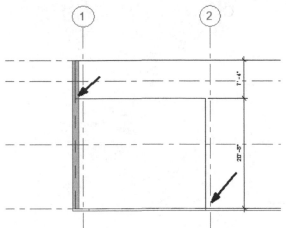

18.   Select the **Green Check** to exit the edit profile mode.

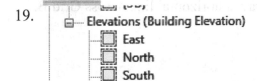

19.   Switch to a **West** elevation.

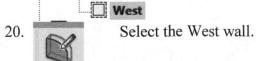

20.   Select the West wall.

Click **Edit Profile** on the ribbon.

21.   Select the **Line** tool from the Draw Panel.

22.

Start the line using the end point located below the Roof level near Grid C.

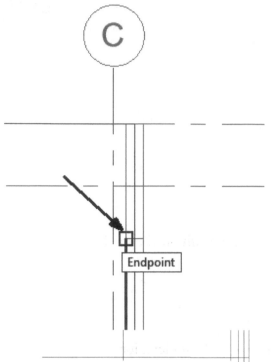

23.

Draw a horizontal line across Grid B.

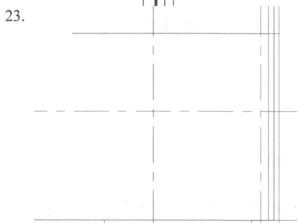

24.

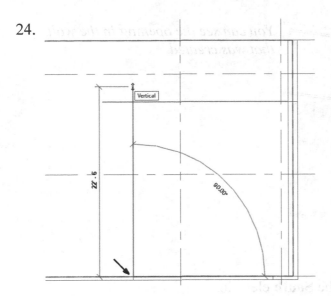

Select the endpoint at Level 1 and to the left of Grid B and draw a vertical line up to intersect with the horizontal line.

25.   Select the **TRIM** tool from the ribbon.

26.

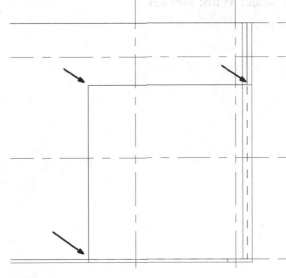

Trim the three corners to create a single continuous wall boundary.

27.   Select the **Green Check** to exit the edit profile mode.

Mode

28. ⊟···· 3D Views   Switch to the **3D** view.
    ⌐···· 🔲 {3D}

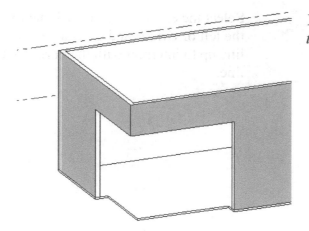

You can see the opening in the wall that was created.

29. Activate the **South** elevation.

30. Under Wall, select **Wall: Reveal**.

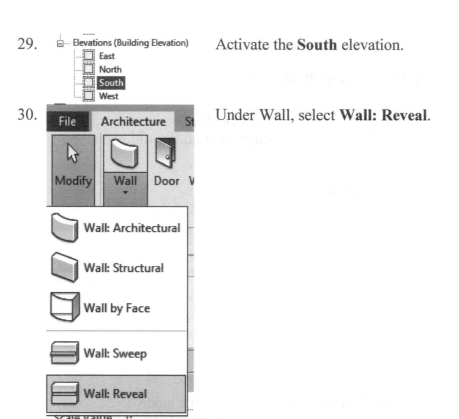

31. Enable **Vertical** on the ribbon.

32.

Click on Grid 3 to place the reveal.

Exit the command.

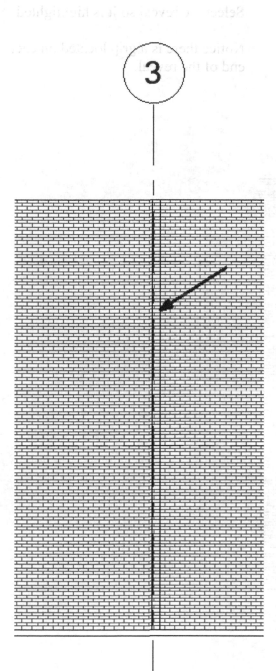

33.

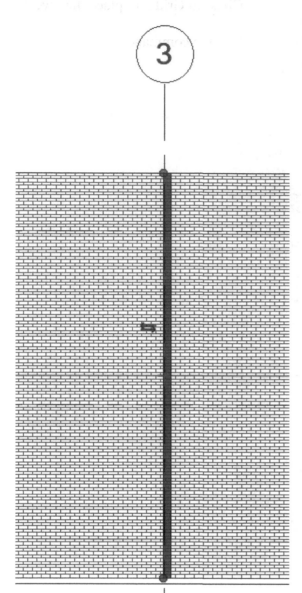

Select the reveal so it is highlighted.

Notice there is a grip located on each end of the reveal.

34.

Drag the top end of the reveal to the intersection of Grid 3 and Roof Level.

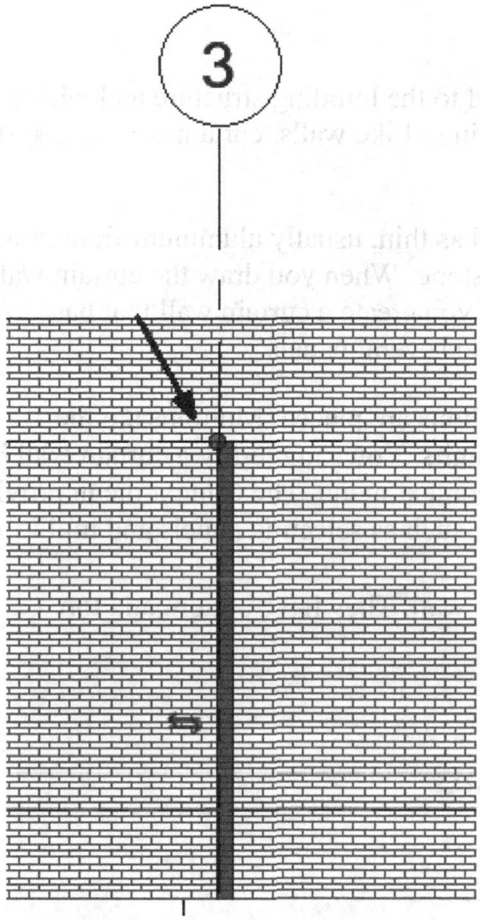

35.

Drag the bottom end of the reveal to the intersection of Grid 3 and Level 2.

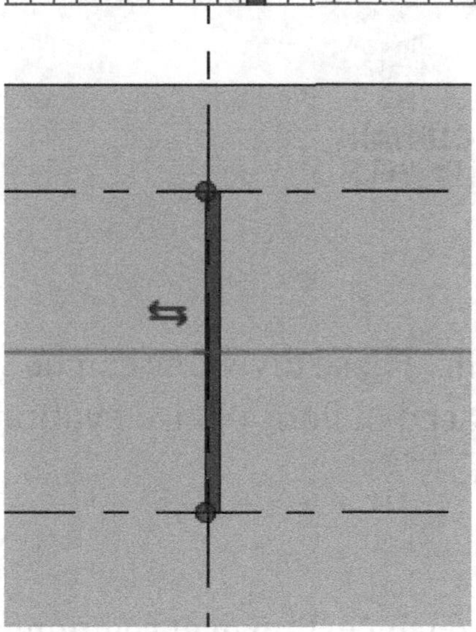

Curtain Walls

A curtain wall is any exterior wall that is attached to the building structure and which does not carry the floor or roof loads of the building. Like walls, curtain walls are system families.

In common usage, curtain walls are often defined as thin, usually aluminum-framed walls containing in-fills of glass, metal panels, or thin stone. When you draw the curtain wall, a single panel is extended the length of the wall. If you create a curtain wall that has automatic curtain grids, the wall is subdivided into several panels.

In a curtain wall, grid lines define where the mullions are placed. Mullions are the structural elements that divide adjacent window units. You can modify a curtain wall by selecting the wall and right-clicking to access a context menu. The context menu provides several choices for manipulating the curtain wall, such as selecting panels and mullions.

Curtain Walls contains most properties of a basic wall. They have bottom and top constraints and their profile can be modified.

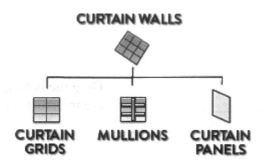

CURTAIN GRIDS
Curtain grids are divisions created on the walls. These divisions can be horizontal or vertical. Curtain grids can be placed in floor plan, elevations, and 3D views.

MULLIONS
Mullions are elements that can be created on each curtain grid segment, as well as on each curtain wall extremity.

CURTAIN PANELS
Curtain panels are rectangular elements located between each curtain grids.

Exercise 1-5

Curtain Walls

Drawing Name: **linear_curtain_wall.rvt**
Estimated Time to Completion: 30 Minutes

Scope

Placing a Curtain Walls
Adding and removing Curtain Wall Grids
Adding a Curtain Wall Door
Adding and removing mullions

Solution

1. Activate **Level 1** Floor Plan.

2. 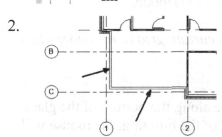 Select the two walls located between Grids 1 and 2 and Grids B and C.

   Hint: *Hold down the CTL key to select more than one element.*

3. 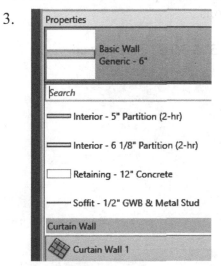 Use the Type Selector to assign **Curtain Wall 1** to the selected walls.

4.  Switch to a **3D** View.

5. 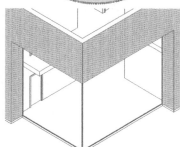 Use the Viewcube to orient the view to a SW perspective view.

6. 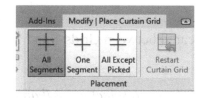 Zoom into the Southwest corner of the building model.

7. Select **Curtain Grid** from the ribbon.

8. Enable **All Segments** on the ribbon.

   All segments applies a curtain grid to the entire glass panel.

9. 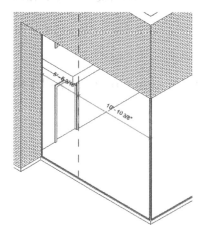 If you hover the mouse along the bottom of the glass panel on the west side of the building, the mouse will snap to different points. These preset points divide the panel into equal sections.

   Click when the value **5' 5 3/16"** is displayed.

10.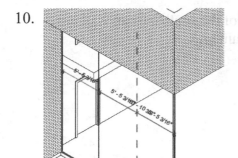

Click a second time to place a curtain grid at the midpoint of the curtain panel.

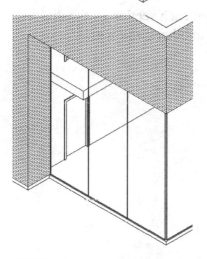

If you cancel out of the command, you see the glass panel has been divided into three equal panes.

11.

Curtain Grid

Select **Curtain Grid** from the ribbon.

12.

Add-Ins	Modify | Place Curtain Grid	
All Segments	One Segment	All Except Picked	Restart Curtain Grid
Placement		

Enable **All Segments** on the ribbon.

All segments applies a curtain grid to the entire glass panel.

13. 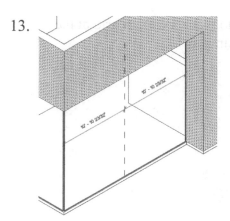 Locate the midpoint of the south curtain wall and select to place a curtain grid.

14. 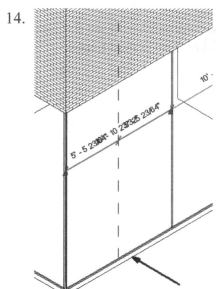 Locate the midpoint of each section of the divided custain wall to place another curtain grid.

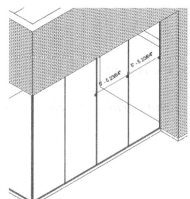 Cancel out of the command so you can inspect your curtain wall.

The south curtain wall should have four vertical glass panels of equal width of 5' 5-23/64".

15. Curtain Grid Select **Curtain Grid** from the ribbon.

16.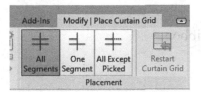

Enable **All Segments** on the ribbon.

All segments applies a curtain grid to the entire glass panel.

17.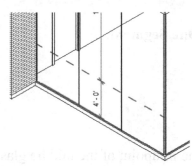

Place a horizontal curtain grid by hovering over the vertical edge of the curtain wall.

Locate the bottom grid 4' above Level 1.

18.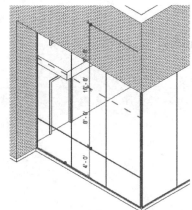

Place a second horizontal grid 8' above first horizontal grid.

19.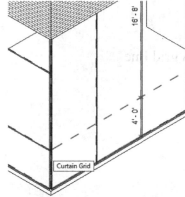

Move over to the west curtain wall.

The curtain grid tool should automatically snap to the horizontal grids placed on the west wall.

Add two horizontal grids at the same distances on the south curtain wall.

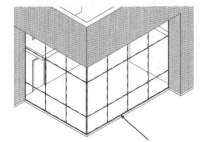

The curtain wall should look as shown.

We will add a door in the area indicated by the arrow, but first we need to define a panel the correct size.

20. 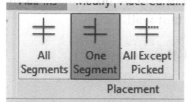 Select the **Curtain Grid** tool from the ribbon.

21. 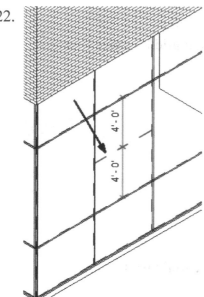 Set the Placement to **One Segment**.

22. 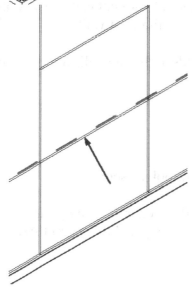 Place the segment at the midpoint of the middle glass panel.

    ESC out of the command.

23. Select the lower curtain grid line.

24. Enable **Add/Remove Segments** from the ribbon.

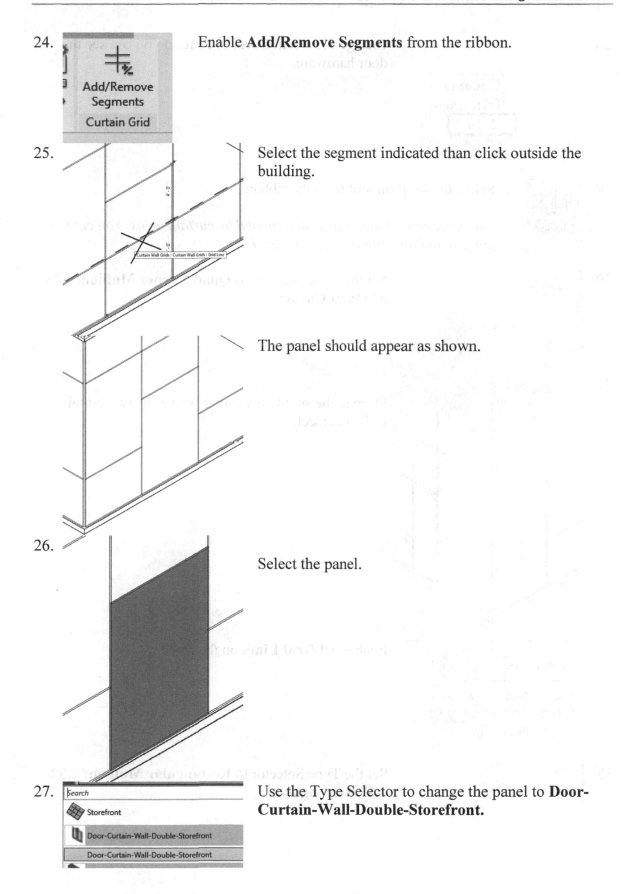

25. Select the segment indicated than click outside the building.

The panel should appear as shown.

26. Select the panel.

27. Use the Type Selector to change the panel to **Door-Curtain-Wall-Double-Storefront.**

28. 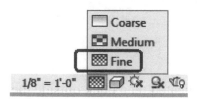 Change the Detail Level to **Fine** so you can see the door hardware.

29. Select the Mullion tool from the ribbon.

    The placement of mullions is determined by curtain grids. You cannot place a mullion without a curtain grid.

30. Set the Type Selector to **Quad Corner Mullion: 5" x 5" Quad Corner**.

31. Place at the southwest corner where the two curtain walls intersect.

32. 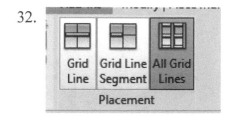 Enable **All Grid Lines** on the ribbon.

33. 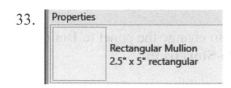 Set the Type Selector to **Rectangular Mullion: 2.5" x 5" Rectangular**.

34.

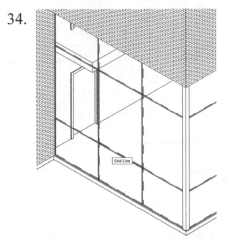

Select the Grid Line on the west curtain wall.

35.

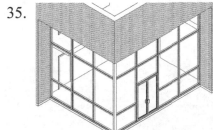

Select the grid lines on the south curtain wall.

ESC out of the command.

36.

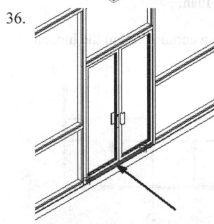

Use the TAB key to cycle through selections until the mullion below the door is selected.

Delete the mullion by clicking the DELETE key on the keyboard or using the Delete tool on the ribbon.

37.

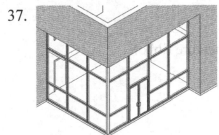

The curtain wall should appear as shown.

Save as *ex1-5.rvt*.

Exercise 1-6

Embedded Curtain Walls

Drawing Name: **embedded curtain wall.rvt**
Estimated Time to Completion: 20 Minutes

Scope
Placing a Curtain Wall inside an existing wall
Adding and removing Curtain Wall Grids
Adding a Curtain Wall Door
Adding and removing mullions

Solution

1. 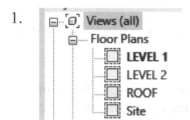 Activate **Level 1** Floor Plan.

    We are going to embed a curtain wall in the indicated south wall.

    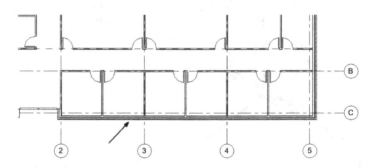

2.  Select the **Wall** tool from the Architecture ribbon.

3. Select the **Curtain Wall: Exterior Glazing** from the Type Selector.

    Click **Edit Type**.

4.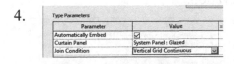

Enable **Automatically Embed**.

Set the Curtain Panel to **System Panel: Glazed**.
Set the Join Condition to **Vertical Grid Continuous**.

5.

Vertical Grid	
Layout	None
Spacing	2' 0"
Adjust for Mullion Size	☐

Horizontal Grid	
Layout	None
Spacing	4' 0"
Adjust for Mullion Size	☐

Vertical Mullions	
Interior Type	None
Border 1 Type	None
Border 2 Type	None

Horizontal Mullions	
Interior Type	None
Border 1 Type	None
Border 2 Type	None

Under Vertical Grid:
Set the Layout to **None**.

Under Horizontal Grid:
Set the Layout to **None**.

Set the Vertical and Horizontal Mullions to **None**.

Click **OK**.

6.

Properties		✕
Curtain Wall Exterior Glazing		
New Walls		Edit Type
Constraints		
Base Constraint	LEVEL 1	
Base Offset	2' 8"	
Base is Attached	☐	
Top Constraint	Up to level: LEVEL 1	
Unconnected Height	6' 0"	
Top Offset	8' 8"	

In the Properties palette:

Set the Base Constraint to **LEVEL 1**.
Set the Base Offset to **2' 8"**.
Set the Top Constraint to **LEVEL 1**.
Set the Top Offset to **8' 8"**.

7.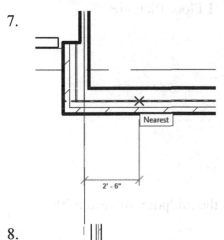

Set the start point at **2' 6"** to the right of Grid 2.

8.

Set the end point at Grid 5.

ESC to exit the command.

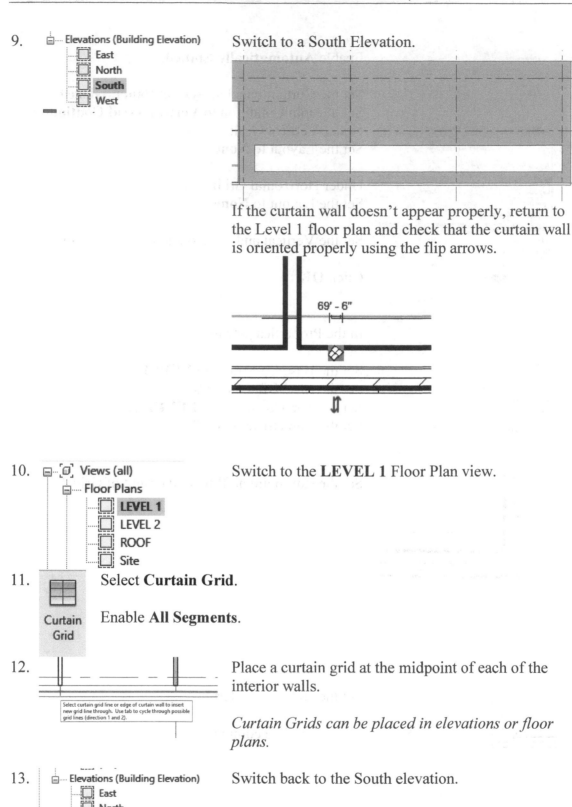

9. Elevations (Building Elevation)
   - East
   - North
   - **South**
   - West

Switch to a South Elevation.

If the curtain wall doesn't appear properly, return to the Level 1 floor plan and check that the curtain wall is oriented properly using the flip arrows.

69' - 6"

10. Views (all)
   - Floor Plans
     - **LEVEL 1**
     - LEVEL 2
     - ROOF
     - Site

Switch to the **LEVEL 1** Floor Plan view.

11. Curtain Grid

Select **Curtain Grid**.

Enable **All Segments**.

12. Select curtain grid line or edge of curtain wall to insert new grid line through. Use tab to cycle through possible grid lines (direction 1 and 2).

Place a curtain grid at the midpoint of each of the interior walls.

Curtain Grids can be placed in elevations or floor plans.

13. Elevations (Building Elevation)
   - East
   - North
   - **South**
   - West

Switch back to the South elevation.

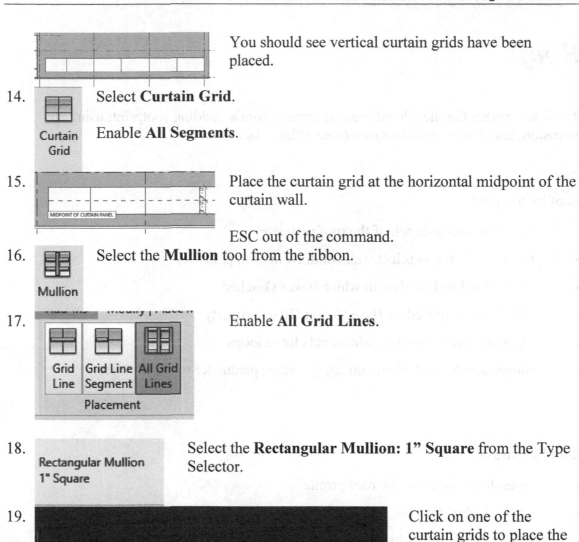

You should see vertical curtain grids have been placed.

14. Select **Curtain Grid**.

Enable **All Segments**.

15. Place the curtain grid at the horizontal midpoint of the curtain wall.

ESC out of the command.

16. Select the **Mullion** tool from the ribbon.

17. Enable **All Grid Lines**.

18. Select the **Rectangular Mullion: 1" Square** from the Type Selector.

19. Click on one of the curtain grids to place the mullions.

Switch to a 3D view to inspect your embedded curtain wall.

20. Save as *ex1-6.rvt*.

Roofs

Roofs are system families. Roofs can be created from a building footprint, using an extrusion, and from a mass instance (converting a face to a roof).

Roof by footprint

- 2D closed-loop sketch of the roof perimeter

- Created when you select walls or draw lines in plan view

- Created at level of view in which it was sketched

- Height is controlled by Base Height Offset property

- Openings are defined by additional closed loops

- Slopes are defined when you apply a slope parameter to sketch lines

Roof by extrusion

- Open-loop sketch of the roof profile

- Created when you use lines and arcs to <u>sketch</u> the profile in an elevation view

- Height is controlled by the location of the sketch in elevation view

- Depth is calculated by Revit based on size of sketch, unless you specify <u>start and end points</u>.

Roofs are defined by material layers, similar to walls.

Exercise 1-7

Creating a Roof by Footprint

Drawing Name: **i_roofs.rvt**
Estimated Time to Completion: 10 Minutes

Scope
Create a roof.

Solution

1. 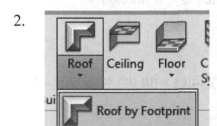 Activate the **T.O. Parapet** floor plan.

2. 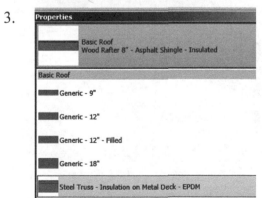 Select **Roof by Footprint** from the tab.

3. Select **Steel Truss - Insulation on Metal Deck - EPDM** using the Type Selector on the Properties panel.

4.  Select **Pick Walls** mode.

5. 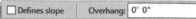 On the Options bar,
uncheck **Defines slope**.
Set the Overhang to **0'-0"**.
Uncheck **Extend to Wall Core**.

6. 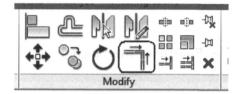 Select all the exterior walls.

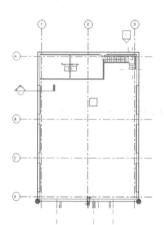

7. 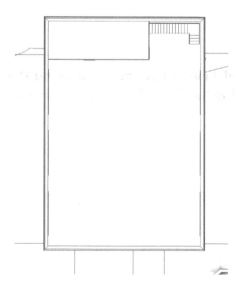 Use the **Trim** tool from the Modify panel to
create a closed boundary, if necessary.

8. Align the boundary lines with the exterior face
of each wall.

9. 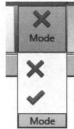 Select the **Green Check** on the Model panel to **Finish Roof**.

10. Close without saving.

Exercise 1-8

Creating a Roof by Extrusion

Drawing Name: **i_roofs_extrusion.rvt**
Estimated Time to Completion: 30 Minutes

Scope
Create a roof by extrusion.
Modify a roof.

Solution

1. 3D Views {3D}   Activate the **3D view.**

2. 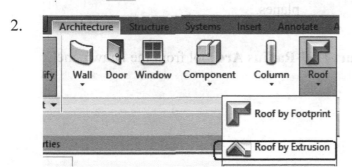 Activate the Architecture tab.
Select **Roof by Extrusion** under the Build panel.

3.  Enable **Name**.
Select **Roof Shape** from the list of reference planes.
Click **OK**.

4.

Roof Reference Level and Offset

Level: Upper Roof

Offset: 0' 0"

OK   Cancel

Select **Upper Roof** from the list.
Click **OK**.

5.

Set Show Ref Viewer
Plane
Work Plane

Select the **Show** tool from the Work Plane panel.
This will display the active work plane.

6.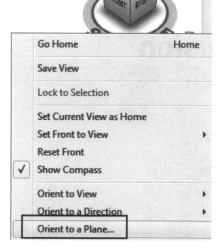

Right click on the ViewCube's ring.
Select **Orient to a Plane**.

Go Home	Home
Save View	
Lock to Selection	
Set Current View as Home	
Set Front to View	▶
Reset Front	
✓ Show Compass	
Orient to View	▶
Orient to a Direction	▶
Orient to a Plane...	

7.

Specify an Orientation Plane

⦿ Name   Reference Plane : Roof Shape

○ Pick a plane

Enable **Name**.
Select **Roof Shape** from the list of reference planes.
Click **OK**.

8.

Draw

Work Pla...

Draw

Select the **Start-End-Radius Arc** tool from the Draw panel.

9.

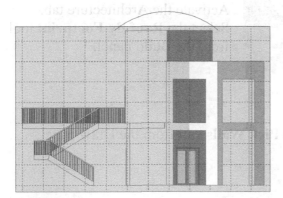

Draw an arc over the building as shown.

10.   Select the **Green Check** under the Mode panel to finish the roof.

11. 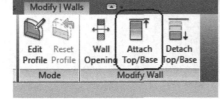   Switch to an isometric 3D view.

12.

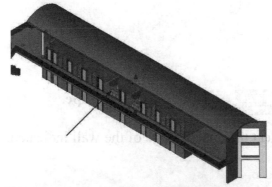

Hover over the wall below the roof.
Click TAB.
All connected walls will be selected.

13. 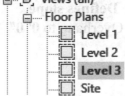   Select **Attach Top/Base** from the Modify Wall panel.
Then select the roof.
The walls will adjust to meet the roof.
Orbit the model to inspect the walls.

14. Views (all)
  Floor Plans
    Level 1
    Level 2
    Level 3
    Site

Activate the **Level 3** floor plan.

15. 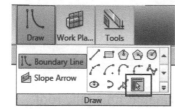 Activate the Architecture tab.
Select the **Roof By Footprint** tool from the Build panel.

16.  Select the **Pick Walls** tool from the Draw panel.

17. On the Options bar: Enable **Defines slope**.
Set the Overhang to **2′ 0″**.
Enable **Extend to wall core**.

18. 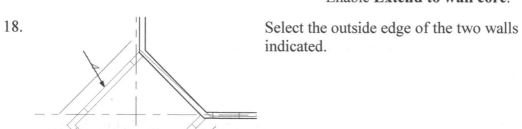 Select the outside edge of the two walls indicated.

19. Disable **Defines slope**.

20. Select the outside edge of the wall indicated.

21.  Disable **Defines slope**.
Set the Overhang to **0′ 0″**.

22. 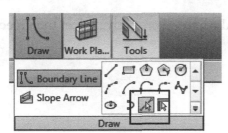 Select the **Pick Line** tool from the Draw panel.

23. 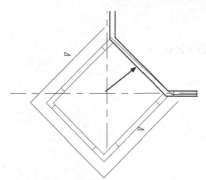 Pick the exterior side of the wall indicated.

24. 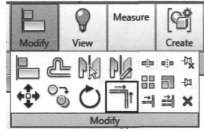 Select the **Trim** tool from the Modify panel.

25. 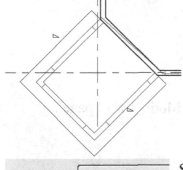 Trim the roof boundaries so they form a closed polygon.

26. 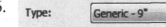 Set the Type to **Generic- 9″**. Click **OK**.

27. 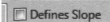 Set the Slope to **6″/12″**.

28. 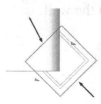 Verify that both the back and front boundary lines have no slope assigned.

    If they have a slope symbol, select the back and front boundary and lines and uncheck **Defines Slope** on the Options bar.

29. 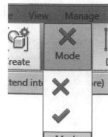 Once the sketch is closed and trimmed properly, select the **Green Check** on the Mode panel.

30. 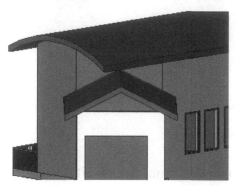 Return to a 3D view.

31.  Hover over the front wall.
Click TAB to select all connected walls.
Left click to select the walls.

32. 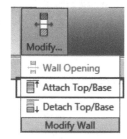 Select Attach Top/Base from the Modify Wall panel.

33.  Select the roof to attach the wall.

34.

| Error - cannot be ignored | — | 1 Error, |

Can't keep wall and target joined

<<   1 of 2   >>   Show

Unjoin Elements

Click **Unjoin Elements**.

Left click in the window to release the selection.

Orbit around the model to inspect the roofs.

35. Close without saving.

Exercise 1-9

Add Split Lines to a Roof

Drawing Name: **Split Lines Roof.rvt**
Estimated Time to Completion: 20 Minutes

Scope
Create a roof by footprint.
Modify a roof using split lines.

Solution

1.
- Views (all)
  - Floor Plans
    - LEVEL 1
    - LEVEL 2
    - **ROOF**
    - Site

Activate the **ROOF** floor plan.

2.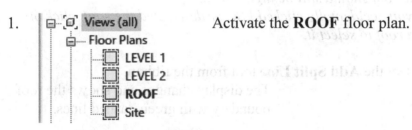

Select the **Roof by Footprint** tool from the ribbon.

3.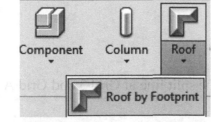

Properties

Basic Roof
Steel Truss - Insulation on Metal Deck - EPDM

Verify that the Type is set to Steel Truss – Insulation on Metal Deck – EPDM.

4. [☐ Defines slope | Overhang: 0' 0" | ☐ Extend to wall core]

Disable **Defines slope**.
Set the Overhang to **0' 0"**.
Disable **Extend to wall core**.

5.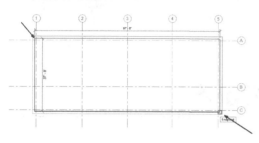

Select the **Rectangle** tool from the Draw panel.

6. Select the two inside corners of the parapet walls to define the rectangle.

7. Select **Green Check** to complete the roof.

8. Add Split Line

*The roof should still be highlighted.
If you accidentally clicked ESC or deselected the roof, click on the roof to select it.*

Select the **Add Split Line** tool from the ribbon.

The display changes and shows the roof boundary with green dashed lines.

9. Start the split line at Grid 3 and Grid A.

10.

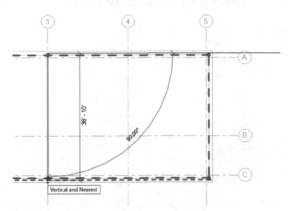

End the split line at Grid 3 and Grid C.

11.

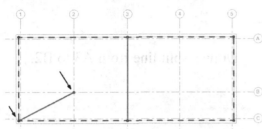

Draw a split line from C1 to B2.

12.

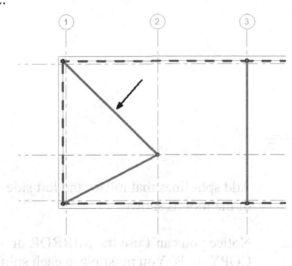

Draw a split line from A1 to B2.

13. 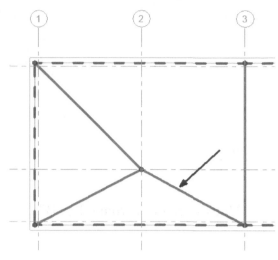 Draw a split line from C3 to B2.

14. 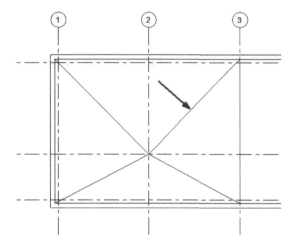 Draw a split line from A3 to B2.

15. 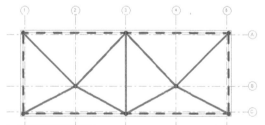 Add split lines that mirror the left side of the roof as shown.

Notice you can't use the MIRROR or COPY tools. You must place each split line individually.

16.   Enable **Modify Sub Elements**.

17. Select the points at B2 and B4 and set them to **-9"**.

18. Select **Edit Type** on the Properties palette.

19. Click **Edit** next to Structure.

20. Note that Layer 2 has Variable enabled.

This is because we added a slope with split lines.
The value of the slope (in this case, 9") must be less than the thickness of the layer.

Click **OK** twice to close the dialog boxes.

21. Click **Modify** on the ribbon to exit the roof editing mode.

22. Switch to a 3D view to see how the roof appears.

Change the display to Realistic to see how the material layers display for the roof.

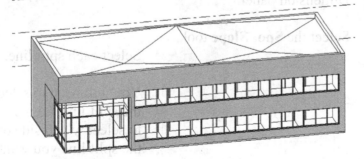

23. Save as *ex1-9.rvt*.

Exercise 1-10
Add Spot Elevation and Slope Annotations

Drawing Name: **spot elevation.rvt**
Estimated Time to Completion: 10 Minutes

Scope
Add a spot slope annotation to a roof.
Add a spot elevation annotation to a roof.

Solution

1. 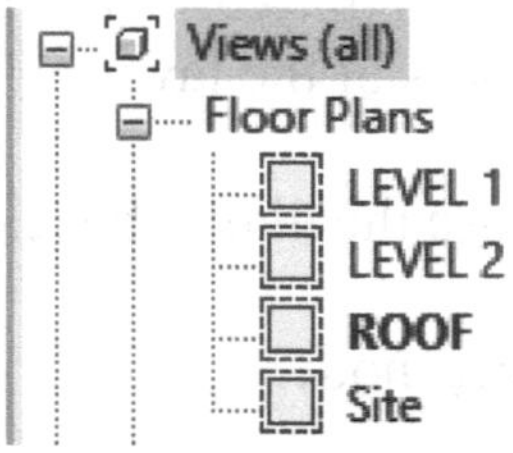
Activate the **ROOF** floor plan.

2. 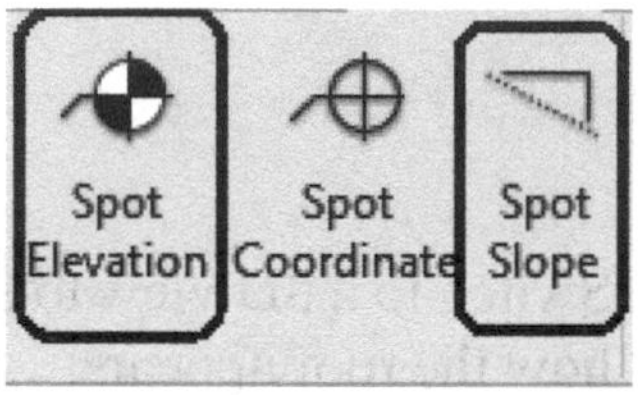
Activate the Annotate ribbon.

Locate the Spot Slope and Spot Elevation tools on the Dimension panel.

Select the **Spot Slope** tool.

3. 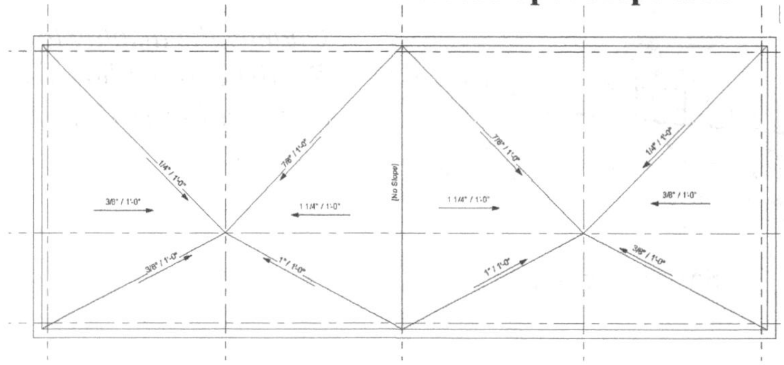
Select each split line.

You will need to select the split line, then click to select which side of the split line you want the label to be placed.

You can also place spot slope labels between the split lines.

4. 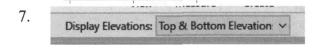 Select the **Spot Elevation** tool.

5. On the Options bar:

   Enable **Actual (Selected) Elevation**.

6. Select the intersection point at B2.

   Click to place the leader and elevation note.

7. On the Options bar:

   Enable**Top & Bottom Elevation**.

8. Select the intersection point at B4.

   Click to place the leader and elevation note.

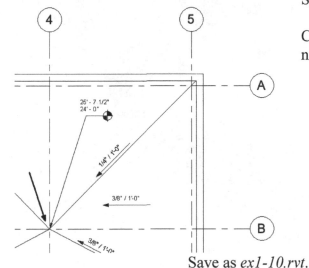

9. Save as *ex1-10.rvt*.

Stairs

Revit stairs are system families, similar to walls, ceilings, and floors. Railings are also system families, but the balusters and the railing profiles are loadable families.

A stair can consist of the following:

- Runs: straight, spiral, U-shaped, L-shaped, custom sketched run

- Landings: created automatically between runs or by picking 2 runs, or by creating a custom sketched landing

- Supports (side and center): created automatically with the runs or by picking a run or landing edge

- Railings: automatically generated during creation or placed later

Be familiar with all the properties that control stairs.

TOP ... BASE ...	**BASE AND TOP LEVELS:** Stairs are based on selected levels that already exist in the project. You can add an offset on these levels if required.
H ...	**DESIRED STAIR HEIGHT:** Total distance between the base and the top of the stairs, including offsets.
19 ...	**DESIRED NUMBER OF RISERS:** Automatically calculated by Revit, dividing Stair Height by Maximum Riser Height. You can change this number, which will modify the stair slope.
11 ...	**ACTUAL NUMBER OF RISERS:** The number of risers you modeled so far.
MAX ...	**MAXIMUM RISER HEIGHT:** Riser height for your stair will never go above this value. This parameter is set on the stair type. Usually on par with code requirements.
a ...	**ACTUAL RISER HEIGHT:** This distance is automatically calculated by Revit, dividing the Stair Height by the Desired Number of Risers.
MIN ...	**MINIMUM TREAD DEPTH:** On the stair type, specify the minimum tread depth. When you start modeling your stair, you can go above this number, but not below.

	ACTUAL TREAD DEPTH: By default, this value is equal to minimum tread depth set in the stair type. However, you can set a bigger value if you want more depth.
	MINIMUM RUN WIDTH: Set on the stair type, you can specify the minimum run width. This does not include support (stringers).
	ACTUAL RUN WIDTH: By default, this will be the same as the minimum run width. You can set a higher value than the minimum, but a lower value will result in a Warning.

Exercise 1-11

Creating Stairs by Sketch

Drawing Name: **i_stairs.rvt**
Estimated Time to Completion: 20 Minutes

Scope
Place stairs using reference work planes.

Solution

1. 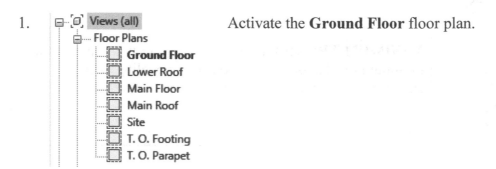   Activate the **Ground Floor** floor plan.

2. 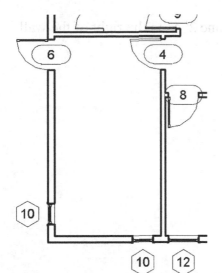 Zoom into the lower left corner of the building.

3. Select the **Reference Plane** tool from the Work Plane panel on the Architecture tab.

v Ref
Plane

4. Offset: 2' 4" Set the Offset to **2' 4"** on the Options bar.

5. Draw Select the **Pick Lines** tool from the Draw panel.

6. Place a vertical reference plane 2' 4" to the left of the wall where Door 4 is placed.

7.   Place a vertical reference plane 2′ 4″ to the right of the wall where Door 6 is placed.

8.   Offset: 4′ 4″ ☐ Lock   Set the Offset to **4′ 4″** on the Options bar.

9.

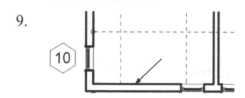

Place a horizontal reference plane 4′ 4″ above the wall where Window 10 is located.

10.   Offset: 7′ 4″ ☐   Set the Offset to **7′ 4″** on the Options bar.

11.

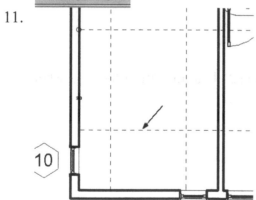

Place a horizontal reference plane 7′ 4″ above the first horizontal reference plane.

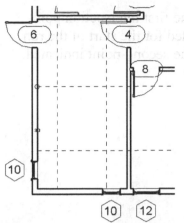

Four reference planes – two horizontal and two vertical should be placed in the room.

12. 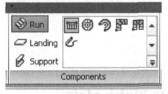 Activate the Architecture tab.
    Select the **Stair** tool from the Circulation panel.

13. Select **Edit Type** from the Properties pane.

14. Change the Maximum Riser Height to **8″**.
    Set the stair width to **4′ 0″**.
    Click **OK**.

Type Parameters	
Parameter	
Calculation Rules	
Maximum Riser Height	0' 8"
Minimum Tread Depth	0' 11"
Minimum Run Width	4' 0"
Calculation Rules	

15. Set the desired number of risers to **16**.

Dimensions	
Desired Number of Risers	16
Actual Number of Risers	1
Actual Riser Height	0' 7 1/2"
Actual Tread Depth	0' 11"
Tread/Riser Start Number	1

16. 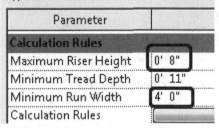 Select the **Run** tool from the Draw panel on the tab.

17. Set the Location line to **Run: Center** on the Options bar.

    Location Line: Run: Center

18.

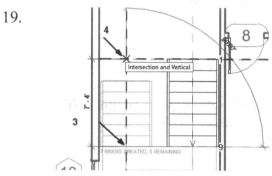

Pick the first intersection point indicated for the start of the run. Pick the second point indicated.

19.

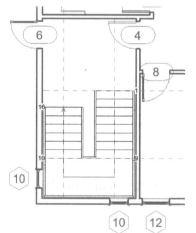

Moving clockwise, select the lower intersection point #3 and then the upper intersection point #4.

U-shaped stairs are placed.

20.  Select the **Green Check** on the Mode panel to **Finish Stairs**.

21. Close without saving.

Exercise 1-12
Creating Stairs by Component

Drawing Name: **i_stairs_component.rvt**
Estimated Time to Completion: 10 Minutes

Scope
Place stairs using the component method.

Solution

1. 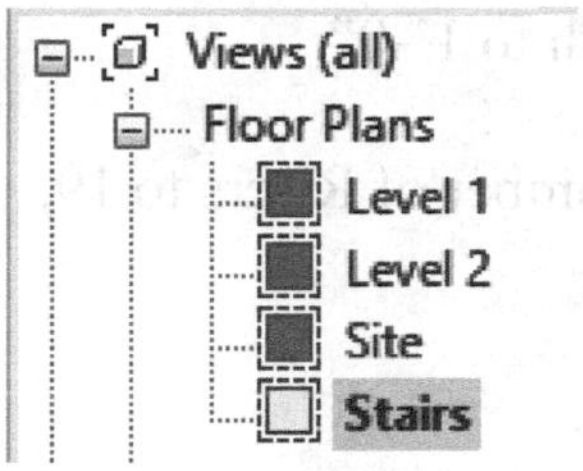   Activate the **Stairs** floor plan.

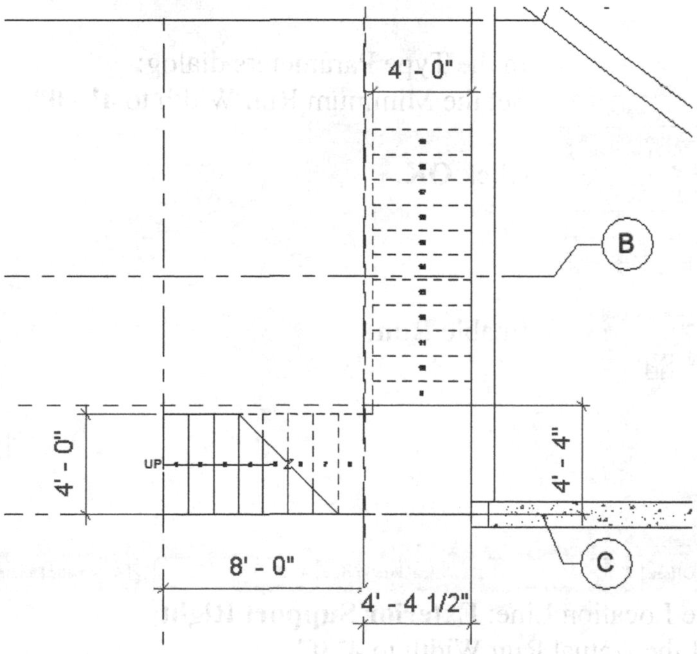

Stairs will be placed using the dimensions shown.

There are eight horizontal risers and 11 vertical risers.

Use the reference planes provided as guides.

Use the Cast in Place Monolithic Stair.

Set the Base Level to Level 1.

Set the Top Level to Level 2.

Set the Tread Depth to 1'-0".

Set the Desired Number of Risers to 19.

2. Activate the Architecture tab.

Select the **Stair** tool.

3. Set the Stair type to Cast-In-Place Stair: Monolithic Stair.

Set the Base Level to Level 1.

Set the Top Level to Level 2.

Stair	
Constraints	
Base Level	Level 1
Base Offset	0' 0"
Top Level	Level 2
Top Offset	0' 0"
Desired Stair Height	11' 0"

4. Scroll down.

Dimensions	
Desired Number of Risers	19
Actual Number of Risers	1
Actual Riser Height	0' 6 243/256"
Actual Tread Depth	1' 0"
Tread/Riser Start Number	1
Identity Data	

Set the Tread Depth to 1'-0".

Set the Desired Number of Risers to 19.

5. 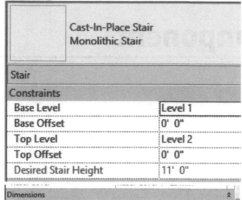 Select **Edit Type** on the Properties panel.

6. In the Type Parameters dialog:
Set the Minimum Run Width to **4' - 0".**

Type Parameters	
Parameter	Value
Calculation Rules	
Maximum Riser Height	0' 7 11/128"
Minimum Tread Depth	0' 11 3/128"
Minimum Run Width	4' 0"
Calculation Rules	Edit...

Click **OK.**

7. Enable **Run**.

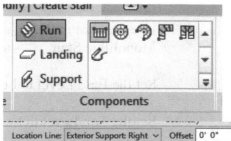

8. On the Options bar: Set the Location Line: **Exterior Support Right**
Set the Offset to **0' 0"**. Set the Actual Run Width to **4' 0"**.

Location Line: Exterior Support: Right	Offset: 0' 0"	Actual Run Width: 4' 0"	☑ Automatic Landing

9. Start the stairs at C6 and end the run at C8.

    This will place 8 risers.

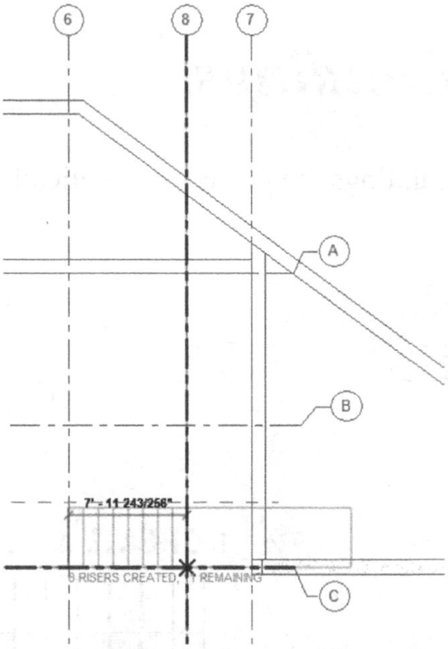

10. 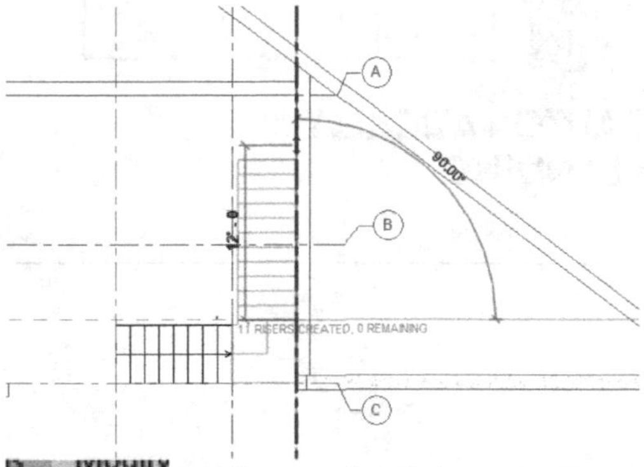 Start the second run at the intersection of the reference plane and Grid line 7.

    Drag the cursor straight up vertically and left click to end when you see 11 risers.

11. 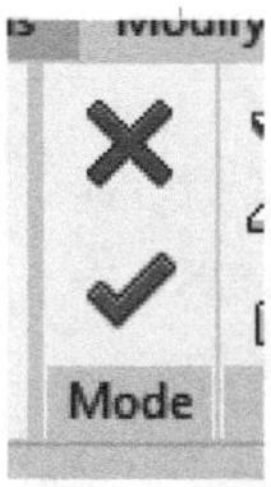 Green check to complete the stairs.

12. Close without saving.

Landings

Landings are placed automatically when you draw more than one run.

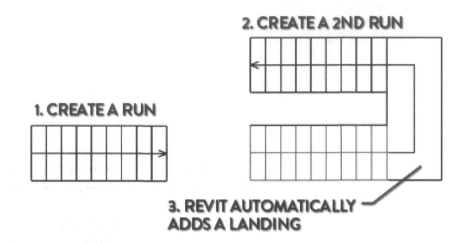

Exercise 1-13

Stair Landings

Drawing Name: **landings.rvt**
Estimated Time to Completion: 10 Minutes

Scope
Create a stair landing

Solution

1. Activate the **Ground Floor – Main Stairs** floor plan.

2. Select **Stair** from the Architecture tab on the ribbon.

3.

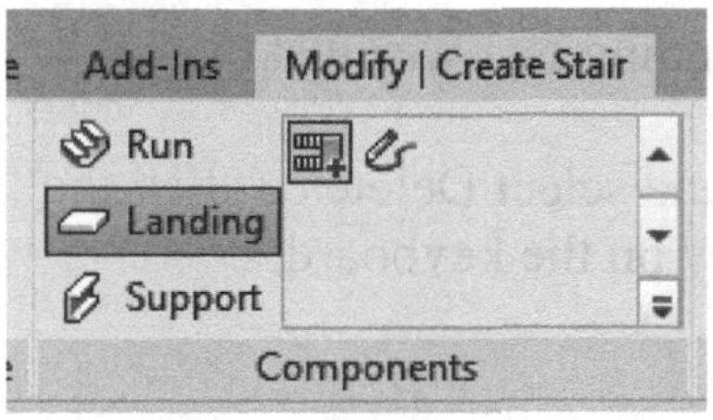

Select **Landing**.

Enable **Select Runs**.

4.

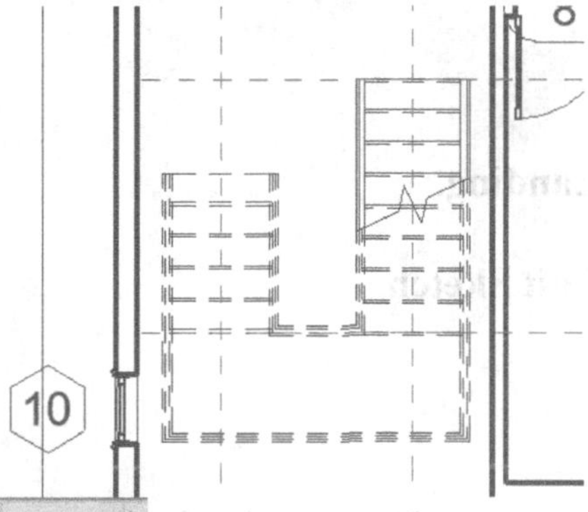

Select the stair on the right and then the stair on the left.

A landing is placed.

5.

Select **Green Check** to exit the stair command.

6.

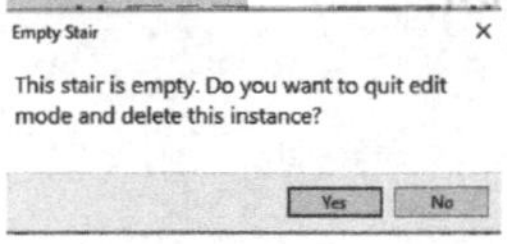

Click **Yes**.

The stair is empty because you only created the landing.

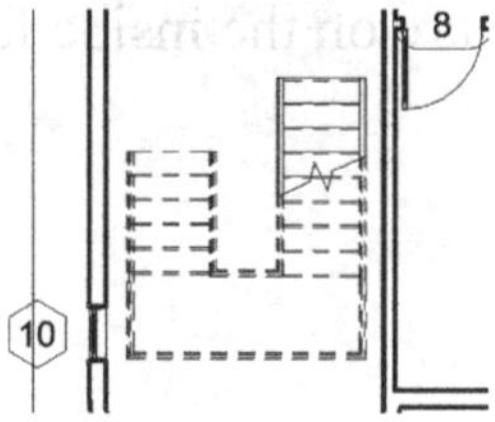

The landing is placed.

7.

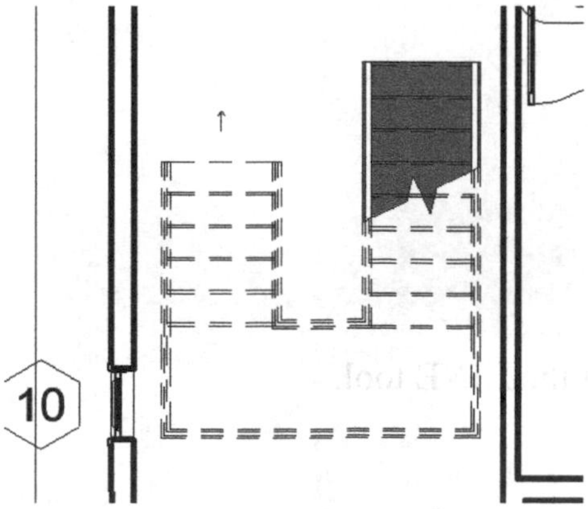

Select the stairs.

Select **Edit Stairs** on the ribbon.

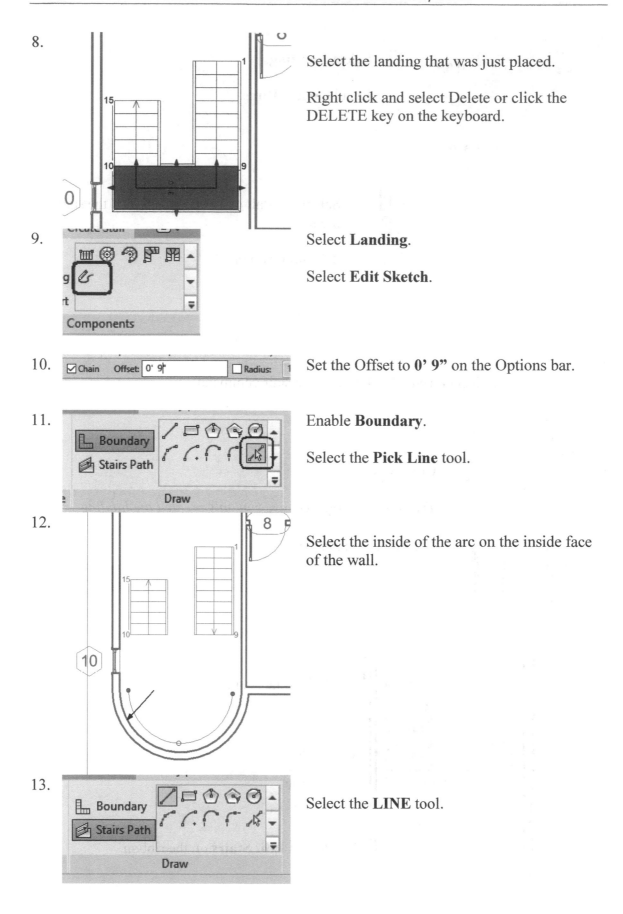

8. Select the landing that was just placed.

Right click and select Delete or click the DELETE key on the keyboard.

9. Select **Landing**.

Select **Edit Sketch**.

10. Set the Offset to **0' 9"** on the Options bar.

11. Enable **Boundary**.

Select the **Pick Line** tool.

12. Select the inside of the arc on the inside face of the wall.

13. Select the **LINE** tool.

14.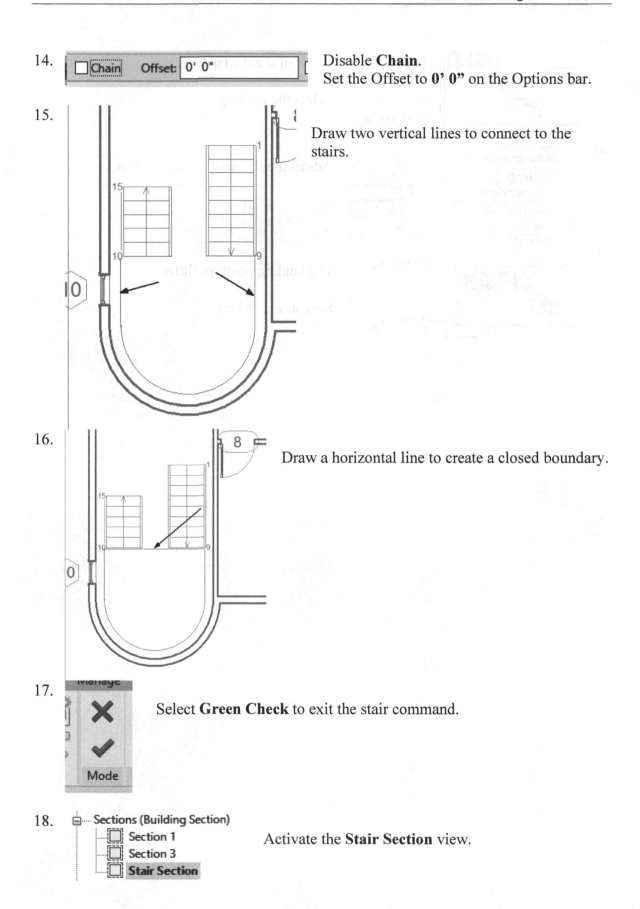
Disable **Chain**.
Set the Offset to **0' 0"** on the Options bar.

15.
Draw two vertical lines to connect to the stairs.

16.
Draw a horizontal line to create a closed boundary.

17.
Select **Green Check** to exit the stair command.

Mode

18. ⊟···· Sections (Building Section)
    └──── ▢ Section 1
    └──── ▢ Section 3
    └──── ▢ **Stair Section**

Activate the **Stair Section** view.

19.

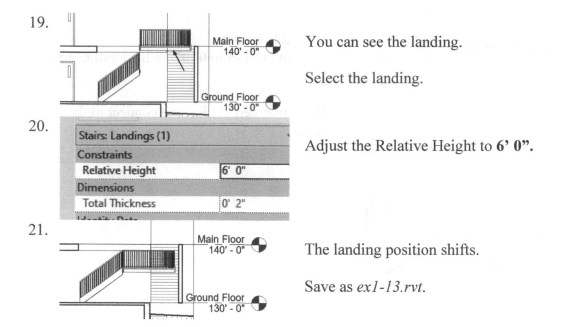

You can see the landing.

Select the landing.

20.

Adjust the Relative Height to **6' 0"**.

21.

The landing position shifts.

Save as *ex1-13.rvt*.

Railings

Railings are system families. Railings can be free-standing or they can be hosted. Railing hosts include stairs, ramps, and floors. (This is a possible exam question.)

You need to understand and identify the different components which make up a railing.

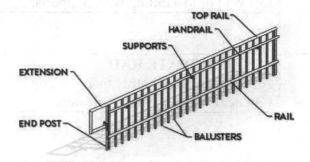

When defining a railing, you should be familiar with the different Type properties.

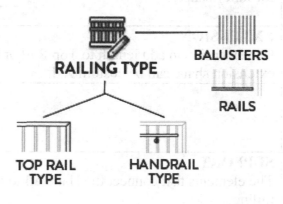

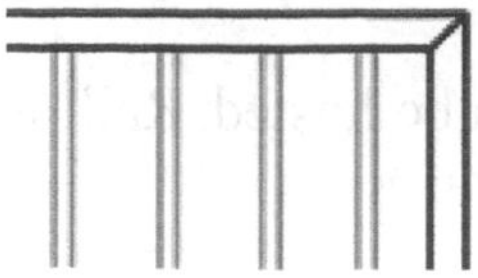	**TOP RAIL** Top rail is the highest horizontal element of a railing. It is created by selecting a 2D profile and a height.
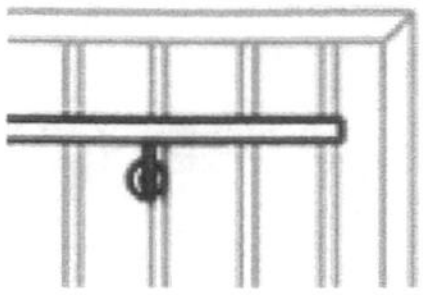	**HANDRAIL** Handrail is an intermediate rail used for hands. They are linked to a wall or to a railing with Supports.
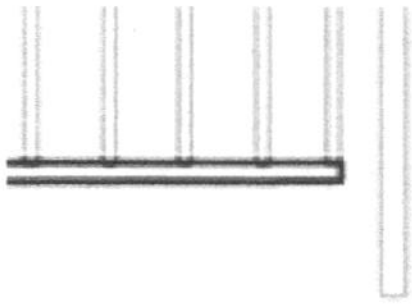	**INTERMEDIATE RAIL** Any horizontal rail other than the Top Rail and the Handrail. Can be used to constrain balusters.
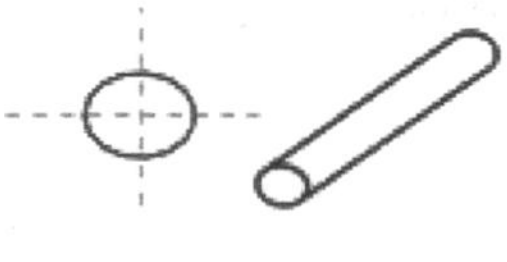	**RAIL 2D PROFILE** Every Rail in Revit is an extrusion from a 2D Profile Family. Use default profiles for simple shapes, or create a custom one for fancy shapes.
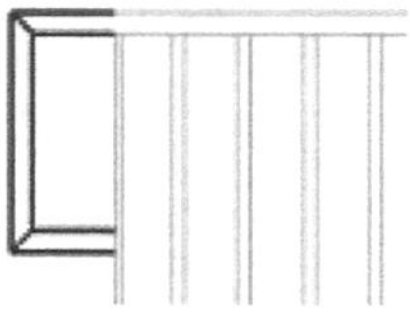	**EXTENSION** Use extension to add length to Top Rail or Handrail. The extension shape can be customized.
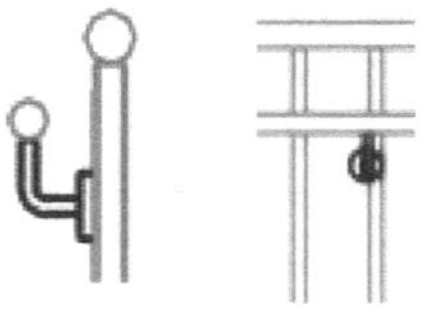	**SUPPORT** The elements that connect the Handrail to the wall or to the railing.

Baluster Elements

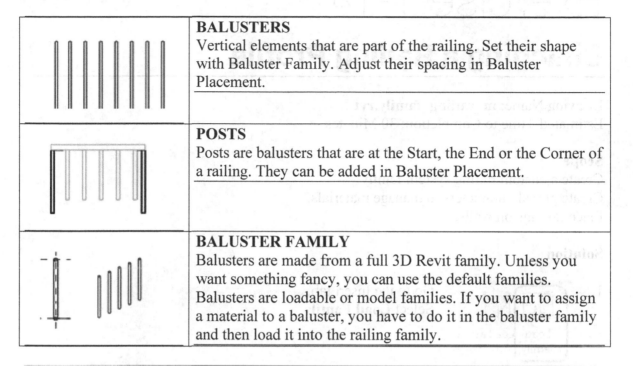	**BALUSTERS** Vertical elements that are part of the railing. Set their shape with Baluster Family. Adjust their spacing in Baluster Placement.
	POSTS Posts are balusters that are at the Start, the End or the Corner of a railing. They can be added in Baluster Placement.
	BALUSTER FAMILY Balusters are made from a full 3D Revit family. Unless you want something fancy, you can use the default families. Balusters are loadable or model families. If you want to assign a material to a baluster, you have to do it in the baluster family and then load it into the railing family.

You should be aware of where you click in the Railing's Type Properties to define different components.

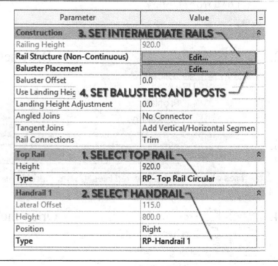

Parameter	Value	=
Construction **3. SET INTERMEDIATE RAILS**		≫
Railing Height	920.0	
Rail Structure (Non-Continuous)	Edit...	
Baluster Placement	Edit...	
Baluster Offset	0.0	
Use Landing Heig **4. SET BALUSTERS AND POSTS**		
Landing Height Adjustment	0.0	
Angled Joins	No Connector	
Tangent Joins	Add Vertical/Horizontal Segmen	
Rail Connections	Trim	
Top Rail **1. SELECT TOP RAIL**		≫
Height	920.0	
Type	RP- Top Rail Circular	
Handrail 1 **2. SELECT HANDRAIL**		≫
Lateral Offset	115.0	
Height	800.0	
Position	Right	
Type	RP-Handrail 1	

Exercise 1-14

Changing a Railing Profile

Drawing Name: **m_railing_family.rvt**
Estimated Time to Completion: 30 Minutes

Scope
Create a custom railing system family.
Create global parameters to manage materials.
Place railings on walls.

Solution

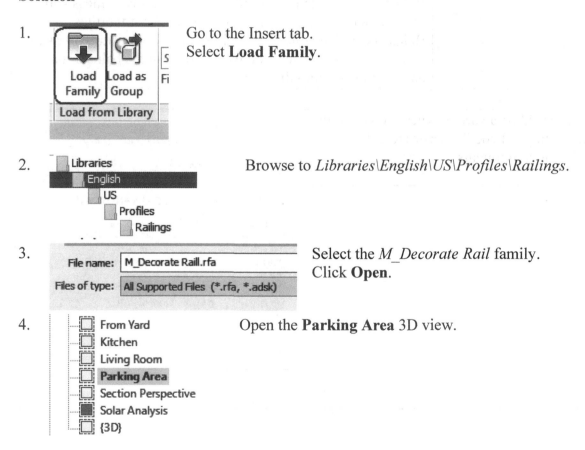

1. Go to the Insert tab.
   Select **Load Family**.

2. Browse to *Libraries\English\US\Profiles\Railings*.

3. Select the *M_Decorate Rail* family.
   Click **Open**.

4. Open the **Parking Area** 3D view.

5.  Select one of the railings placed on the concrete walkway.

6. 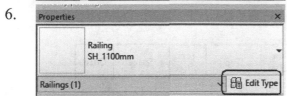 Select **Edit Type** on the Properties panel.

7. Select **Duplicate**.

Duplicate...

8. Type **Railing_Crescent_City**.
Click **OK**.

Name: Railing_Crescent_City

9. Select **Edit** next to Rail Structure.

Type Parameters			
Parameter		Value	=
Construction			
Railing Height		800.0	
Rail Structure (Non-Continuous)		Edit...	
Baluster Placement		Edit...	
Baluster Offset		0.0	
Use Landing Height Adjustment		No	
Landing Height Adjustment		0.0	
Angled Joins		Add Vertical/Horizontal Segmen	
Tangent Joins		Extend Rails to Meet	
Rail Connections		Trim	

10. Delete all the rails except for **New Rail (1)**. Set the Height to **250.0**. Set the Profile to **M_Rectangular HandRail: 50 x 50 mm.** Set the Material to **Iron, Wrought.**
You will have to import the material into the document.
Click **OK.**

Rails					
	Name	Height	Offset	Profile	
1	New Rail(1)	250.0	0.0	M_Rectangular Handrail : 50 x 50mm	Iron, Wrought

11. Select **Edit** next to Baluster Placement.

Construction		
Railing Height	800.0	
Rail Structure (Non-Continuous)	Edit...	
Baluster Placement	Edit...	
Baluster Offset	0.0	
Use Landing Height Adjustment	No	
Landing Height Adjustment	0.0	
Angled Joins	Add Vertical/Horizontal Segments	
Tangent Joins	Extend Rails to Meet	
Rail Connections	Trim	
Top Rail		

12. Set the Baluster Family to **Baluster –Square: 25 mm**. Set the Host to **Top Rail Element**. Set the Top Offset to **50.0**. Set Distance from Previous to **200.00.**

	Name	Baluster Family	Base	Base offset	Top	Top offset	Dist. from previous	Offset
1	Pattern start	N/A	N/A	N/A	N/A	N/A	N/A	N/A
2	Regular balust	M_Baluster - Square : 25mm	Host	0.0	Top Rail Eleme	50.0	200	0.0
3	Pattern end	N/A	N/A	N/A	N/A	N/A	0.0	N/A

13. Set the Start Post to **None**. Set the Corner Post to **None**. Set the End Post to **None**. Click **OK**.

Posts

	Name	Baluster Family	Base	Base offset	Top	Top offset	Space	Offset
1	Start Post	None	Host	-247.0	Top Rail Ele	0.0	0.0	0.0
2	Corner Post	None	Host	-247.0	Top Rail Ele	0.0	0.0	0.0
3	End Post	None	Host	-247.0	Top Rail Ele	0.0	0.0	0.0

14.

Rail Connections   mm

Top Rail	
Use Top Rail	☑
Height	800.0
Type	Rectangular - 50x50mm

Set the Top Rail Height to **800.00**. Click **OK**.

15.

Inspect the railing.

16.

Global Parameters   Transfer Project Standards   Purge Unused   Project Units

Settings

Select the **Manage** tab.
Select **Global Parameters**.

17. ✏️ 📄 📄 ↑ ↓ A↓ A↑

Select **New Parameter**.

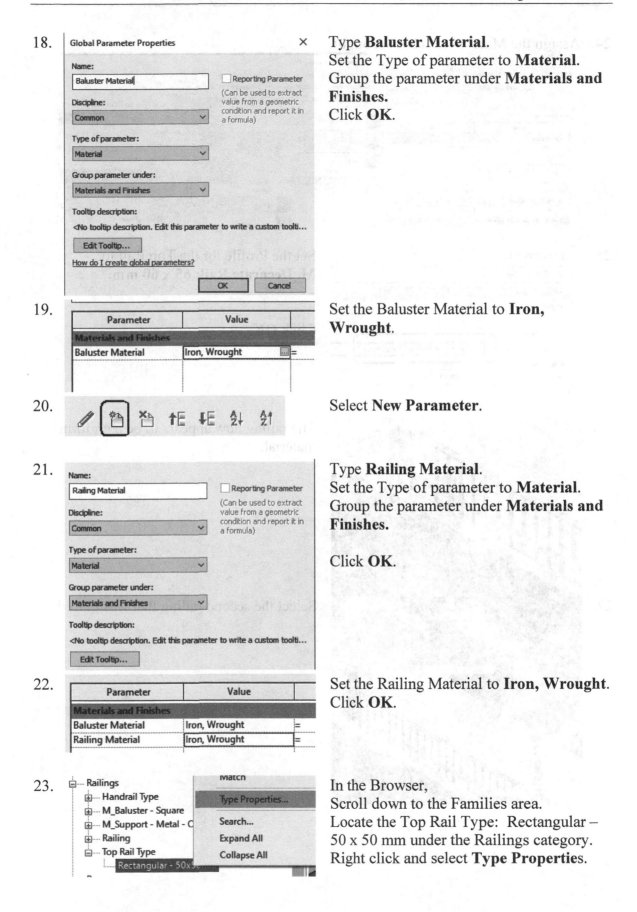

18. Type **Baluster Material**.
Set the Type of parameter to **Material**.
Group the parameter under **Materials and Finishes.**
Click **OK**.

19. Set the Baluster Material to **Iron, Wrought**.

20. Select **New Parameter**.

21. Type **Railing Material**.
Set the Type of parameter to **Material**.
Group the parameter under **Materials and Finishes.**

Click **OK**.

22. Set the Railing Material to **Iron, Wrought**.
Click **OK**.

23. In the Browser,
Scroll down to the Families area.
Locate the Top Rail Type: Rectangular –
50 x 50 mm under the Railings category.
Right click and select **Type Properties**.

24. Assign the Material to **Iron, Wrought**.

25.

Set the Profile for the Top Rail to: **M_Decorate Rail: 65 x 60 mm.**

Click **OK**.

26. The railing now appears to be a uniform material.

27. Select the second railing that was placed.

28. 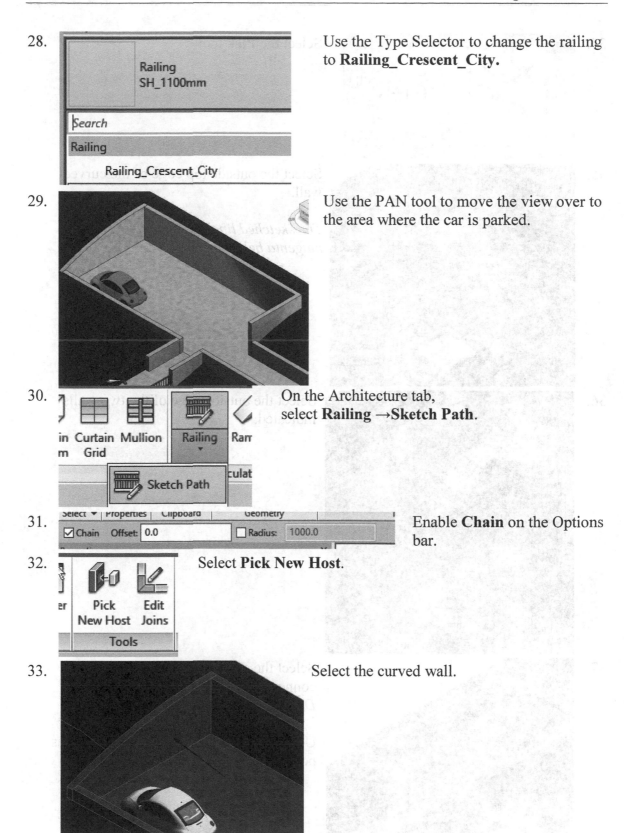 Use the Type Selector to change the railing to **Railing_Crescent_City.**

Railing
SH_1100mm

Search

Railing

Railing_Crescent_City

29. Use the PAN tool to move the view over to the area where the car is parked.

30. On the Architecture tab,
select **Railing →Sketch Path.**

in Curtain Mullion   Railing   Ram
m   Grid

Sketch Path

31. Enable **Chain** on the Options bar.

Select ▼ | Properties | Clipboard   Geometry

☑ Chain   Offset: 0.0   ☐ Radius: 1000.0

32. Select **Pick New Host.**

er   Pick   Edit
     New Host Joins

Tools

33. Select the curved wall.

34. 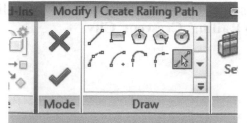 Select the **Pick** tool.

35.  Select the outside top edge of the curved wall.

The sketched line will be displayed in magenta below the wall.

36.  Select the outside edge of the two walls indicated.

37. 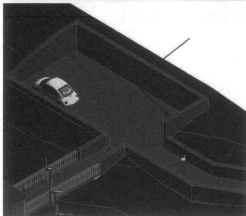 Select the outside edge of the remaining connecting walls.
Do not select the lower walls.

Use the TRIM tool to create a continuous polyline.

38.

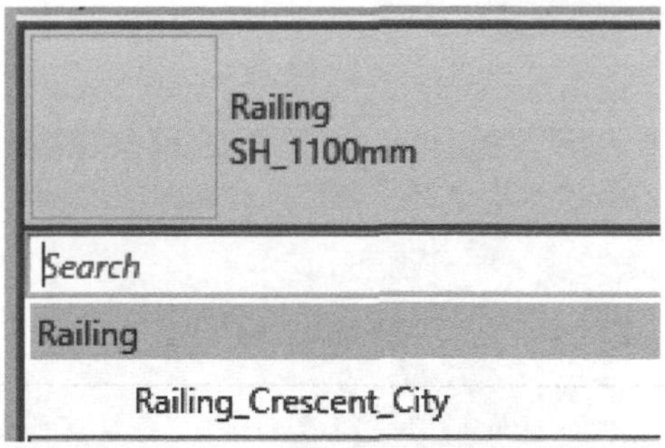

Railing
SH_1100mm

Search

Railing

Railing_Crescent_City

Set the Railing Type to
Railing_Crescent_City.

39.  Click the green check to finish.

Mode

40.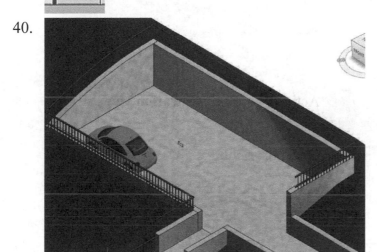

The railing is placed.

Save as *ex1-14.rvt*.

Exercise 1-15

Modify a Railing

Drawing Name: **railing.rvt**
Estimated Time to Completion: 15 Minutes

Scope
Modify an existing railing.

Solution

1. Sections (Building Section)
   Stair Elevation

   Activate the **Stair Elevation** view.

2.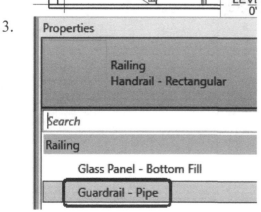

   Select the railing.

3.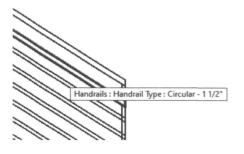

   Use the Type Selector to change the railing type to **Guardrail – Pipe**.

4. Hover your mouse over the second from the top railing.

   Click the TAB key to cycle through selections until the handrail highlights.

   Click to select the handrail.

5. 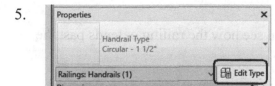 Select **Edit Type** from the Properties palette.

6. 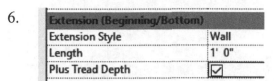 Under Extension (Beginning/Bottom):
Set the Extension Style to **Wall**.
Set the Length to **1'-0"**.
Enable **Plus Tread Depth**.

7. 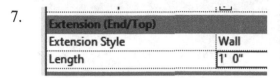 Under Extension (End/Top):
Set the Extension Style to **Wall**.
Set the Length to **1'-0"**.

To preview, click **Apply**.

Try changing the extension style to Post and then Floor to see how it looks.

The railing will adjust to the new settings.

Click **OK**.

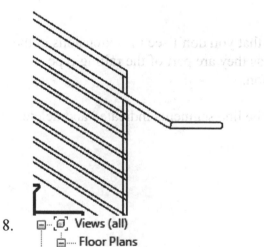

8. Activate the **LEVEL 1** floor plan view.

9.

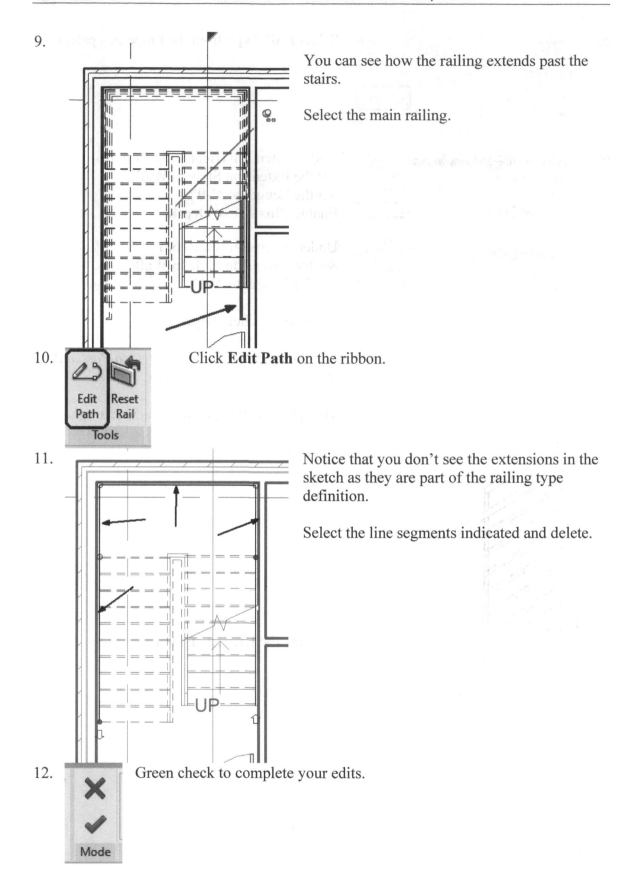

You can see how the railing extends past the stairs.

Select the main railing.

10. Click **Edit Path** on the ribbon.

11. Notice that you don't see the extensions in the sketch as they are part of the railing type definition.

Select the line segments indicated and delete.

12. Green check to complete your edits.

13.

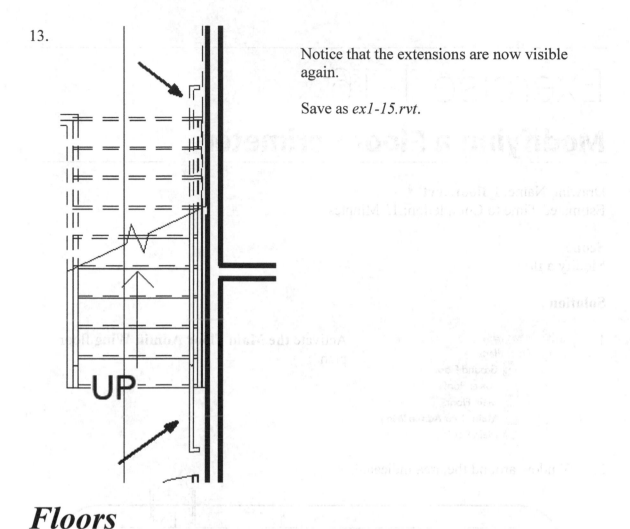

Notice that the extensions are now visible again.

Save as *ex1-15.rvt*.

Floors

Floors are system families. They have layers, just like walls. You create floors by defining their boundaries, either by picking walls or using drawing tools. Floors are placed relative to levels. You do not need walls or a building pad to place a floor element.

Typically, you sketch a floor in a plan view, although you can use a 3D view if the work plane of the 3D view is set to the work plane of a plan view.

Floors are offset downward from the level on which they are sketched.

Exercise 1-16

Modifying a Floor Perimeter

Drawing Name: **i_floors.rvt**
Estimated Time to Completion: 15 Minutes

Scope
Modify a floor.

Solution

1. 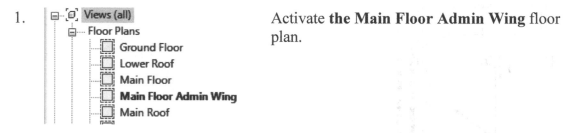 Activate **the Main Floor Admin Wing** floor plan.

2. Window around the area indicated.

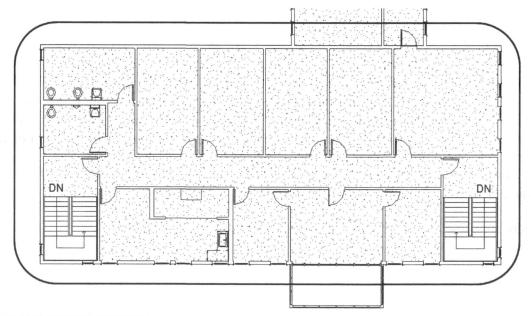

3. Select the **Filter** tool located in the lower right corner of the window.

4.

Category:	Count:
☐ <Room Separation>	1
☐ Casework	5
☐ Curtain Panels	12
☐ Curtain Wall Grids	5
☐ Curtain Wall Mullions	16
☐ Doors	15
☑ Floors	1
☐ Lines (Lines)	16
☐ Plumbing Fixtures	7
☐ Railings	4
☐ Rooms	13
☐ Stairs	2
☐ Walls	27
☐ Windows	18

Total Selected Items: 1

Select **Check None** to disable all the checks.

Then check only the **Floors**.

Click **OK**.

5.

Dimensions	
Slope	
Perimeter	254' 8"
Area	3254.30 SF
Volume	3457.69 CF
Thickness	1' 0 3/4"

In the Properties pane,
Note that the floor has a perimeter of 254' 8".

6.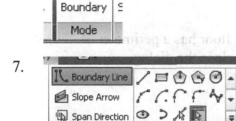

Select **Edit Boundary** under the Mode panel.

7.

Select **Pick Walls** mode under the Draw panel.

8. Offset: 0' 0"   ☐ Extend into wall (to core)

Uncheck the Extend into wall (to core) option.

9.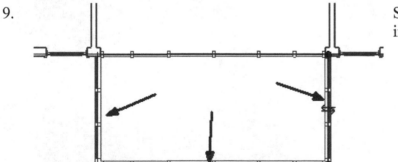

Select the three walls indicated.

10.

Select the **Trim** tool on the Modify panel from the tab.

11. 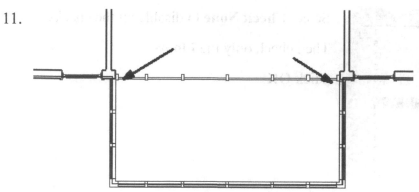 Trim the two corners indicated so that there is no wall in the section between the two vertical walls on the upper ends.

12. Select the **Green Check** to **Finish Floor** from the tab.

13. Attaching to floor ×

Would you like walls that go up to this floor's level to attach to its bottom?

☐ Do not show me this message again   Attach   Don't attach

Select **Don't attach**.

14.

Dimensions	
Slope	
Perimeter	270' 8"
Area	3390.71 SF
Volume	3602.63 CF
Thickness	1' 0 3/4"

Note that the floor has a perimeter of 270′ 8″. Click **OK** to close the dialog.

15. Close without saving.

Exercise 1-17
Modifying Floor Properties

Drawing Name: **floors_i.rvt**
Estimated Time to Completion: 5 Minutes

Scope
Modify a floor using Type Properties
Determine the floor's Elevation at Bottom

Solution

1. 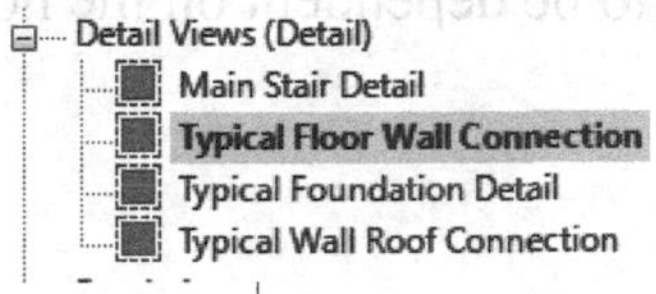 Activate the **Typical Floor Wall Connection** Detail view.

2. 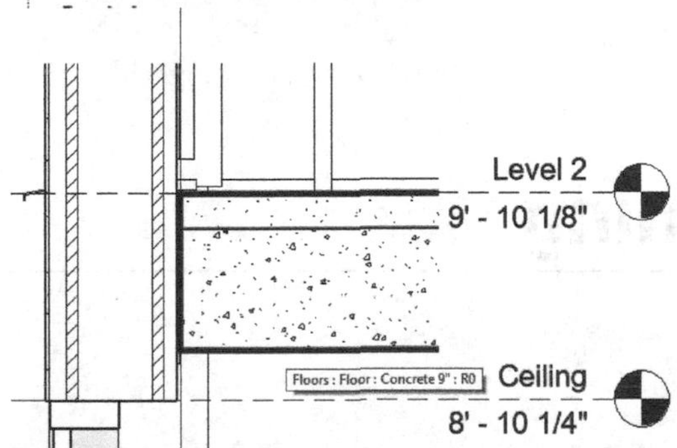 Select the floor.

   If you hover over an element, you will see some text providing element information.

3. 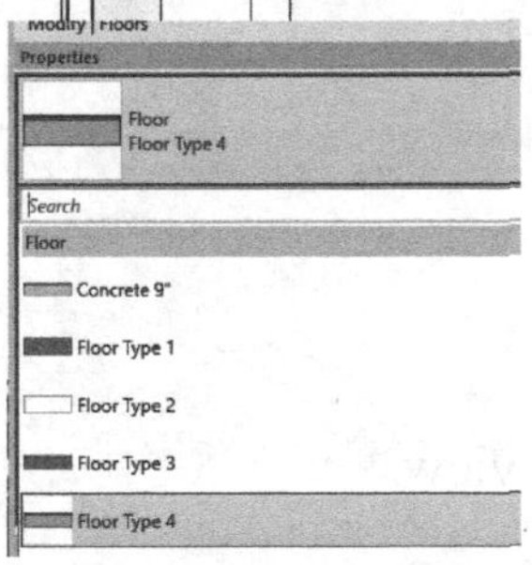 Using Type Properties change the floor to Floor Type 4.

4. 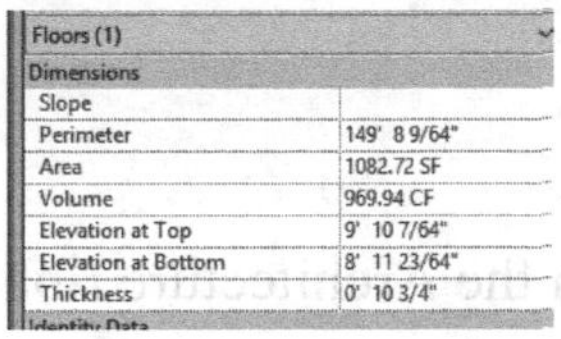 What is the Elevation at Bottom of the floor?

   It should read 8' 11 23/64"

Vertical Openings

The three methods for creating a vertical opening are:

- Shaft
- Vertical
- Boundary Edit

You should be familiar with the different methods of creating vertical openings, such as shafts and vertical as well as by modifying floor or ceiling boundaries. There may be a question to determine if you understand best practices and when which tool should be applied. In general, shafts are used when the openings go through more than one level/floor/ceiling. Shafts are independent elements. Vertical openings are hosted by floors, ceilings, or slabs. Vertical openings only create an opening on the selected level. Boundary edit can also be used when you want the opening to be dependent on the host and specific to a level.

Exercise 1-18

Place a Vertical Opening

Drawing Name: **floor_openings.rvt**
Estimated Time to Completion: 20 Minutes

Scope
Modify a floor using Shaft
Modify a floor using Vertical Opening

Solution

1. 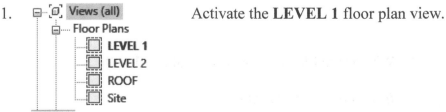 Activate the **LEVEL 1** floor plan view.

2. Select the **Shaft** tool from the Opening panel on the Architecture ribbon.

3.

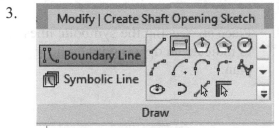

Enable **Boundary Line**.

Select the **Rectangle** tool from the Draw panel.

4.

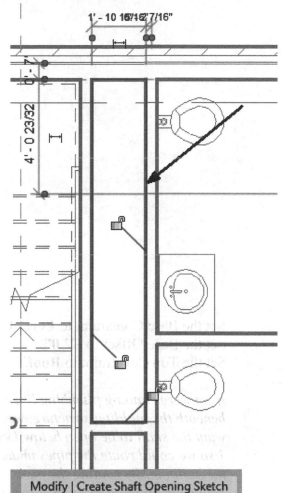

Place the rectangle flush to the finish face of the walls behind the top bathroom area.

This will be a shaft used for routing plumbing lines. We are using a shaft instead of an opening to ensure that the plumbing can travel the entire vertical distance from Level 1 to the roof.

5.

Enable **Symbolic Line**.

Select the **Line** tool from the Draw panel.

6.

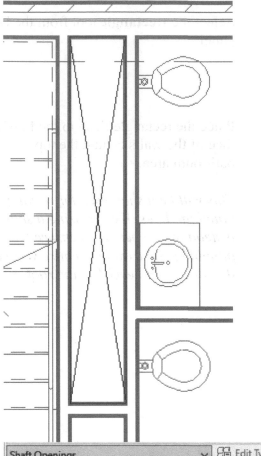

Place an X using the symbolic lines to indicate an opening in the floor plan.

7.

Shaft Openings		⊟📋 Edit Type
Constraints		⊗
Base Constraint	LEVEL 1	
Base Offset	-2' 0"	
Top Constraint	Up to level: ROOF	
Unconnected Height	26' 0"	
Top Offset	0' 0"	
Phasing		⊗
Phase Created	New Construction	
Phase Demolished	None	

Set the Base Constraint to **Level 1**.
Set the Base Offset to **-2' 0"**.
Set the Top Constraint to **Roof.**

If we were running plumbing lines beneath the building structure, we want the shaft to be open below Level 1 so we could route the pipes under the building. By routing the pipes below the building, we can connect to the city water supply as well as to any sewer or septic lines.

8.

✕	⎪Ӏ Boundary Line
✓	🗐 Symbolic Line
Mode	

Green check to complete the shaft.

9.   ⊟ Sections (Building Section)
        🔲 **Floor Opening**
        ___

Switch to the **Floor Opening** Section view.

10.

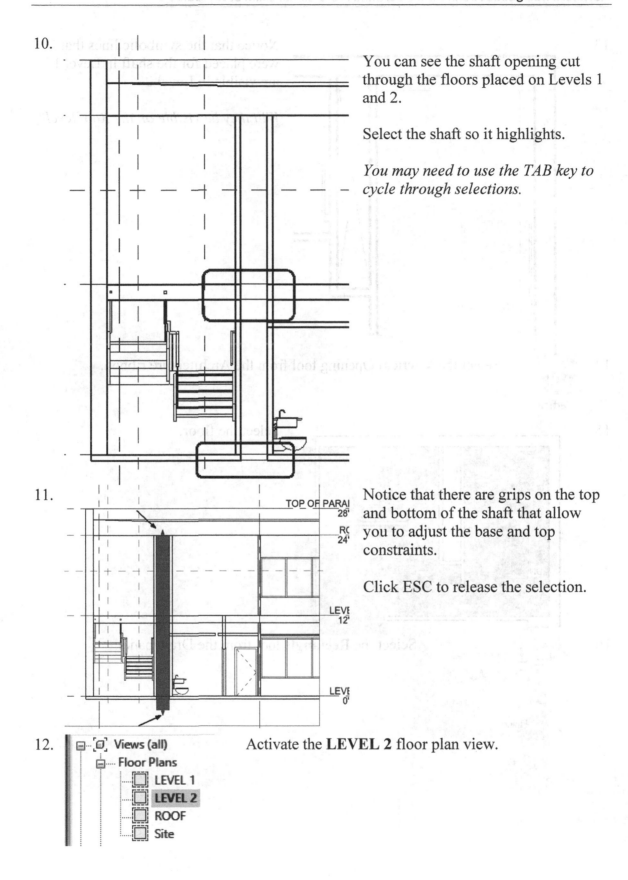

You can see the shaft opening cut through the floors placed on Levels 1 and 2.

Select the shaft so it highlights.

You may need to use the TAB key to cycle through selections.

11.

Notice that there are grips on the top and bottom of the shaft that allow you to adjust the base and top constraints.

Click ESC to release the selection.

12.

Activate the **LEVEL 2** floor plan view.

13.

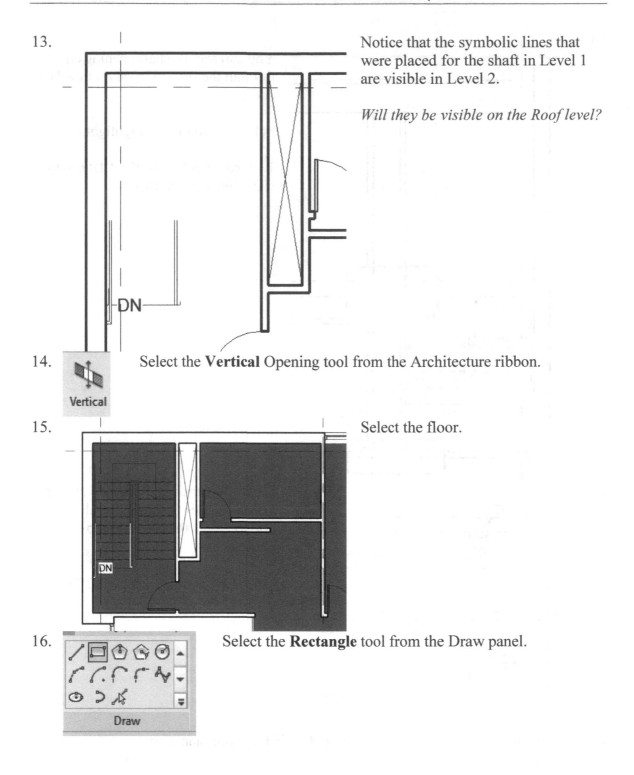

Notice that the symbolic lines that were placed for the shaft in Level 1 are visible in Level 2.

Will they be visible on the Roof level?

14. Select the **Vertical** Opening tool from the Architecture ribbon.

15. Select the floor.

16. Select the **Rectangle** tool from the Draw panel.

17.

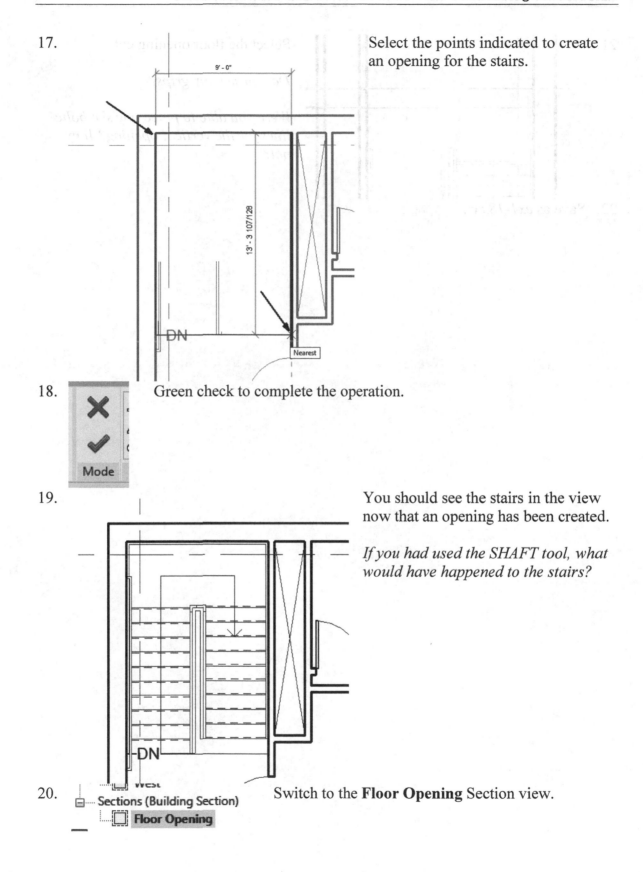

Select the points indicated to create an opening for the stairs.

18. Green check to complete the operation.

Mode

19. You should see the stairs in the view now that an opening has been created.

If you had used the SHAFT tool, what would have happened to the stairs?

20. Switch to the **Floor Opening** Section view.

Sections (Building Section)
    Floor Opening

21. 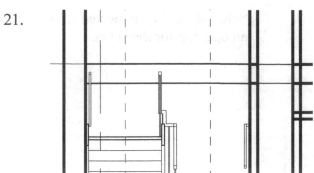   Select the floor opening cut.

Do you see any grips?

Were you able to place any symbolic lines for the vertical opening? Why not?

22. Save as *ex1-18.rvt*.

Placing Columns

Although structural columns share many of the same properties as architectural columns, structural columns have additional properties defined by their configuration and industry standards which provide different behaviors.

Structural elements such as beams, braces, and isolated foundations join to structural columns; they do not join to architectural columns.

In addition, structural columns have an analytical model that is used for data exchange.

Typically, drawings or models received from an architect may contain a grid and architectural columns. You create structural columns by manually placing each column or by using the At Grids tool to add a column to selected grid intersections. In most cases, it is helpful to set up a grid before adding structural columns, as they snap to grid lines.

Structural columns can be created in plan or 3D views.

You can place a slanted structural column, but you cannot place a slanted architectural column.

You can use architectural columns to model column box-outs around structural columns and for decorative applications. Architectural columns inherit the material of other elements to which they are joined.

By default, structural columns extend downward when you place them. For example, if you place a structural column on Level 2 of your project, it will extend from Level 2 down to Level 1.

Architectural columns behave in the opposite manner. If you place one on Level 1 of your project, it will extend up to Level 2.

To extend a structural column up to the next level

1. On the Structural tab of the Design bar, click Structural Column.

2. On the Options bar, select Height from the drop-down menu and then select a Level from the drop-down menu. If you are placing the structural column on Level 1, select Level 2 for the height. The structural column will extend from Level 1 to Level 2.

Exercise 1-19
Placing Columns

Drawing Name: **columns.rvt**
Estimated Time to Completion: 30 Minutes

Scope
Place structural columns at grid lines.
Place an architectural column.
Insert a structural column into an architectural column.

Solution

1.
  Views (all)
  Floor Plans
  Level 1
  Level 2
  Site

  Activate the **Level 1** floor plan.

2.
  Column Roof Ceiling

  Structural Column

  Select **Structural Column** from the drop-down on the Architecture tab.

3.
  W Shapes-Column
  W10X33
  W10X49

  Select **W10X33** using the Type Selector.

4.
  Modify   Vie

  Height ∨ Level 2 ∨ 9' 0"

  On the Options bar:
  Specify Height.
  Specify Level 2.

  Remember structural columns extend DOWN not up.

5.

  Left click on **At Grids**.

6.

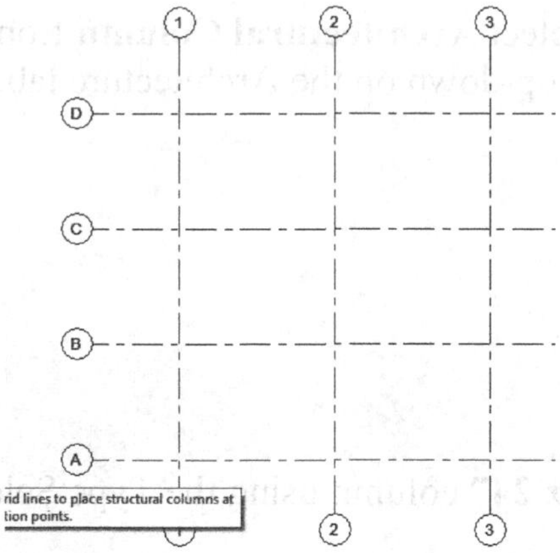

Hold down the CTL key.

Select Grids 1,2, 3, A,B,C, and D.

Finish

Select Finish on the tab.

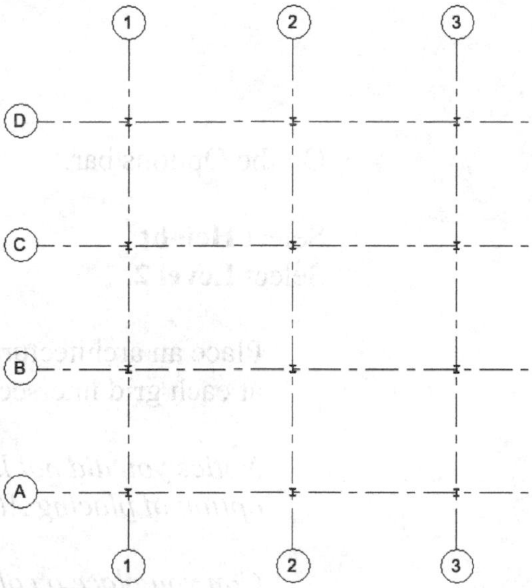

A structural column was placed at each grid intersection.

Click Cancel to exit the command.

7. Elevations (Building Elevation)

☐ **East**
☐ North
☐ South
☐ West

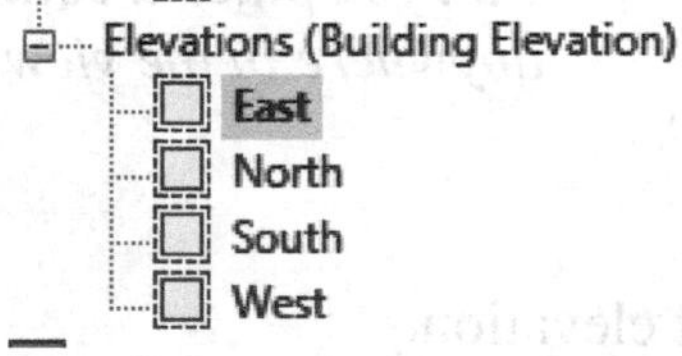

Activate the East elevation.

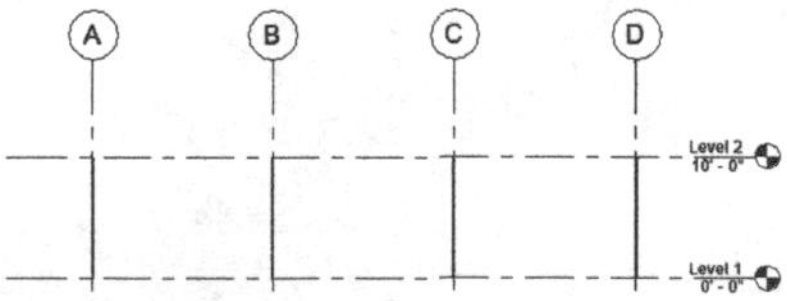

Notice that the structural columns were placed between Level 1 and Level 2.

8. Views (all)

Floor Plans

☐ **Level 1**
☐ Level 2
☐ Site

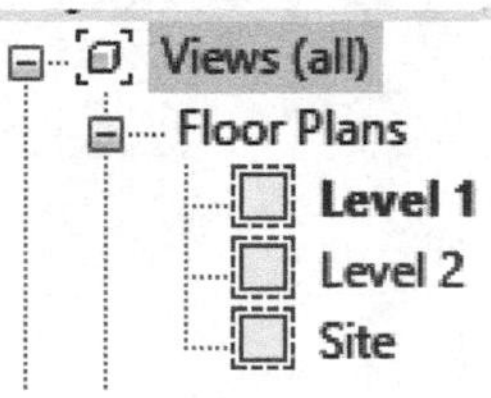

Activate the **Level 1** floor plan.

9. 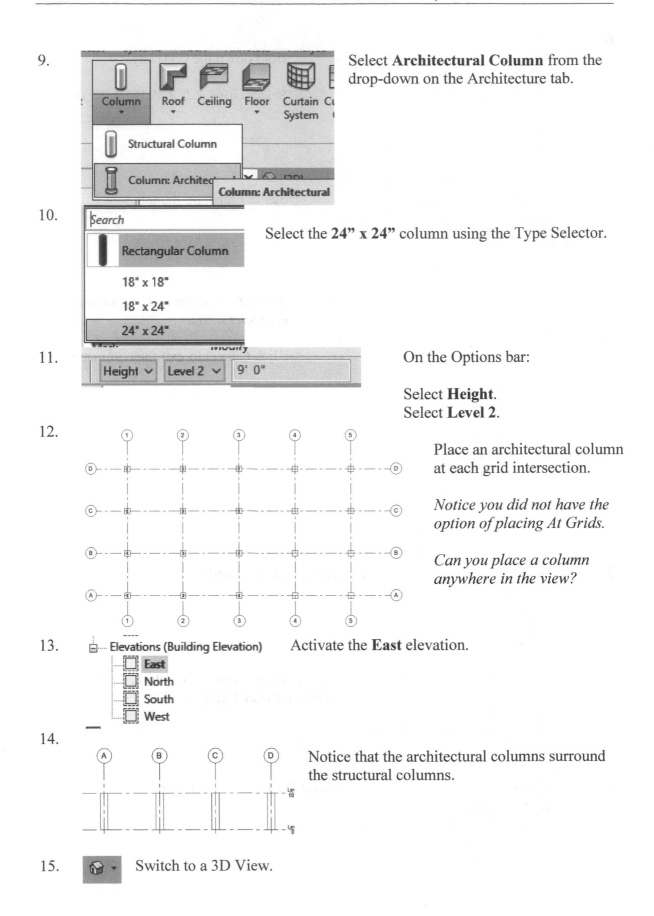 Select **Architectural Column** from the drop-down on the Architecture tab.

10. Select the **24" x 24"** column using the Type Selector.

11. On the Options bar:

Select **Height**.
Select **Level 2**.

12. Place an architectural column at each grid intersection.

Notice you did not have the option of placing At Grids.

Can you place a column anywhere in the view?

13. Activate the **East** elevation.

14. Notice that the architectural columns surround the structural columns.

15. Switch to a 3D View.

16.

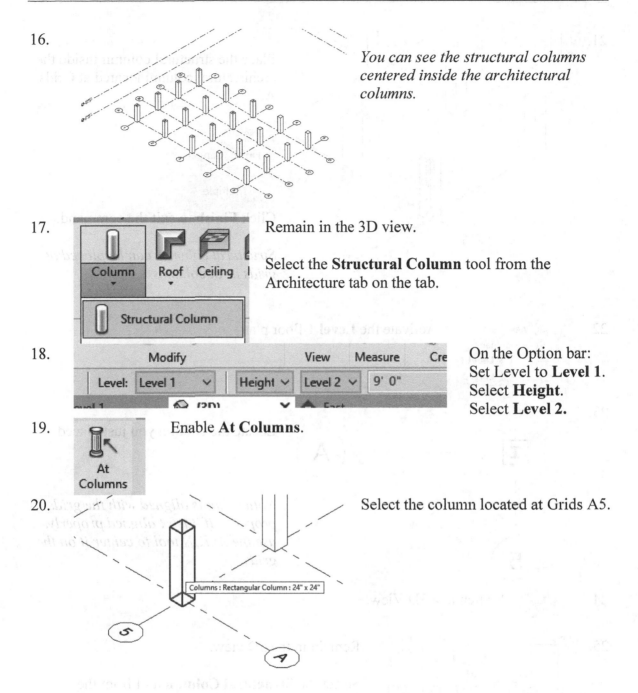

You can see the structural columns centered inside the architectural columns.

17. Remain in the 3D view.

Select the **Structural Column** tool from the Architecture tab on the tab.

18. On the Option bar:
Set Level to **Level 1**.
Select **Height**.
Select **Level 2.**

19. Enable **At Columns**.

20. Select the column located at Grids A5.

21.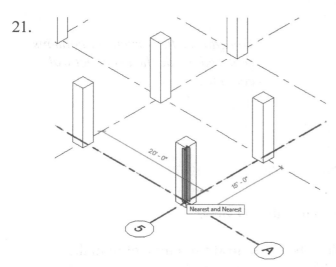

Place the structural column inside the architectural column located at Grids A5.

Click **Finish** to exit the command.

Structural columns can be placed at grids or at columns.

22.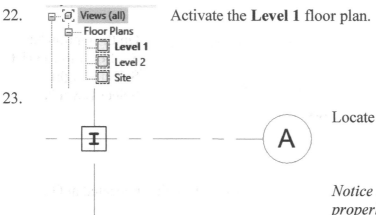

Activate the **Level 1** floor plan.

23.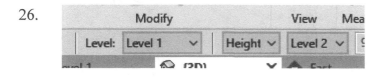

Locate the column you just placed.

Notice that is aligned with the grids properly. If it isn't aligned properly, use the ALIGN tool to center it on the grids.

24. Switch to a 3D View.

25.

Remain in the 3D view.

Select the **Structural Column** tool from the Architecture tab on the tab.

26.

On the Option bar:
Set Level to **Level 1**.
Select **Height**.
Select **Level 2.**

27. 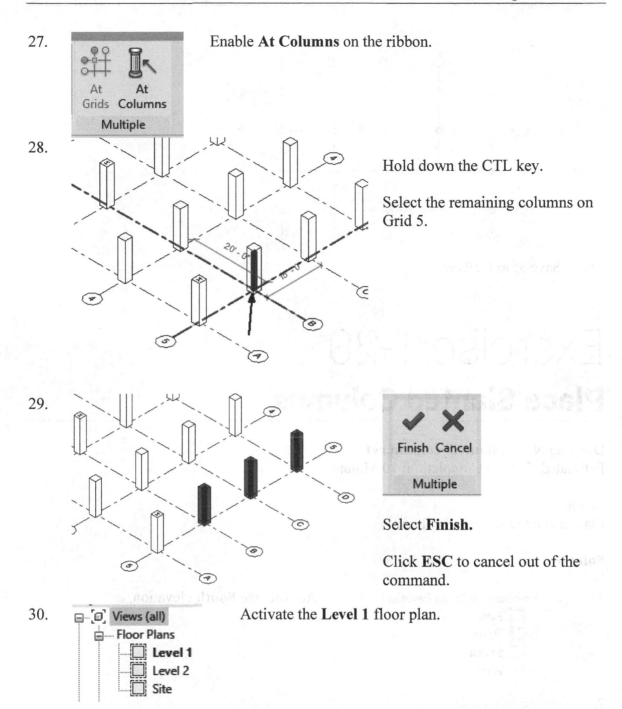 Enable **At Columns** on the ribbon.

28. Hold down the CTL key.

Select the remaining columns on Grid 5.

29. Select **Finish.**

Click **ESC** to cancel out of the command.

30. Activate the **Level 1** floor plan.

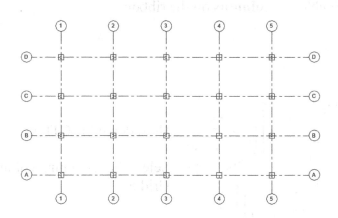

31.   Save as *ex1-19.rvt*.

Exercise 1-20
Place Slanted Columns

Drawing Name: **slanted_column.rvt**
Estimated Time to Completion: 10 Minutes

Scope
Place a slanted structural column

Solution

1.   Elevations (Building Elevation)   Activate the **South** elevation.
         East
         North
         South
         West

2.   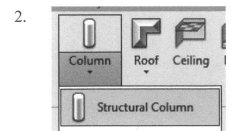   Select the **Structural Column** tool from the
                                      Architecture ribbon.

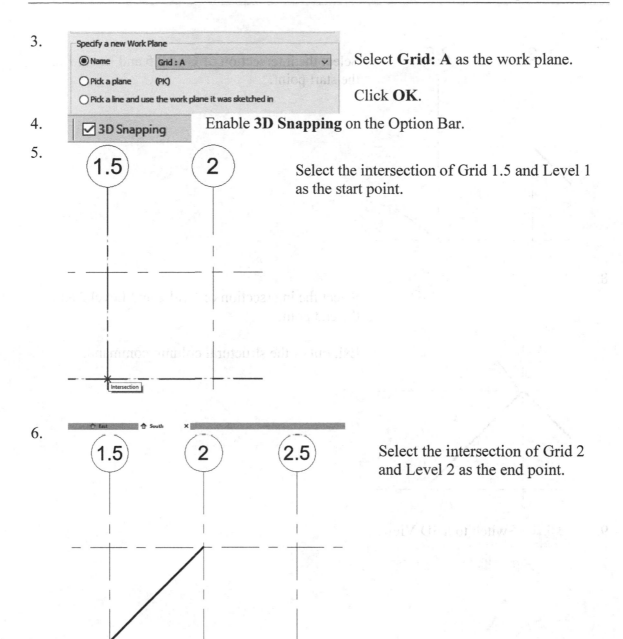

3. Select **Grid: A** as the work plane.

   Click **OK**.

4. Enable **3D Snapping** on the Option Bar.

5. Select the intersection of Grid 1.5 and Level 1 as the start point.

6. Select the intersection of Grid 2 and Level 2 as the end point.

7.

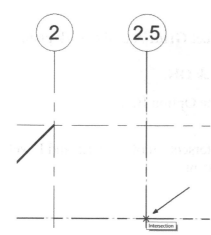

Select the intersection of Grid 2.5 and Level 1 as the start point.

8.

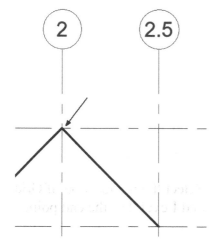

Select the intersection of Grid 2 and Level 2 as the end point.

Exit out of the structural column command.

9. 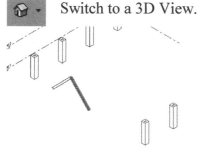 Switch to a 3D View.

10. Save as *ex1-20.rvt*.

Exercise 1-21

Columns and Materials

Drawing Name: **columns_materials.rvt**
Estimated Time to Completion: 20 Minutes

Scope
Explore how architectural and structural columns manage materials when
intersecting with a wall

Solution

1.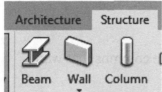
    Activate the **LEVEL 1 Framing
    Plan** floor plan.

2.
    Activate the Structure ribbon.

    Select the **Column** tool.

    *Note that the Structure ribbon only allows
    you to place structural columns while the
    Architecture ribbon allows you to place
    structural OR architectural columns.*

3.
    Select the **W10X33** column family
    using the Type Selector.

4. On the Option bar:
    Height is enabled.
    LEVEL 2 is selected.
    9'-0" is displayed as the default.

5.

Enable **Vertical Column**.

Enable **At Grids**.

6.

Use a crossing to select Grids 5 to Grids 1. Click a point near point 1 and then click a point near point 2.

7.

Hold down the CTL key and select Grid A.

8.

Click **Finish** to complete the command.

You don't see the columns that were placed because they are not in the view range.

9.

Switch to the LEVEL 2 floor plan.

10.

Locate the columns that were placed.

11.   Set the Detail Level of the view to **Fine**.

12.

Switch to the Architecture ribbon.

Select **Column: Architectural**.

13.

Set the Type to **24" x 24"**.

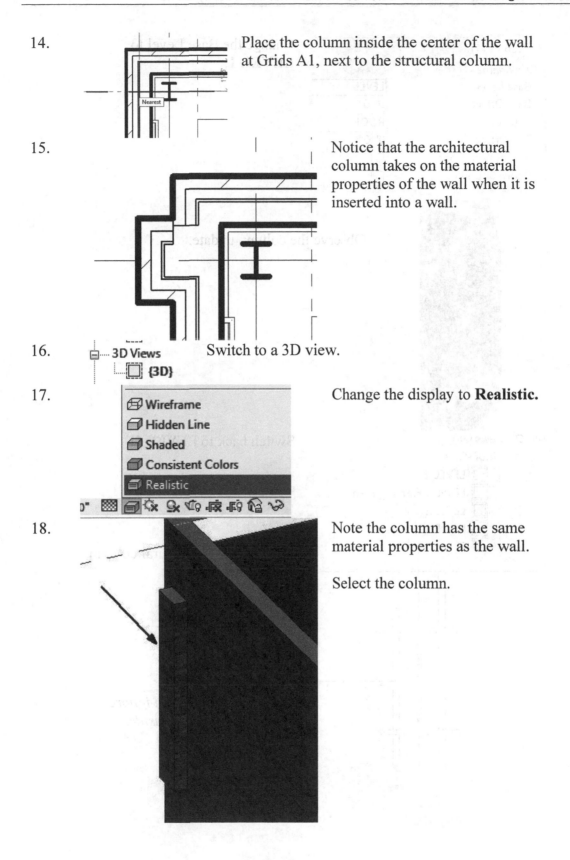

14. Place the column inside the center of the wall at Grids A1, next to the structural column.

15. Notice that the architectural column takes on the material properties of the wall when it is inserted into a wall.

16. Switch to a 3D view.

17. Change the display to **Realistic.**

18. Note the column has the same material properties as the wall.

Select the column.

19.

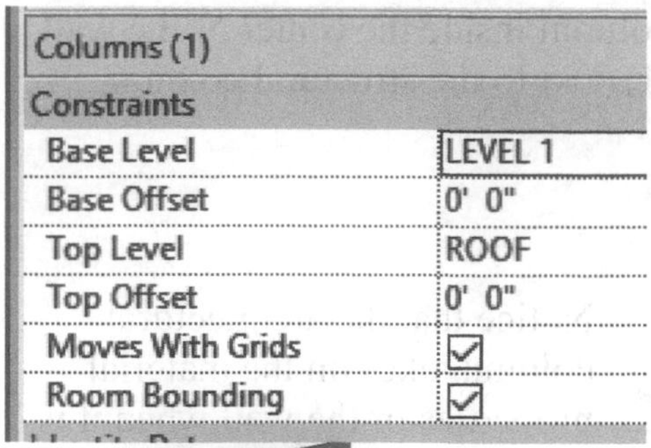

Columns (1)	
Constraints	
Base Level	LEVEL 1
Base Offset	0' 0"
Top Level	ROOF
Top Offset	0' 0"
Moves With Grids	☑
Room Bounding	☑

Change the Base Level to **LEVEL 1**.

20.

Observe the column update.

21.

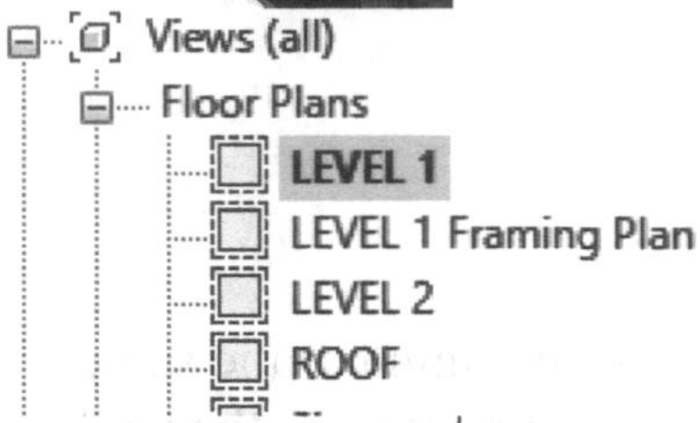

```
□ [◌] Views (all)
   □ Floor Plans
        ☐ LEVEL 1
        ☐ LEVEL 1 Framing Plan
        ☐ LEVEL 2
        ☐ ROOF
```

Switch back to LEVEL 1.

22.

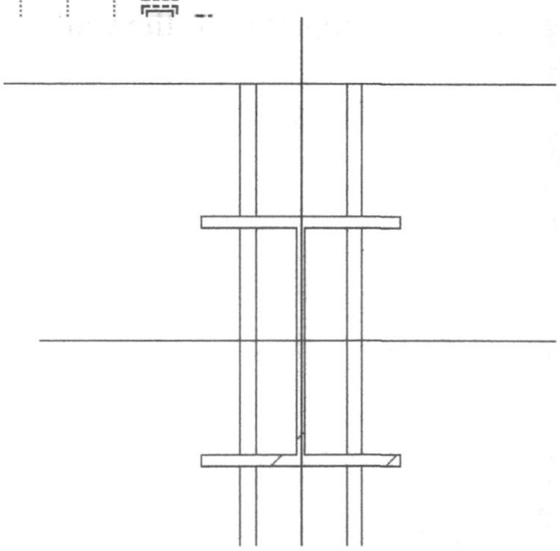

Zoom into the structural column located at A3.

Enable **Thin Lines**.

The structural column ignores the wall it is placed inside.

23.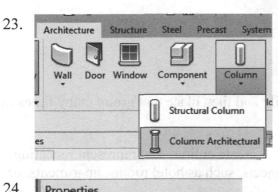

Switch to the Architecture ribbon.

Select **Column: Architectural**.

24.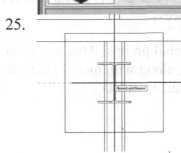

Set the Type to **24" x 24"**.

25.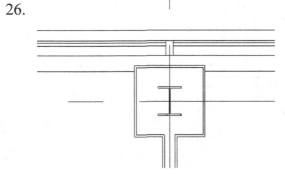

Place the column inside the center of the wall at Grids A3, on top of the structural column.

26.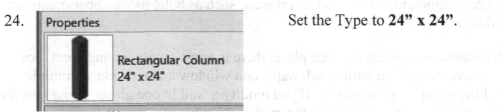

Note that the architectural column joins with the wall and adopts the material properties of the wall.

Save as *ex1-21.rvt*.

Groups

You can group elements in a project or family and then place that group many times in a project or family.

Grouping elements is useful when you need to create entities that represent repeating layouts or are common to many building projects, such as hotel rooms, apartments, or repeating floors.

With each instance of a group that you place, there is associativity among them. For example, you create a group with a bed, walls, and window and then place multiple instances of the group in your project. If you modify a wall in one group, it changes for all instances of that group, simplifying the modification process.

You can create model groups, detail groups, and attached detail groups. Attached detail groups contain view-specific elements, such as annotations, along with model elements. You cannot create a group that includes both model and detail elements.

Exercise 1-22

Model Groups

Drawing Name: **model group.rvt**
Estimated Time to Completion: 15 Minutes

Scope
Create a model group.
Place a model group.
Edit a model group.

Solution

1.

Activate the **GROUND FLOOR – Exam Rooms** floor plan.

2.

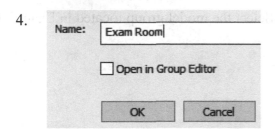

Exam room 2-4 has a patient bed, a chair, a desk, a rolling chair and some casework.

We are going to create a model group of all the furniture in the exam room.

Select all the items in the room except for the casework.

3.

Select the Create Group tool from the ribbon.

4.

Name: Exam Room

☐ Open in Group Editor

OK   Cancel

Type **Exam Room** for the name.

Click **OK**.

5.

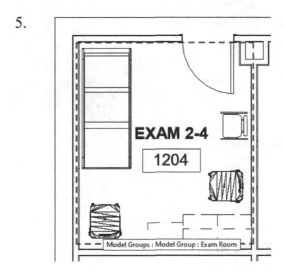

If you hover over the furniture, you should see the Model Group displayed.

Click to select the model group.

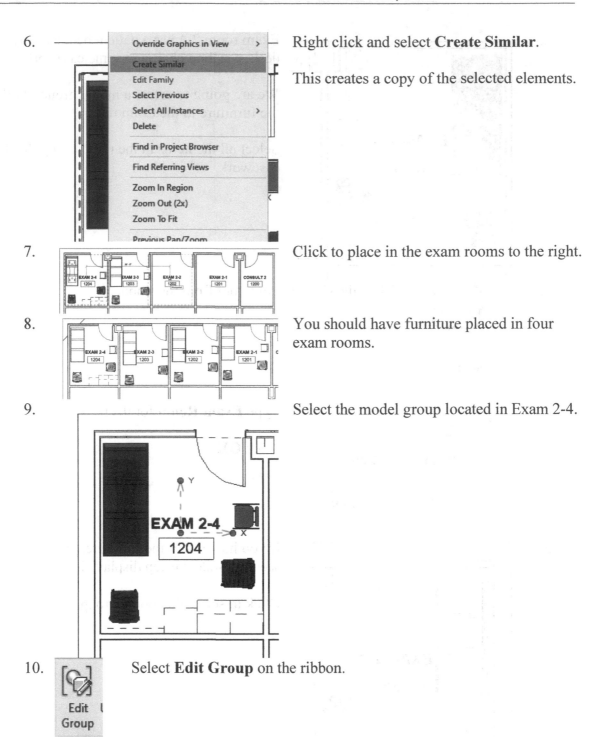

6. Right click and select **Create Similar**.

This creates a copy of the selected elements.

7. Click to place in the exam rooms to the right.

8. You should have furniture placed in four exam rooms.

9. Select the model group located in Exam 2-4.

10. Select **Edit Group** on the ribbon.

11. Select **Add.**

Notice that elements that are already part of the model group are grayed out.

12. 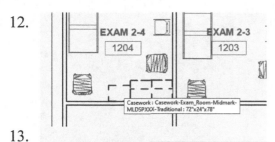 Select the casework in Exam 2-4.

Right click and select **Cancel**.

13. 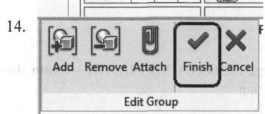 Adjust the position of the patient bed so it is below the column located in the upper right of the room.

14. 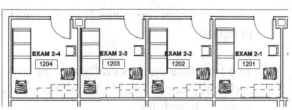 Click **Finish**.

15. Notice that all the model groups update.

Adjust the position of the model groups so they fit properly in each exam room.

16. Save as *ex1-22.rvt*

Exercise 1-23
Detail Groups

Drawing Name: **detail groups.rvt**
Estimated Time to Completion: 30 Minutes

Scope
Create an attached detail group.
Create a model group.
Modify a model group.
Mirror an attached detail group.

Solution

1. Working Ground Floor
    SITE PLAN
    GROUND FLOOR
    GROUND FLOOR -Exam Rooms

    Activate the **GROUND FLOOR – Exam Rooms** floor plan.

    The first exam room has furniture and casework installed.

    The furniture and specialty equipment (patient bed) have been tagged to make it easier to manage schedules.

    We want to create an attached detail group that includes the furniture, casework, and specialty equipment as well as the tags.

2.

    On the Architecture ribbon:

    Select **Model Group→Create Group**.

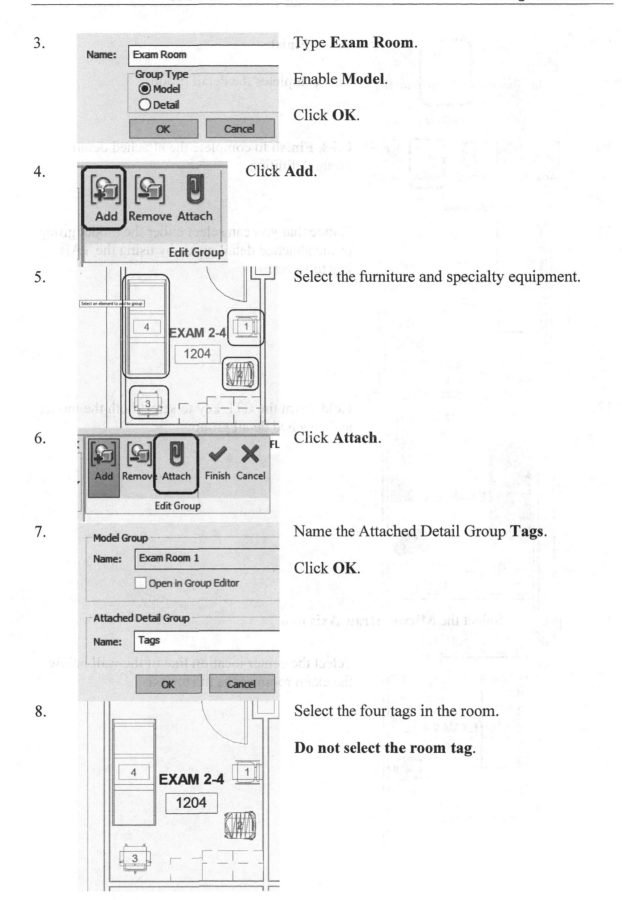

3.   Type **Exam Room**.

      Enable **Model**.

      Click **OK**.

4.   Click **Add**.

5.   Select the furniture and specialty equipment.

6.   Click **Attach**.

7.   Name the Attached Detail Group **Tags**.

      Click **OK**.

8.   Select the four tags in the room.

      Do not select the room tag.

9. 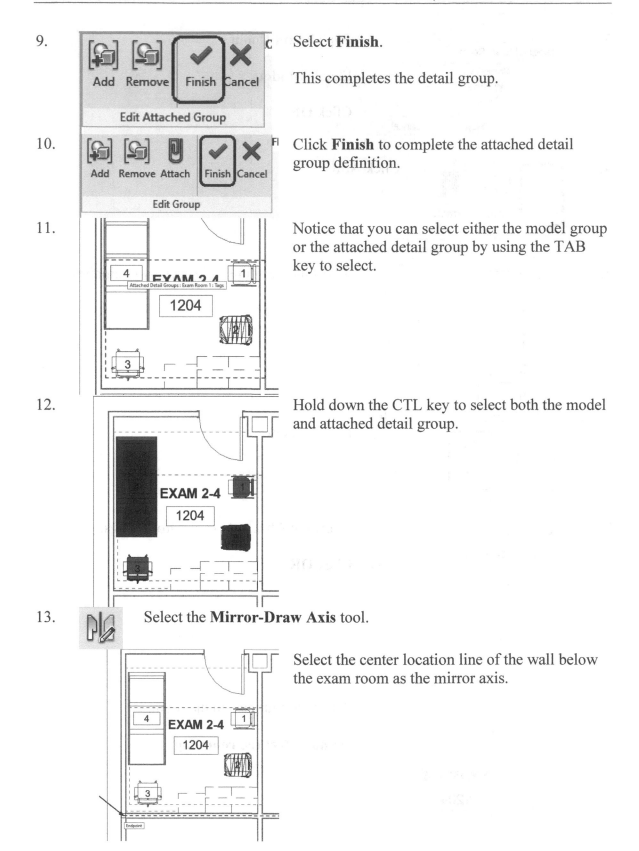 Select **Finish**.

This completes the detail group.

10. Click **Finish** to complete the attached detail group definition.

11. Notice that you can select either the model group or the attached detail group by using the TAB key to select.

12. Hold down the CTL key to select both the model and attached detail group.

13. Select the **Mirror-Draw Axis** tool.

Select the center location line of the wall below the exam room as the mirror axis.

14.

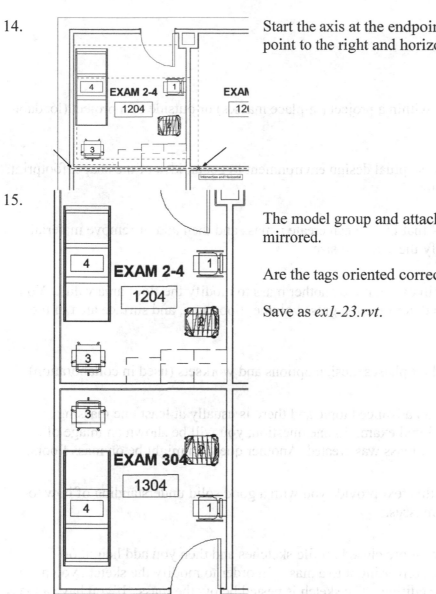

Start the axis at the endpoint and then select a point to the right and horizontal to the first point.

15.

The model group and attached detail group were mirrored.

Are the tags oriented correctly?

Save as *ex1-23.rvt*.

Masses

You can create masses within a project (in-place masses) or outside of a project (loadable mass families).

Masses are used in a conceptual design environment to get an idea of the shape, footprint, and volume of a building.

Masses can be nested – that is, you can create a mass and then add or remove material from the mass to modify the shape or size.

In a project, you can join one mass to another mass to modify the floor area value. You can create schedules to determine the gross volume, floor area, and surface area of the building model.

Masses can be affected by phases, design options and worksets (used in collaboration).

Masses are considered an advanced topic and there is usually at least one massing question on the professional exam. In one question, you will be shown an image of a mass and asked how that mass was created. Another question might be on mass floors.

The Mass exercises in this text provide you with a good solid understanding of how to create and manipulate masses.

Masses start with one or more closed profile sketches and then you add height or thickness to the sketch, converting it to a mass. In order to modify the sketch, you have to open the mass up for editing. The sketch is nested below the mass. If you have a mass which consists of more than one mass, such as a base mass, then a void/hole, you have to open the mass for editing, then select the mass element to be edited and open that feature. This can be confusing for students as they have to go down the nested elements to get to the one they want to change.

Exercise 1-24
Create a Mass Using Revolve

Drawing Name: **mass_revolve.rvt**
Estimated Time to Completion: 20 Minutes

Scope
Create a mass using revolve.

Solution

1. Open *mass_revolve.rvt.*

2.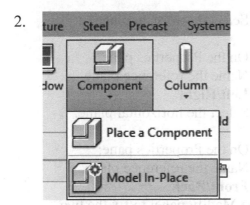
On the Architecture tab:

Select Model In-Place.

Model In-Place allows you to create a mass object on the fly. This is a system family.

3.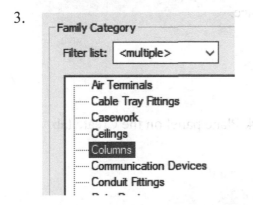
Select **Columns**.

Click **OK**.

Note that architectural columns have a separate category from structural columns.

4.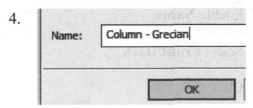
Type **Column – Grecian**.

Click **OK**.

5.    Select **Reference Plane** on the tab.

6. 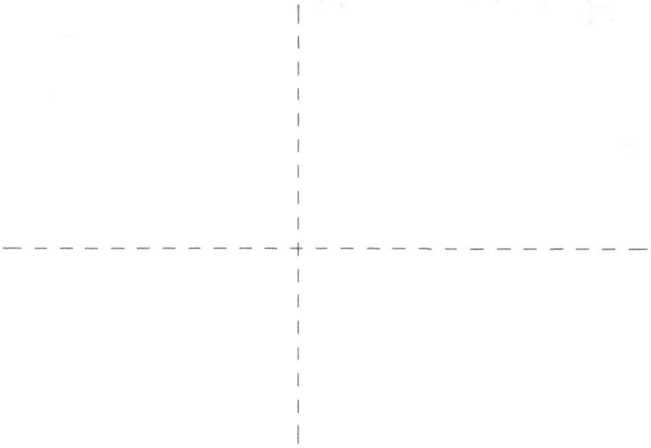   Draw a horizontal and a vertical plane.

7. 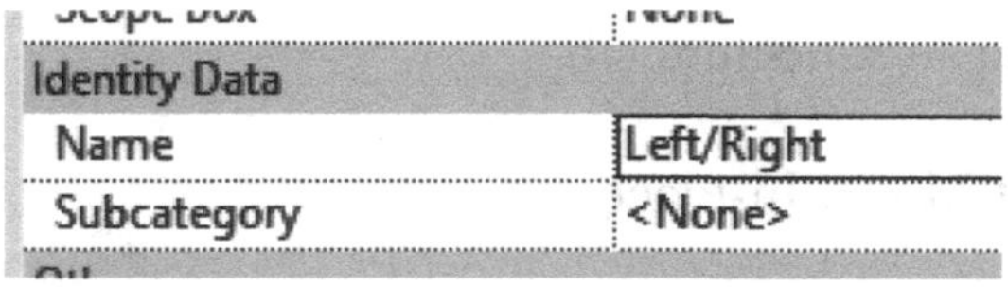   Select the vertical plane.

On the Properties pane:
Name the reference plane
Left/Right.

8. 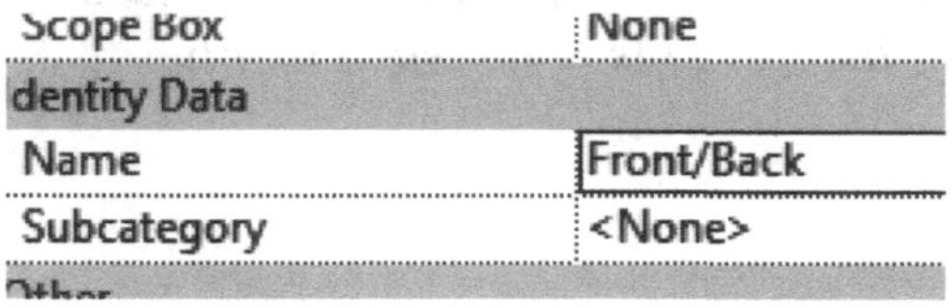   Select the horizontal plane.

On the Properties pane:
Name the reference plane
Front/Back.

9. 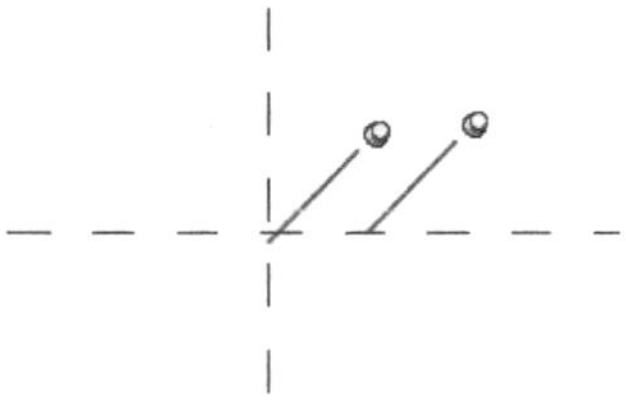   Use the **PIN** tool on the Modify panel to fix the two reference planes in place.

10. 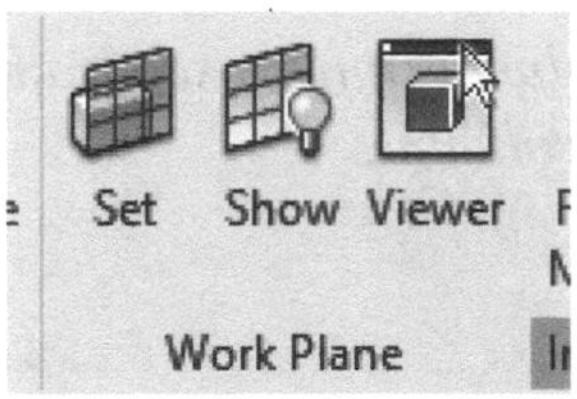 Select **Set** on the Work Plane panel on the Create tab.

11. 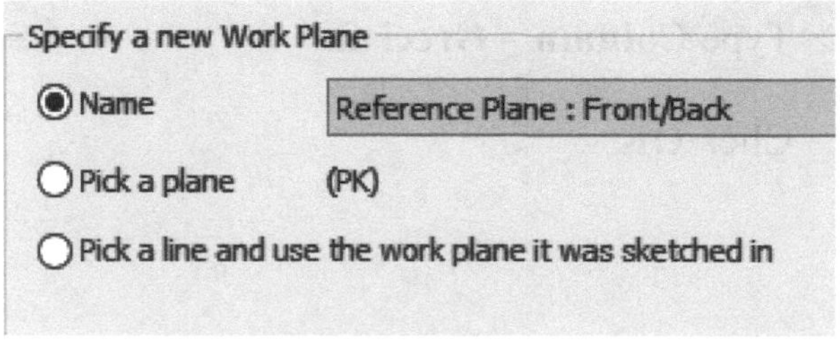 Enable **Name**.

Select **Front/Back**.

Click **OK**.

12.

| Elevation: North |
| Elevation: South |

Highlight **South**.

Click **Open View.**

13. Select **Revolve**.

14. Select **Import CAD** from the Insert tab on the ribbon.

15.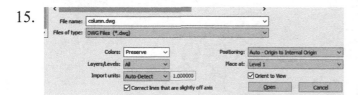
Select *column.dwg*.
Set colors to **Preserve**.
Set Layers to **All.**
Set Positioning to **Auto – Origin to Internal Origin**
Enable **Orient to View**
Enable **Correct lines that are slightly off axis**.
Click **Open**.

16.

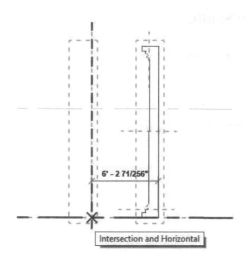

Use the **MOVE** tool to position the profile's right edge aligned with the vertical reference plane.

17.

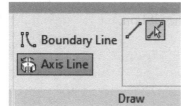

Highlight **Axis Line**.

Select the **Pick Line** tool.

18.

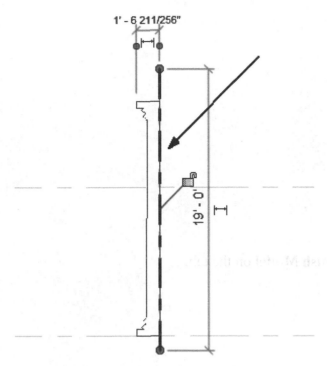

Select the vertical reference plane to place the axis line.

19.  Switch to a **3D** view.

20. In the Properties pane:

Materials and Finishes	
Material	Concrete, Cast-in-Place gray

Set the Material to **Concrete, Cast-in-Place gray.**

21.

Click the **Green Check** to complete the revolve.

22.

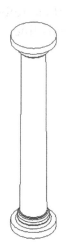

The column is created.

23.

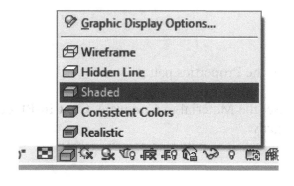

Select **Finish Model** on the tab.

24.

Change the display to **Shaded**.

If you insert this column into a wall how will that affect the materials that were applied?

25.

Save as *ex1-24.rvt*.

Exercise 1-25

Create a Mass Using Blend

Drawing Name: **mass_blend.rvt**
Estimated Time to Completion: 15 Minutes

Scope
Create a mass using blend.

Solution

1. Open *mass_blend.rvt*.

2. 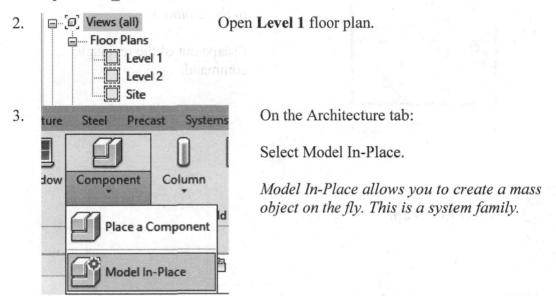 Open **Level 1** floor plan.

3. On the Architecture tab:

    Select Model In-Place.

    Model In-Place allows you to create a mass object on the fly. This is a system family.

4. 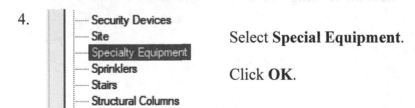 Select **Special Equipment**.

    Click **OK**.

5.  Type **Range Hood**.

    Click **OK**.

6. Select the **Blend** tool on the tab.

7.  Select the **Modify|Create Blend Base Boundary** tab on the tab.

Select the **rectangle** tool.

8.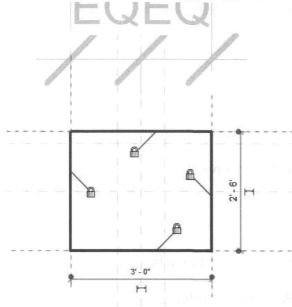

Draw the rectangle so it lies on top of the outside planes.

Enable the locks for each side of the rectangle.

This constrains the bottom profile to those planes.

Escape out of the rectangle command.

9. 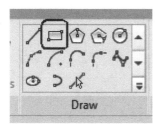 Select **Edit Top**.

10. Select the **rectangle** tool.

11. 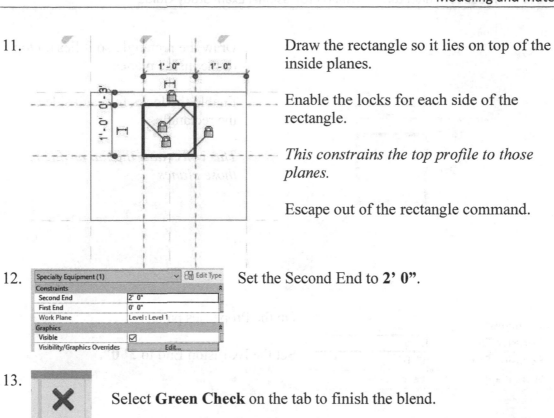 Draw the rectangle so it lies on top of the inside planes.

Enable the locks for each side of the rectangle.

This constrains the top profile to those planes.

Escape out of the rectangle command.

12. Set the Second End to **2' 0"**.

13. 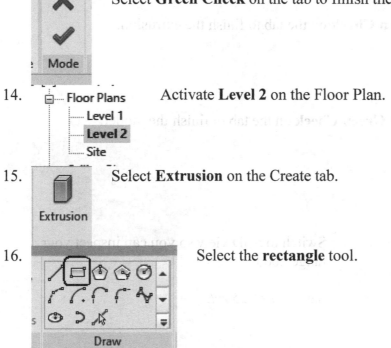 Select **Green Check** on the tab to finish the blend.

14. Floor Plans — Level 1 — **Level 2** — Site   Activate **Level 2** on the Floor Plan.

15. Extrusion   Select **Extrusion** on the Create tab.

16. Select the **rectangle** tool.

17.

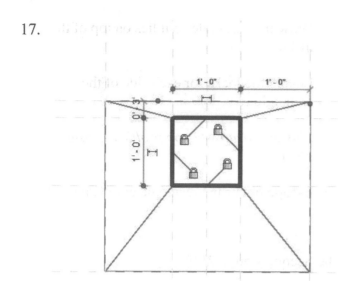

Draw the rectangle so it lies on top of the inside planes.

Enable the locks for each side of the rectangle.

This constrains the top profile to those planes.

18.

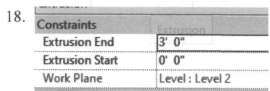

On the Properties pane:

Set the Extrusion End to **3' 0"**.

19.

Select **Green Check** on the tab to finish the extrusion.

20.

Select **Green Check** on the tab to finish the model.

21.

Switch to a 3D view so you can inspect your range hood.

Save as *ex1-25.rvt*.

Exercise 1-26

Create a Mass Using a Swept Blend

Drawing Name: **mass_swept_blend.rvt**
Estimated Time to Completion: 20 Minutes

Scope
Create a mass using a swept blend.

Solution

1. Open *mass_swept_blend.rvt*.

2. 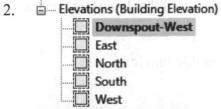 Open the **Downspout-West** elevation view.

3. 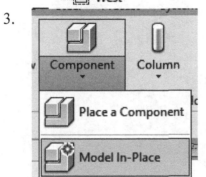 Select **Model In-Place** from the Architecture tab on the tab.

4. 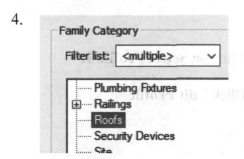 Highlight **Roofs**.

    Click **OK**.

5. Name: Downspout-End | Type **Downspout-End** in the Name field.

    Click **OK**.

6.  Select **Swept Blend** from the Create tab on the tab.

7.  Select **Sketch Path**.

8. 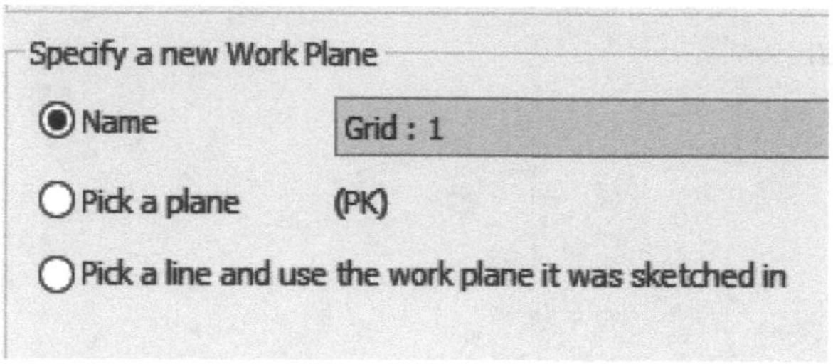 Enable **Name**.

Select **Grid: 1** from the drop-down list.

Click **OK**.

9. 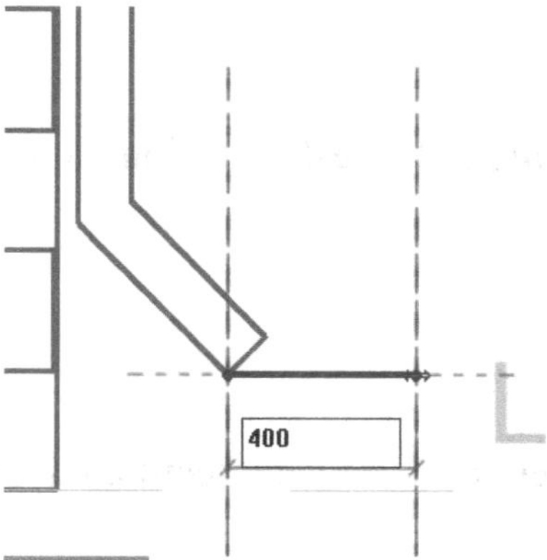 Select the lower end point of the downspout as the start point.

Draw a horizontal line 400 mm long.

10.  Select **Green Check** to complete the path.

11. 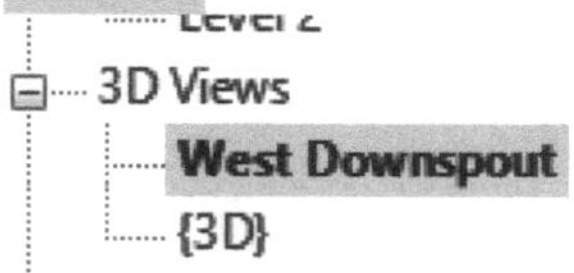 Switch to the **West Downspout 3D** view.

12. 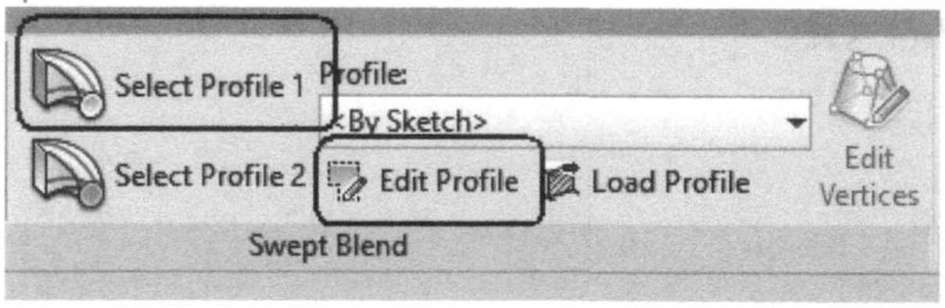 Click on **Select Profile 1**.

Click **Edit Profile**.

13. 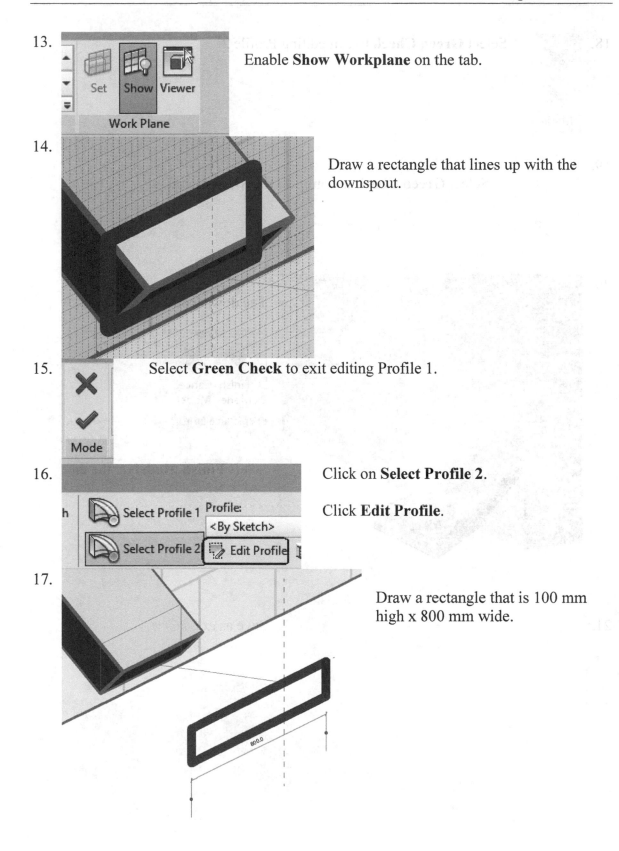 Enable **Show Workplane** on the tab.

14. Draw a rectangle that lines up with the downspout.

15. Select **Green Check** to exit editing Profile 1.

16. Click on **Select Profile 2**.

   Click **Edit Profile**.

17. Draw a rectangle that is 100 mm high x 800 mm wide.

18. Select **Green Check** to exit editing Profile 2.

19. Select **Green Check** to complete the Swept Blend.

20.

The downspout is completed.

Select **Finish Model** on the tab.

21. Save as *ex1-26.rvt*.

Exercise 1-27

Placing a Mass

Drawing Name: **new**
Estimated Time to Completion: 15 Minutes

Scope
Placing a conceptual mass

Solution

1. Start a new project using the *Metric-Architectural* template.

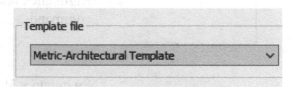

2.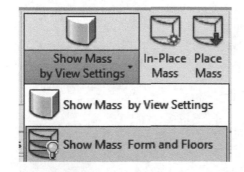

    Activate the Massing & Site ribbon.

    Enable **Show Mass Form and Floors** from the Conceptual Mass panel.

3.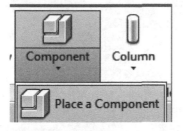

    Activate the Architecture tab.

    Select the **Component→Place a Component** tool on the Build panel.

4.

    Select **Load Family** from the Mode panel.

5. 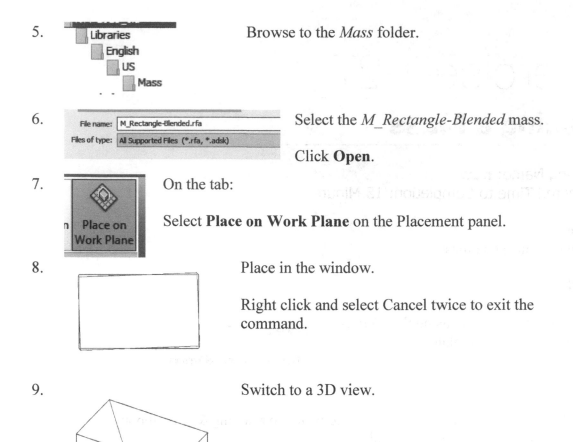   Browse to the *Mass* folder.

6.   Select the *M_Rectangle-Blended* mass.

       Click **Open**.

7.   On the tab:

       Select **Place on Work Plane** on the Placement panel.

8.   Place in the window.

       Right click and select Cancel twice to exit the command.

9.   Switch to a 3D view.

10. Close without saving.

Exercise 1-28
Placing Mass Floors

Drawing Name: mass_floors.rvt
Estimated Time to Completion: 5 Minutes

Scope
Placing mass floors

Solution

1.   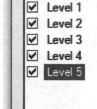   Activate the Massing & Site ribbon.

     Enable **Show Mass Form and Floors** from the
     Conceptual Mass panel.

2.   Select the mass so that it highlights.

     Select the **Mass Floors** tool.

3.   Place a check on all the levels.

     Click **OK**.

     Mass Floors

     ☑ Level 1
     ☑ Level 2
     ☑ Level 3
     ☑ Level 4
     ☑ Level 5

4.
Mass Floor (1)	
Dimensions	
Floor Perimeter	159' 2 57/64"
Floor Area	1485.80 SF
Exterior Surface Area	2302.81 SF
Floor Volume	7462.14 CF
Level	Level 4

Select the mass floor for Level 4.
Use TAB to help select.
What is the floor area?

You should see a floor area of 1485.80 SF.
Close without saving.

Things to remember about Masses:

- The default material for a mass is 5 percent transparent.
- Masses will not print unless the category is enabled in Visibility/Graphics Overrides.
- Masses are created from a single closed profile.
- Masses are a nested entity. In order to modify the profile, you have to open the mass up for editing and then open the desired form component up for editing.
- Masses can be comprised of multiple forms, a combination of voids and solids.

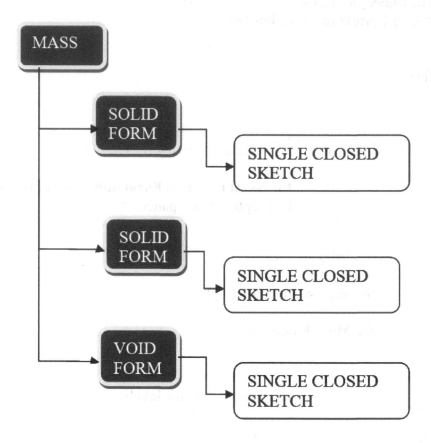

Toposurfaces

A topographical surface (a toposurface) is created using points or imported data. You can create toposurfaces in 3D views or site plans.

You can modify toposurfaces by adding building pads and creating subregions.

Subregions allow you to assign different materials to a toposurface.

When viewing a toposurface, consider the following:

- **Visibility**. You can control the visibility of topographic points. There are 2 topographic point subcategories: Boundary and Interior. Revit classifies points automatically.
- **Triangulation edges**. Triangulation edges for toposurfaces are turned off by default. You can turn them on by selecting them from the Model Categories/Topography category in the Visibility/Graphics dialog.

Exercise 1-29
Creating a Toposurface

Drawing Name: topo_1.rvt
Estimated Time to Completion: 5 Minutes

Scope
Importing a cad file to create a toposurface.

Solution

1. Views (all)
   Floor Plans
   LEVEL 1
   LEVEL 2
   ROOF
   Site

   Activate the **Site** floor plan.

2.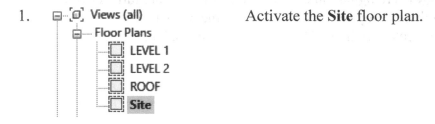

   Go to the Insert ribbon.

   Select **Import CAD**.

3.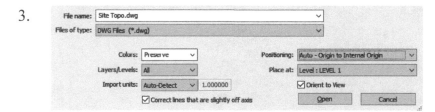

   Locate the *Site Topo.dwg* file.
   Set colors to **Preserve**.
   Set Layers to **All**.
   Set Positioning to **Auto – Origin to Internal Origin**.
   Enable **Orient to View**.
   Enable **Correct lines that are slightly off axis**.
   Click **Open**.

4.

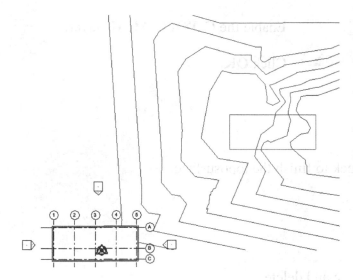

The topo file is imported but it is not aligned to the building.

Unpin the imported file and align it to the building using the green rectangle to position the cad file.

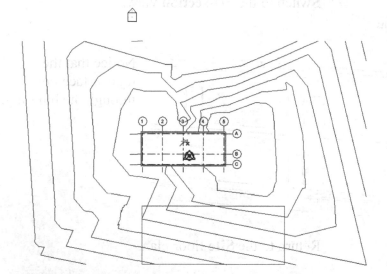

When it is realigned, it should look like this.

5.

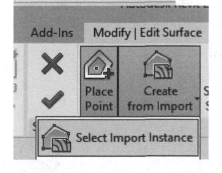

Switch to the Massing & Site ribbon.

Select the **Toposurface** tool.

6.

Select **Create from Import→Select Import Instance**.

Select the imported file.

7. 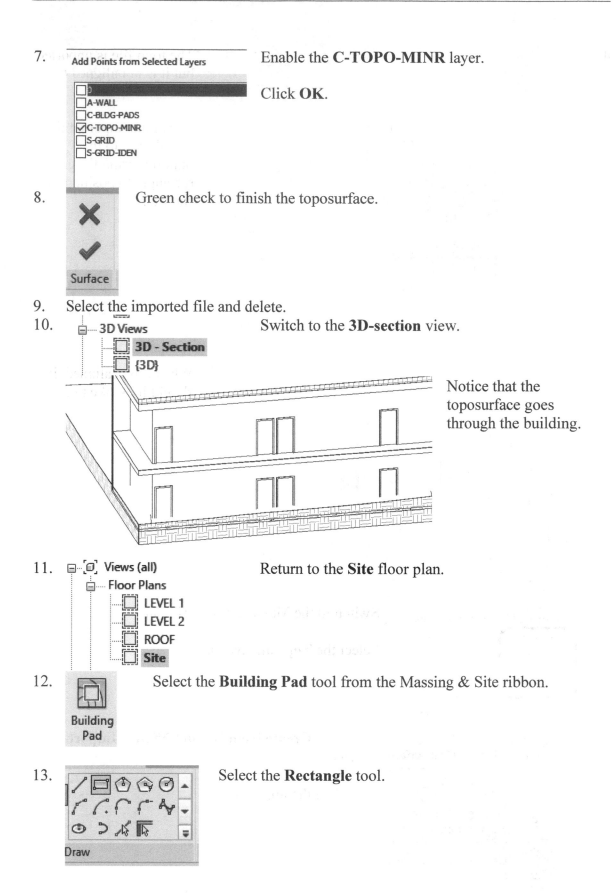 Enable the **C-TOPO-MINR** layer.

Click **OK**.

8. Green check to finish the toposurface.

9. Select the imported file and delete.

10. Switch to the **3D-section** view.

Notice that the toposurface goes through the building.

11. Return to the **Site** floor plan.

12. Select the **Building Pad** tool from the Massing & Site ribbon.

13. Select the **Rectangle** tool.

14.

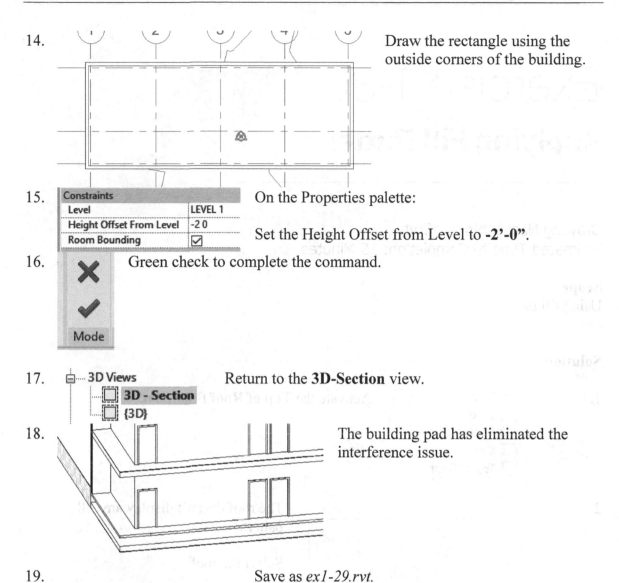

Draw the rectangle using the outside corners of the building.

15. On the Properties palette:

Constraints	
Level	LEVEL 1
Height Offset From Level	-2 0
Room Bounding	☑

Set the Height Offset from Level to **-2'-0"**.

16. Green check to complete the command.

17. Return to the **3D-Section** view.

3D Views
  3D - Section
  {3D}

18. The building pad has eliminated the interference issue.

19. Save as *ex1-29.rvt*.

Exercise 1-30
Applying Fill Patterns

Drawing Name: hip_roof.rvt
Estimated Time to Completion: 15 Minutes

Scope
Using fill patterns.

Solution

1. 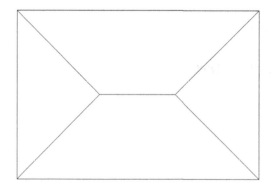   Activate the **Top of Roof** floor plan.

2.   The roof doesn't display any fill pattern.

    Select the roof.

3. 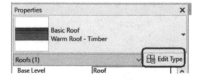   Select **Edit Type**.

4. 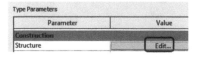   Click **Edit** next to Structure.

5. 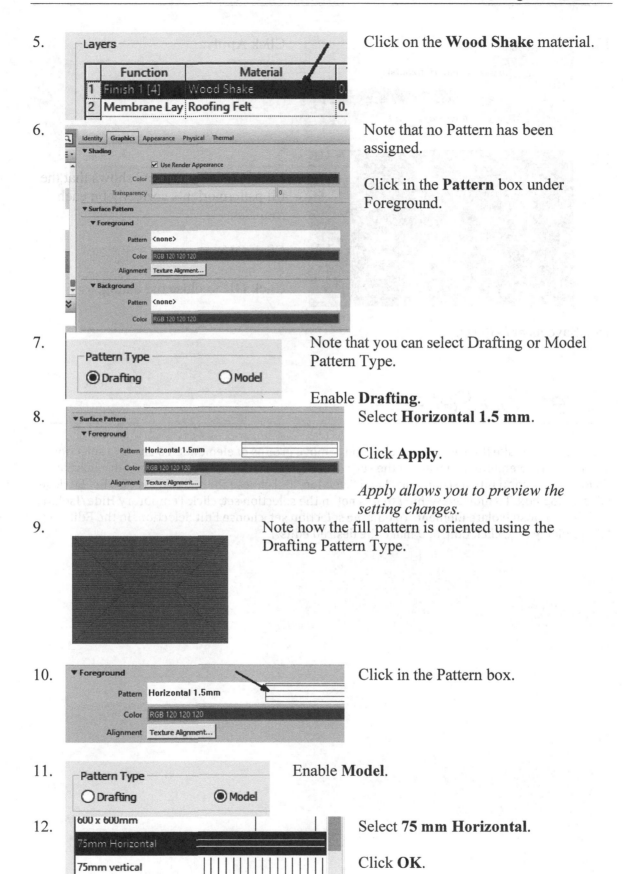 Click on the **Wood Shake** material.

6. Note that no Pattern has been assigned.

Click in the **Pattern** box under Foreground.

7. Note that you can select Drafting or Model Pattern Type.

Enable **Drafting**.

8. Select **Horizontal 1.5 mm**.

Click **Apply**.

Apply allows you to preview the setting changes.

9. Note how the fill pattern is oriented using the Drafting Pattern Type.

10. Click in the Pattern box.

11. Enable **Model**.

12. Select **75 mm Horizontal**.

Click **OK**.

13.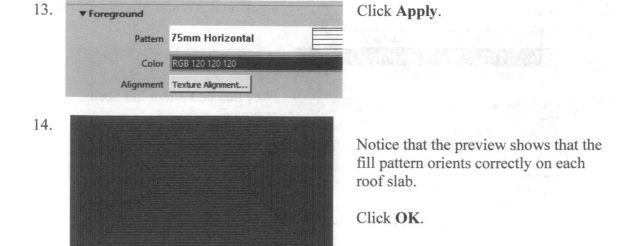

Click **Apply**.

14.

Notice that the preview shows that the fill pattern orients correctly on each roof slab.

Click **OK**.

Click **OK** to close the dialog.

15. Save as *ex1-30.rvt*.

Selection Sets

Selection sets make it easier to edit, add, and remove groups of elements. To create a selection set, select the elements you want in the set, choose Modify, then on the Selection tab, click Save. To load a selection into a scene, click Load Selection, and choose the selection set in the Retrieve Files dialog box. To hide everything that is not in the selection set, click Temporary Hide/Isolate, and then choose Isolate Element. To edit the selection set, choose Edit Selection. In the Edit Filters dialog box, click Edit, and make the desired edits.

Exercise 1-31
Selection Sets

Drawing Name: selection sets.rvt
Estimated Time to Completion: 20 Minutes

Scope
Using selection sets.

Solution

1. 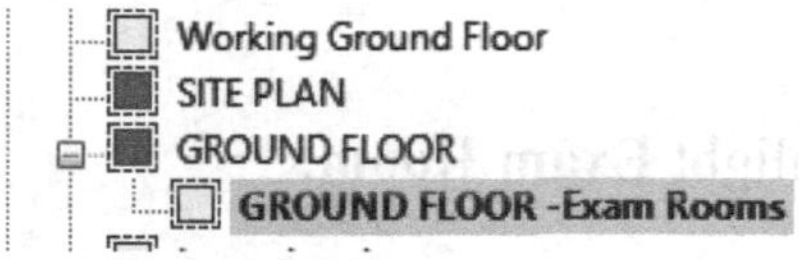 Activate the **GROUND FLOOR – Exam Rooms** floor plan.

2. 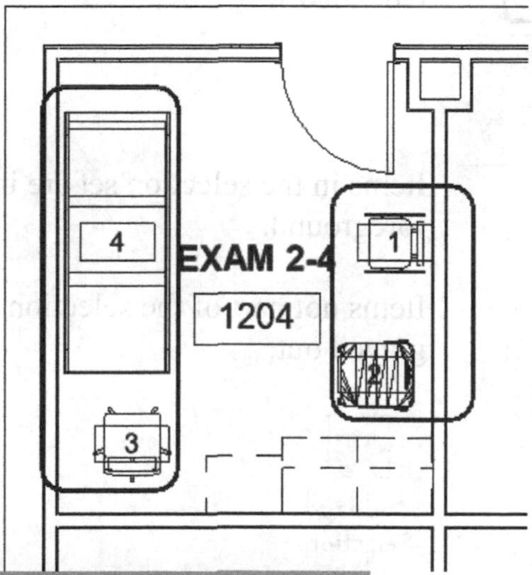 Select the patient bed, the two chairs, and the desk.
Select the tags as well.

Select using window and holding down the CTL key.

3. 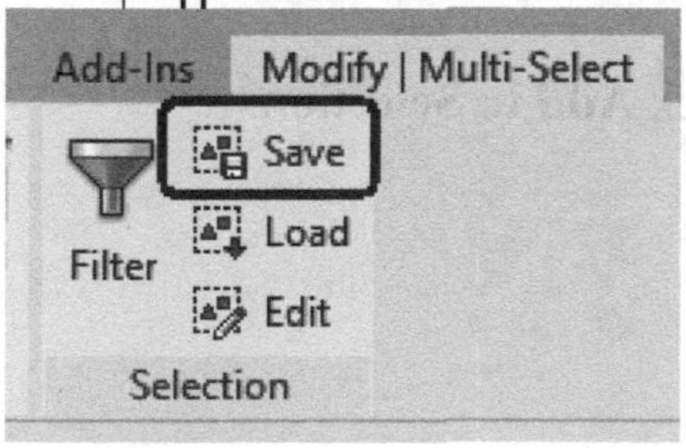 On the ribbon:

Click **Save** on the Selection panel.

4. Save Selection

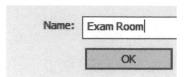

Type **Exam Room** for the name.

Click **OK**.

Left click anywhere in the window to release the selection.

5.

Click on the chair with the tag labeled 3.

Notice that when you create a model group and you select an item in the group, the entire group highlights. When you create a selection set, the elements remain independent.

Also, you were able to create a selection set of model and annotation elements without having to add the annotations as an attachment.

6.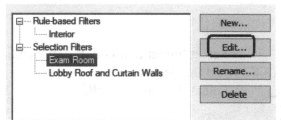

Switch to the Manage ribbon.

Click **Edit** on the Selection panel.

7.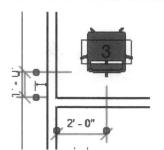

Highlight **Exam Room**.

Click **Edit.**

8.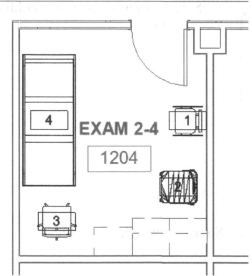

Items in the selection set are in the foreground.

Items not part of the selection set are grayed out.

Click **Add to Selection**.

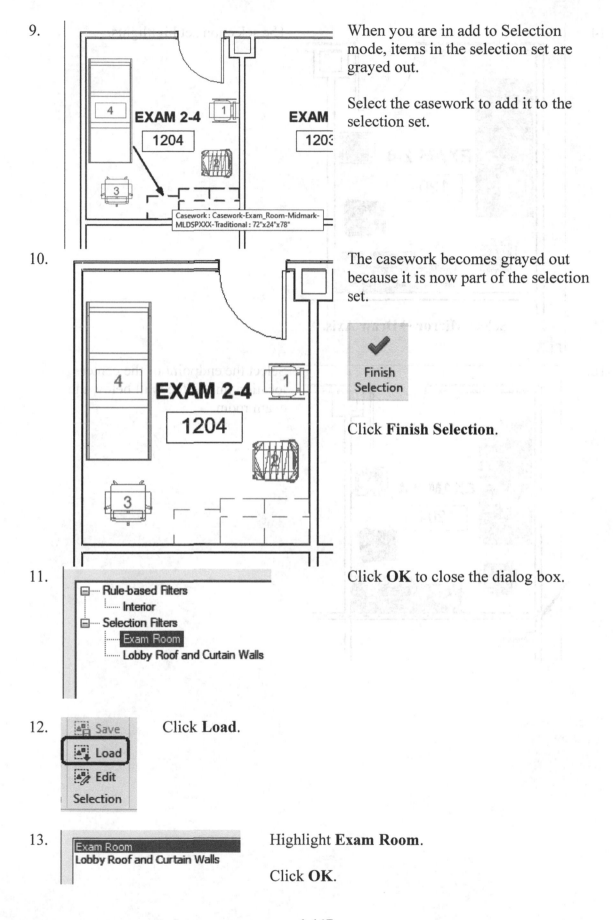

9.

When you are in add to Selection mode, items in the selection set are grayed out.

Select the casework to add it to the selection set.

Casework : Casework-Exam_Room-Midmark-MLDSPXXX-Traditional : 72"x24"x78"

10.

The casework becomes grayed out because it is now part of the selection set.

Finish Selection

Click **Finish Selection**.

11.

⊟ Rule-based Filters
       Interior
⊟ Selection Filters
       Exam Room
       Lobby Roof and Curtain Walls

Click **OK** to close the dialog box.

12.

Save
Load
Edit
Selection

Click **Load**.

13.

Exam Room
Lobby Roof and Curtain Walls

Highlight **Exam Room**.

Click **OK**.

14.

The selection set highlights.

15. Select **Mirror →Draw Axis**.

16.

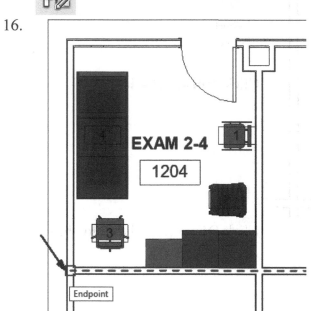

Select the endpoint of the center location line of the wall below the exam room.

17.

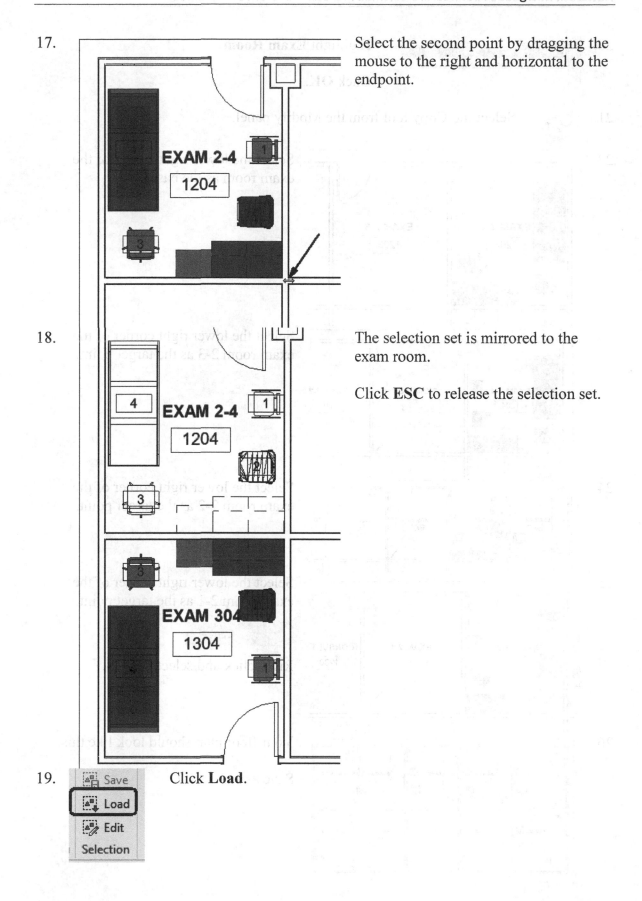

Select the second point by dragging the mouse to the right and horizontal to the endpoint.

18.

The selection set is mirrored to the exam room.

Click **ESC** to release the selection set.

19.

Click **Load**.

20.   Exam Room
       Lobby Roof and Curtain Walls

       Highlight **Exam Room**.

       Click **OK**.

21.   Select the **Copy** tool from the Modify panel.

22.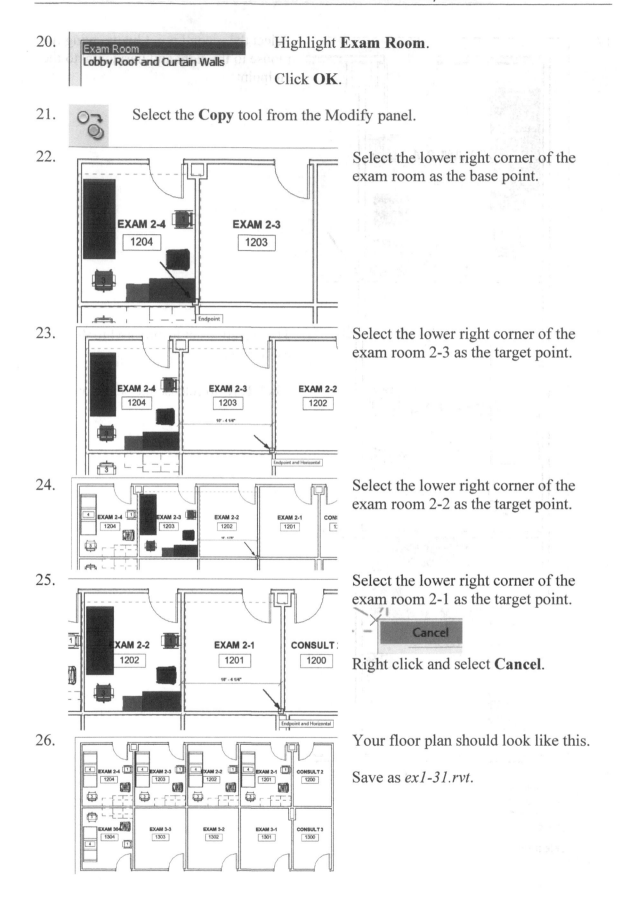

       Select the lower right corner of the exam room as the base point.

23.   Select the lower right corner of the exam room 2-3 as the target point.

24.   Select the lower right corner of the exam room 2-2 as the target point.

25.   Select the lower right corner of the exam room 2-1 as the target point.

       Cancel

       Right click and select **Cancel**.

26.   Your floor plan should look like this.

       Save as *ex1-31.rvt*.

Rooms

A room is a subdivision of space within a building model, based on elements such as walls, floors, roofs, and ceilings.

Walls, floors, roofs, and ceilings are defined as room-bounding. Revit uses these room-bounding elements when computing the perimeter, area, and volume of a room.

You can turn on/off the Room Bounding parameter of many elements. You can also use room separation lines to further subdivide space where no room-bounding elements exist. When you add, move, or delete room-bounding elements, the room's dimensions update automatically.

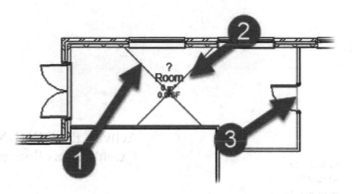

1: The Room Reference Crosshair- The intersection of the two lines represents the center of the room object. This is not necessarily the "geographic" center of the room or space.

2: Room Tag – This is a text annotation that takes the data from the room object paramenters and displays it in the tag. This particular tag displays the Room Number, Room Name, and Area. You can create your own custom room tags or use the Type Selector to display different room tag types.

3: Room Boundary Line – When the room is selected, the boundary of the room is displayed as a light blue line. If you have an open floor plan, you can add room boundary lines to designate different rooms/spaces.

Exercise 1-32
Rooms

Drawing Name: rooms_1.rvt
Estimated Time to Completion: 10 Minutes

Scope
Creating rooms.
Modify room tags.

Solution

1. 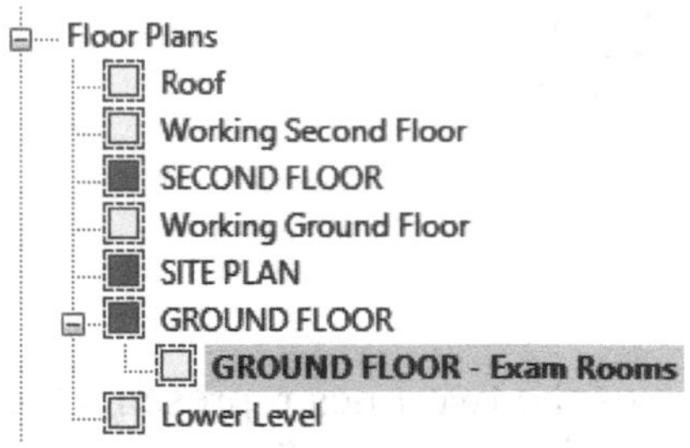 Floor Plans
    ☐ Roof
    ☐ Working Second Floor
    ■ SECOND FLOOR
    ☐ Working Ground Floor
    ■ SITE PLAN
    ■ GROUND FLOOR
       ☐ **GROUND FLOOR - Exam Rooms**
    ☐ Lower Level

    Activate the **GROUND FLOOR –
    Exam Rooms** floor plan.

2.  Activate the Architecture ribbon.

    Select the **Room** tool.

3. Room: New
       New
       1008 NURSE STATION
       1204 EXAM 2-4

    From the Options bar:

    Select **1204 EXAM 2-4** from the
    drop-down list for rooms.

4.

    Verify that **Tag on Placement** is enabled on the ribbon.

5.

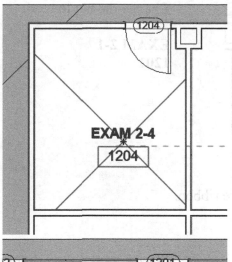

Place the room in the room with the door tagged 1204.

6.

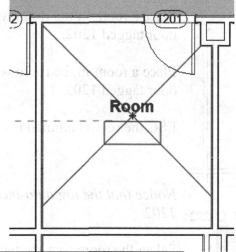

Move the mouse to the right and place a room in the room with the door tagged 1201.

ESC the room command.

7.

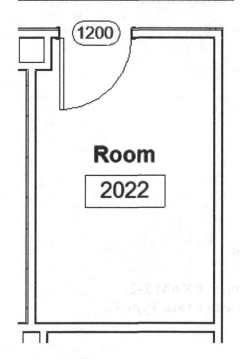

Notice that the room is placed.

The tag defaulted to the default values.

Click on the tag to edit.

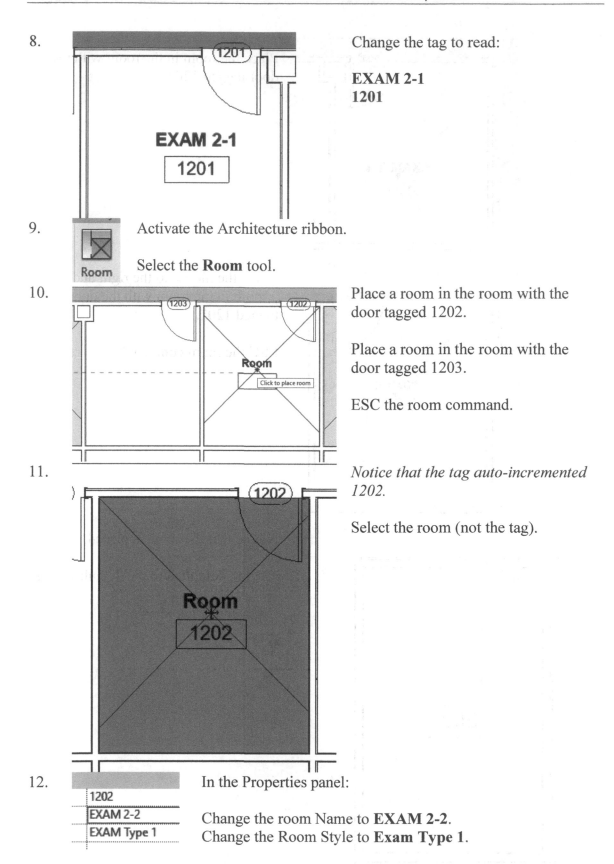

8.   Change the tag to read:

EXAM 2-1
1201

9.   Activate the Architecture ribbon.

Select the **Room** tool.

10.   Place a room in the room with the door tagged 1202.

Place a room in the room with the door tagged 1203.

ESC the room command.

11.   *Notice that the tag auto-incremented 1202.*

Select the room (not the tag).

12.   In the Properties panel:

Change the room Name to **EXAM 2-2**.
Change the Room Style to **Exam Type 1**.

13.

Identity Data	
Number	1203
Name	EXAM 2-3
Room Style	EXAM Type 1
Image	

Select the room tagged 1203.

In the Properties panel:

Change the room Name to **EXAM 2-3**.
Change the Room Style to **Exam Type 1**.

14.

Save as *ex1-32.rvt*.

Exercise 1-33

Room Separators

Drawing Name: rooms_2.rvt
Estimated Time to Completion: 5 Minutes

Scope
Creating rooms using a room separator.
Modify room tags.

Solution

1.

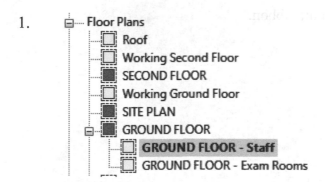

Activate the **GROUND FLOOR –
Staff** floor plan.

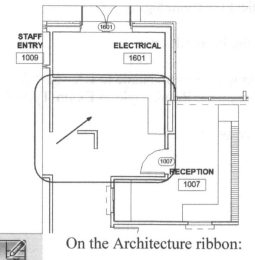

The area indicated is a nurse's station, but it is missing necessary room-bounding elements as it has openings between the walls.

We need to use Room Separators to define the room.

2. 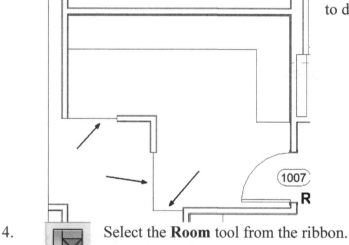 On the Architecture ribbon:

Select **Room Separator**.

3. Using the Line tool, place three lines to define the room.

4. Select the **Room** tool from the ribbon.

5.

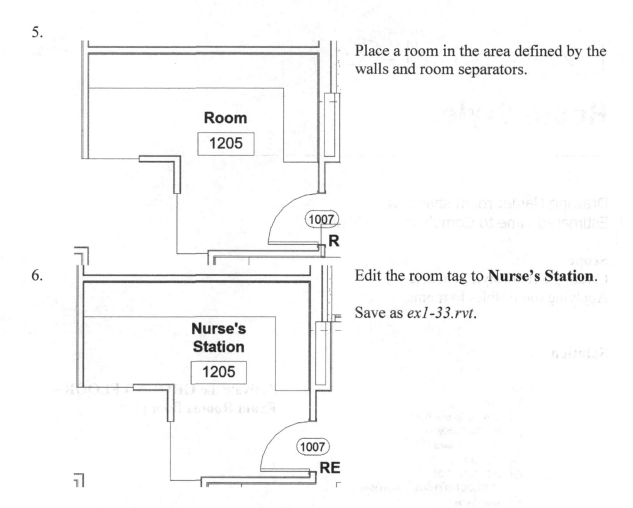

Place a room in the area defined by the walls and room separators.

6.

Edit the room tag to **Nurse's Station**.

Save as *ex1-33.rvt*.

Room Styles

Room styles are created using a room style schedule. Once you have defined your room style, you can then apply it to new or existing rooms.

Exercise 1-34

Room Styles

Drawing Name: room styles.rvt
Estimated Time to Completion: 15 Minutes

Scope
Creating a room style schedule.
Applying room styles to rooms.

Solution

1.

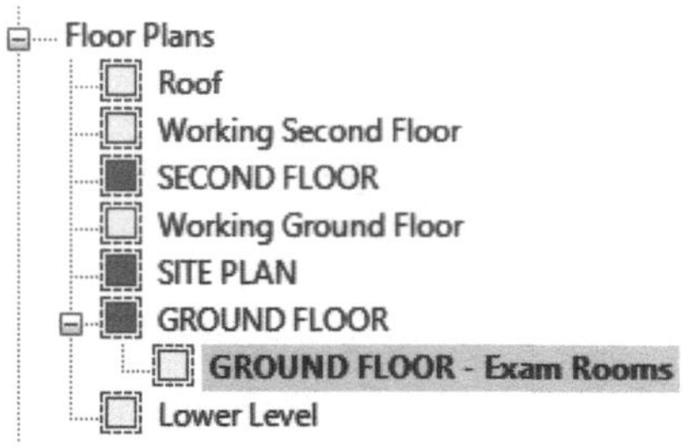

Floor Plans
   Roof
   Working Second Floor
   SECOND FLOOR
   Working Ground Floor
   SITE PLAN
   GROUND FLOOR
      GROUND FLOOR - Exam Rooms
   Lower Level

Activate the **GROUND FLOOR – Exam Rooms** floor plan.

2.

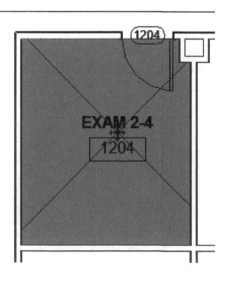

Select Exam 2-4 room.

3.

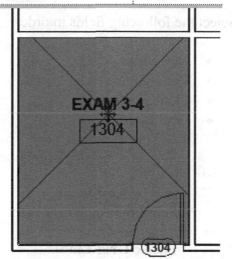

Note that in the Identity Data the finishes have been filled in.

Click ESC to release the selection.

4.

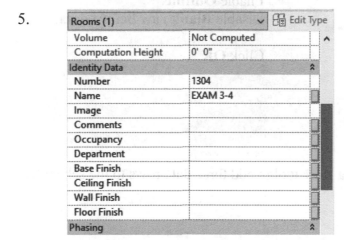

Select EXAM 3-4.

5.

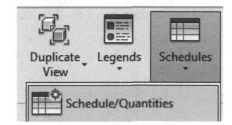

Note that the finish data has not been filled in.

Click ESC to release the selection.

6.

Activate the View ribbon.

Select **Schedules→Schedule/Quantities**.

7. 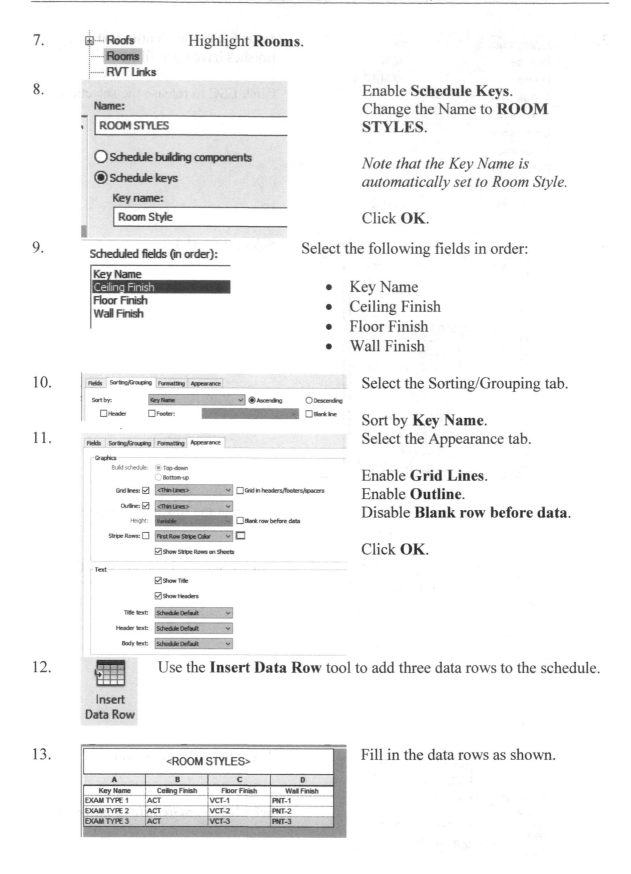 Highlight **Rooms**.

8. Enable **Schedule Keys**.
Change the Name to **ROOM STYLES**.

Note that the Key Name is automatically set to Room Style.

Click **OK**.

9. Select the following fields in order:

- Key Name
- Ceiling Finish
- Floor Finish
- Wall Finish

10. Select the Sorting/Grouping tab.

Sort by **Key Name**.

11. Select the Appearance tab.

Enable **Grid Lines**.
Enable **Outline**.
Disable **Blank row before data**.

Click **OK**.

12. Use the **Insert Data Row** tool to add three data rows to the schedule.

13. Fill in the data rows as shown.

<ROOM STYLES>

Key Name	Ceiling Finish	Floor Finish	Wall Finish
EXAM TYPE 1	ACT	VCT-1	PNT-1
EXAM TYPE 2	ACT	VCT-2	PNT-2
EXAM TYPE 3	ACT	VCT-3	PNT-3

14.

Open the **GROUND FLOOR – Exam Rooms** floor plan view.

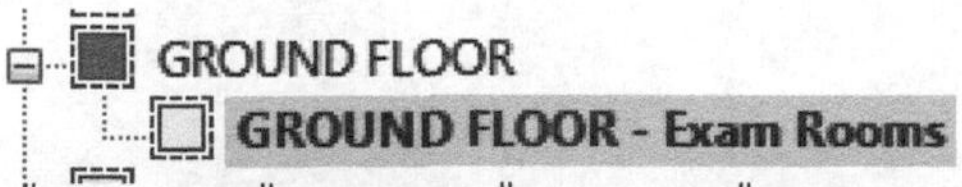

15.

Hold down the CTL key and select the four exam rooms tagged 3-x.

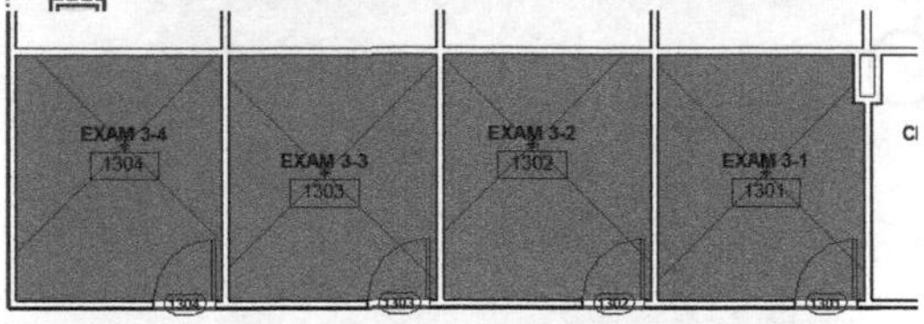

16.

In the Properties panel:

Room Style is now available in the Identity Data.

Select **EXAM TYPE 3** to assign that room style to the selected rooms.

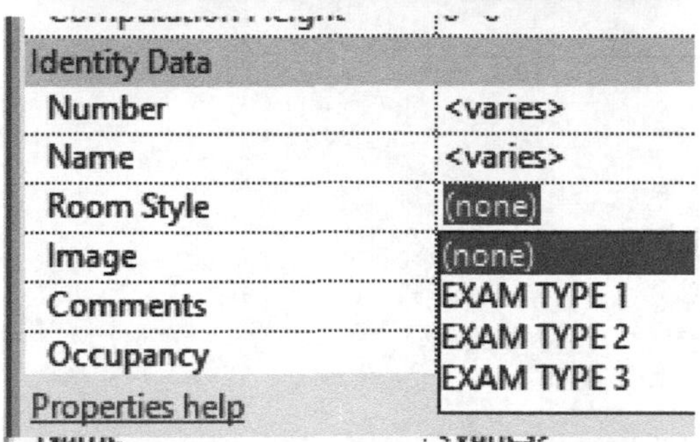

17.

Scroll down and you will see the finish values have been filled in based on the room style that was assigned.

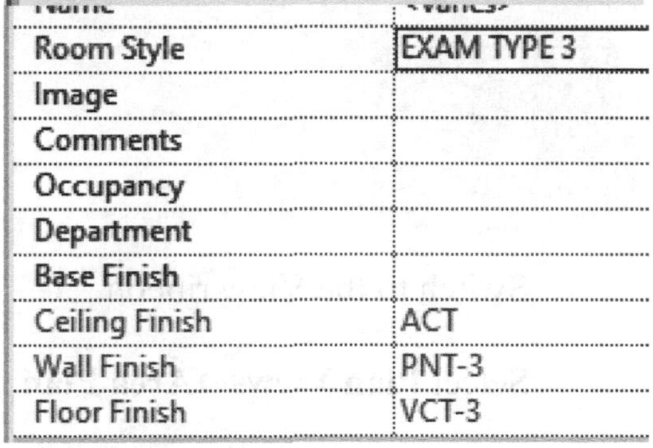

Ceiling Finish	ACT
Wall Finish	PNT-3
Floor Finish	VCT-3

18.

Save as *ex1-34.rvt*.

Exercise 1-35

Area Plans

Drawing Name: area plans.rvt
Estimated Time to Completion: 20 Minutes

Scope
Creating an area plan view.
Tag an existing area.
Create area boundaries.
Change area types.

Solution

1.

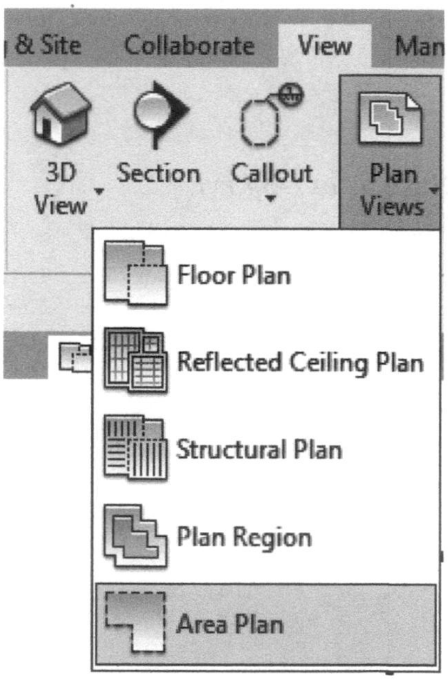

Switch to the View ribbon.

Select **Plan Views→Area Plan**.

2.

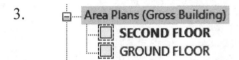

Select **Gross Building**.

Highlight **SECOND FLOOR**.

Click **OK**.

3.

Area Plans (Gross Building)
    ☐ **SECOND FLOOR**
    ☐ GROUND FLOOR

Note that the view was created and organized in the Project Browser under Area Plans (Gross Building).

You should be able to see some blue boundaries that define different areas in the building.

4.

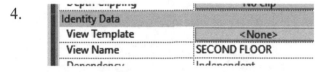

On the Properties palette:

Click **<None>** next to View Template.

5.

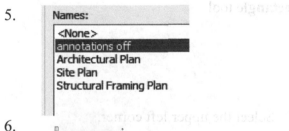

Highlight **annotations off.**

Click **OK**.

This turns off visibility of elevations, sections, and reference planes.

6.

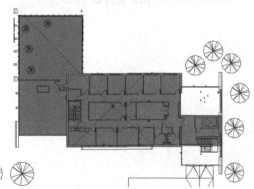

Click inside the building to select the area.

The area is previewed, but you can see that some areas should be added or separated.

This area was generated automatically when you created the area plan using the bounding elements.

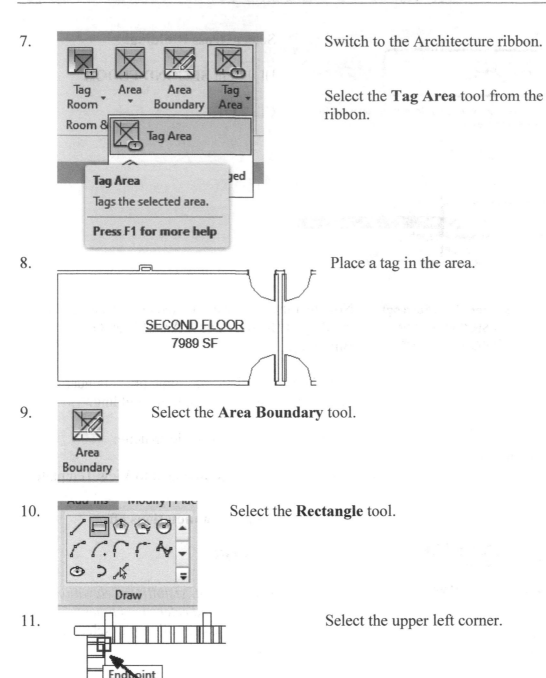

7. Switch to the Architecture ribbon.

Select the **Tag Area** tool from the ribbon.

8. Place a tag in the area.

SECOND FLOOR
7989 SF

9. Select the **Area Boundary** tool.

10. Select the **Rectangle** tool.

11. Select the upper left corner.

Endpoint

12.

Select the lower right corner.

13.

Place a second rectangle as indicated.

You will get an error message about overlapping lines.

14.

Place a third rectangle.

15.

Delete the overlapping lines and trim the corners as needed.

16.

Select the **Area** tool.

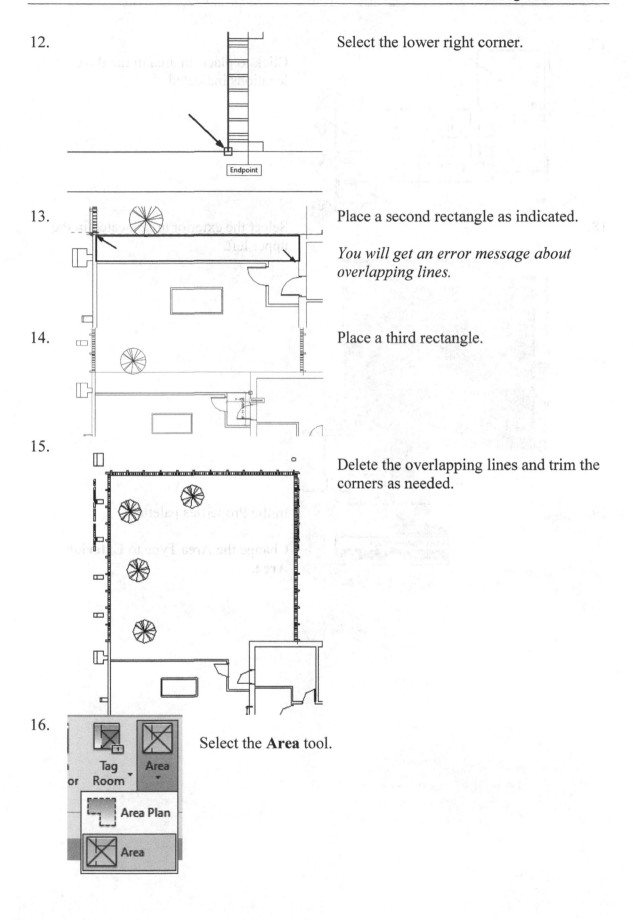

17.

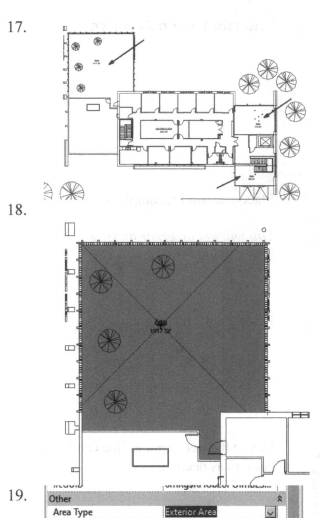

Click to place an area in the three locations indicated.

18.

Select the exterior area located in the upper left.

19.

Other	
Area Type	Exterior Area
	Gross Building Area
	Exterior Area
Properties help	Apply

In the Properties palette:

Change the Area Type to **Exterior Area.**

20.   Hold down the CTL key and select these two areas.

In the Properties palette:

Change the Area Type to **Exterior Area.**

21.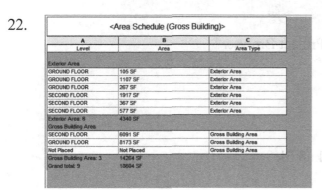

Locate the Area Schedule (Gross Building) in the Project Browser.

Open the view.
Save as *ex1-35.rvt*

22.

<Area Schedule (Gross Building)>		
A	B	C
Level	Area	Area Type
Exterior Area		
GROUND FLOOR	105 SF	Exterior Area
GROUND FLOOR	1107 SF	Exterior Area
GROUND FLOOR	267 SF	Exterior Area
SECOND FLOOR	1917 SF	Exterior Area
SECOND FLOOR	367 SF	Exterior Area
SECOND FLOOR	577 SF	Exterior Area
Exterior Area: 6	4340 SF	
Gross Building Area		
SECOND FLOOR	6091 SF	Gross Building Area
GROUND FLOOR	8173 SF	Gross Building Area
Not Placed	Not Placed	Gross Building Area
Gross Building Area: 3	14264 SF	
Grand total: 9	18604 SF	

Practice Exam

1. Select the answer which is NOT an example of bidirectional associativity:

    A. Flip a section line and all views update.
    B. Draw a wall in plan view and it appears in all other views.
    C. Change an element type in a schedule and the change is displayed in the floor plan view as well.
    D. Flip a door orientation so the door swing is on the exterior of the building.

2. Select the answer which is NOT an example of a parametric relationship:

    A. A floor is attached to enclosing walls. When a wall moves, the floor updates so it remains connected to the walls.
    B. A series of windows are placed along a wall using an EQ dimension. The length of the wall is modified, and the windows remain equally spaced.
    C. A door is placed in a wall. The wall is moved, and the door remains constrained in the wall.
    D. A shared parameter file is loaded to the server.

3. Which tab does NOT appear on Revit's ribbon?

    A. Architecture
    B. Basics
    C. Insert
    D. View

4. Which item does NOT appear in the Project Browser?

    A. Families
    B. Groups
    C. Callouts
    D. Notes

5. Which is the most recently saved backup file?

    A. office.0001
    B. office.0002
    C. office.0003
    D. office.0004

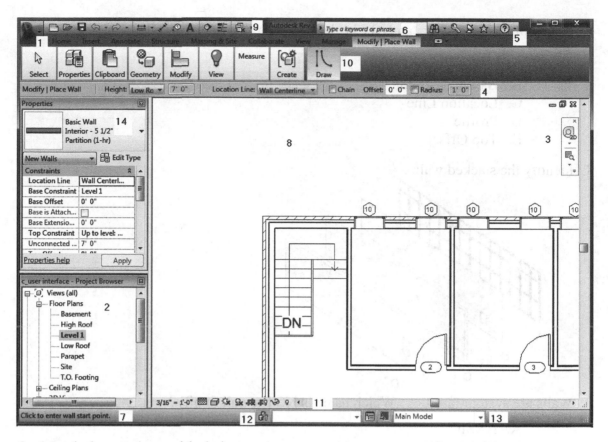

6. Match the numbers with their names.

View Control Bar	InfoCenter
Project Browser	Status Bar
Navigation Bar	Properties Pane
Options Bar	Application Menu
Design Options	Drawing Area
Help	Quick Access Toolbar
Ribbon	Worksets

7. Which of the following can NOT be defined prior to placing a wall?

    A. Unconnected Height
    B. Base Constraint
    C. Location Line
    D. Profile
    E. Top Offset

8. Identify the stacked wall.

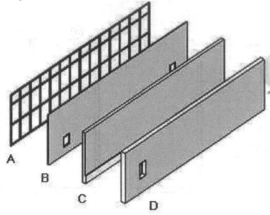

9. Walls are system families. Which name is NOT a wall family?

    A. BASIC
    B. STACKED
    C. CURTAIN
    D. COMPLICATED

10. Select the TWO that are wall type properties:

    A. COARSE FILL PATTERN
    B. LOCATION LINE
    C. TOP CONSTRAINT
    D. FUNCTION
    E. BASE CONSTRAINT

11. Use this key to cycle through selections:

    A. TAB
    B. CTRL
    C. SHIFT
    D. ALT

12. The construction of a stacked wall is defined by different wall _____.

    A. Types
    B. Layers
    C. Regions
    D. Instances

13. To change the structure of a basic wall you must modify its:

    A. Type Parameters
    B. Instance Parameters
    C. Structural Usage
    D. Function

14. If a stacked wall is based on Level 1 but one of its subwalls is on Level 7, the base level for the subwall is Level _____.

    A. 7
    B. 1
    C. Unconnected
    D. Variable

15. A designer wants to adjust the horizontal rail indicated in the image. Where should the designer click in the Type Properties dialog?

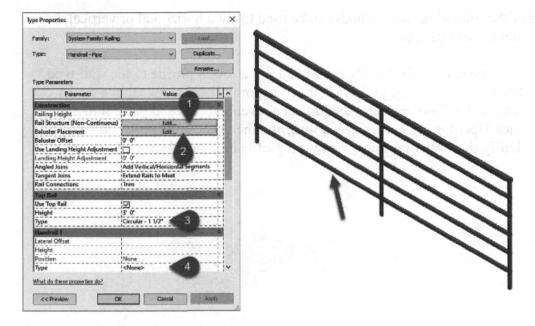

16. Which of the following methods is used to join a wall to a roof and have the wall shape conform to the underside of the roof?

    A. Select the wall, click Attach Top/Base on the ribbon, and then select the roof.
    B. Edit the profile of the wall, select the top boundary, click Attach Top/Base, and then select the roof.
    C. Use the Wall by Face tool and create a wall based on the underside face of the roof.
    D. Select the roof, click Attach Top/Base, and then select the wall.

17. A designer is working with a room schedule. The designer only wants to see rooms that are offices in the schedule, and not restrooms, break rooms, etc. How should this be accomplished?

    A. Delete the non-office room rows in the schedule.
    B. Only add the office rooms to the list of fields.
    C. Sort the schedule to only show office rooms.
    D. Add a filter to filter out the non-office rooms.

18. Which of the following best describes the hierarchy of elements in a Revit project?

    A. Roofs > Walls > Floors > Ceilings
    B. Category > Family > Type > Instance
    C. Category > Type > Family > Instance
    D. Family > Category > Instance > Type

19. Which of the following two methods can be used to cut a horizontal or vertical projection from a wall? (Select 2)

    A. Open the wall in the Family Editor and add a reveal profile to the wall type.
    B. Use the Wall: Reveal tool to add a vertical or horizontal reveal in a wall.
    C. Activate the Cut Geometry tool, select a reveal profile, and place it in a wall.
    D. Click Opening By Face, select a wall, and then sketch the projection to cut.
    E. Modify the wall type and add a reveal profile to the wall structure.

Answers:
    1) D; 2) D; 3) B; 4) D; 5) D; 6) 1- Application Menu, 2- Project Browser, 3- Navigation Bar, 4- Options Bar, 5- Help, 6- InfoCenter, 7- Status Bar, 8- Drawing Area, 9- Quick Access Toolbar, 10- Ribbon, 11- View Control Bar,12- Worksets, 13- Design Options; 7) D ; 8) C; 9) D; 10) A & D; 11) D; 12) A; 13) A; 14) B; 15) 1; 16) A; 17) D; 18) B; 19) B & E

02

Families

This lesson addresses the following Professional exam questions:

- Managing Family Types
- Identifying a family
- Working with hosted and non-hosted elements
- Family parameters
- Using Symbolic Lines in Families
- Controlling the display of families
- Detail Component Families
- System and Component families
- Edit Room-Aware Families
- Identifying family properties

Family Types

Using the Family Types tool, you can create many types (sizes) for a family.

To do this, you need to have labeled the dimensions and created the parameters that are going to vary.

Each family type has a set of properties (parameters) that includes the labeled dimensions and their values. You can also add values for standard parameters of the family (such as Material, Model, Manufacturer, Type Mark, and others).

To manage families and family types, use the shortcut menu in the Project Browser.

1.   In the Project Browser, under Families, locate the desired family or type.

2.   To manage families and family types, do the following:

If you want to...	then...
change type properties	right-click a type, and click Type Properties. As an alternative, double-click a type, or select it and Click Enter.
rename a family or type	right-click the family or type, and click Rename. As an alternative, select the type and Click F2.
add a type from an existing type	right-click a type, and click Duplicate. Enter a name for the type. The new type displays in the type list. Double-click the new type to open the Type Properties dialog, and change properties for the new type.
add a new type	right-click a family, and click New Type. Enter a name for the type. The new type displays in the list. Double-click the new type to open the Type Properties dialog, and define properties for the new type.
copy and paste a type into another project	right-click a type, and click Copy to Clipboard. Open the other project and, in the drawing area, Click Ctrl+V to paste it.
reload a family	right-click a family, and click Reload. Navigate to the location of the updated family, select it, and click Open.
edit a family in the Family Editor	right-click a family, and click Edit. The family opens in the Family Editor.
delete a family or type	right-click a family or type, and click Delete from Project. In addition to deleting the family or type from the project, this function deletes instances of matching types that exist in the model. Note: You cannot delete the last type in a system family.

Exercise 2-1

Create a New Family Type

Drawing Name: **family_type.rvt**
Estimated Time to Completion: 5 Minutes

Scope
Create a new family type.

Solution

1. 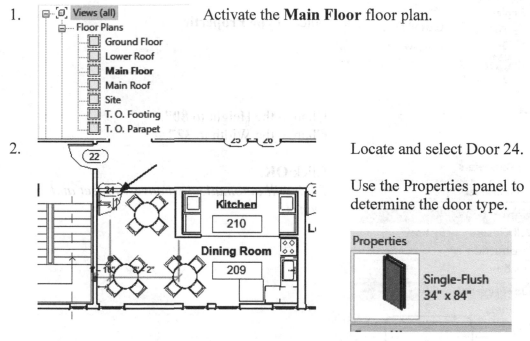 Activate the **Main Floor** floor plan.

2. Locate and select Door 24.

   Use the Properties panel to determine the door type.

3. 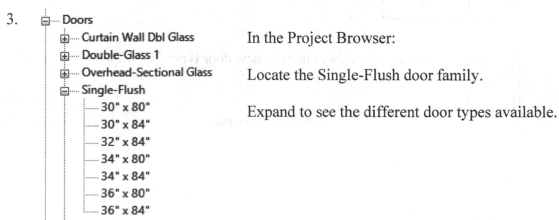 In the Project Browser:

   Locate the Single-Flush door family.

   Expand to see the different door types available.

4. 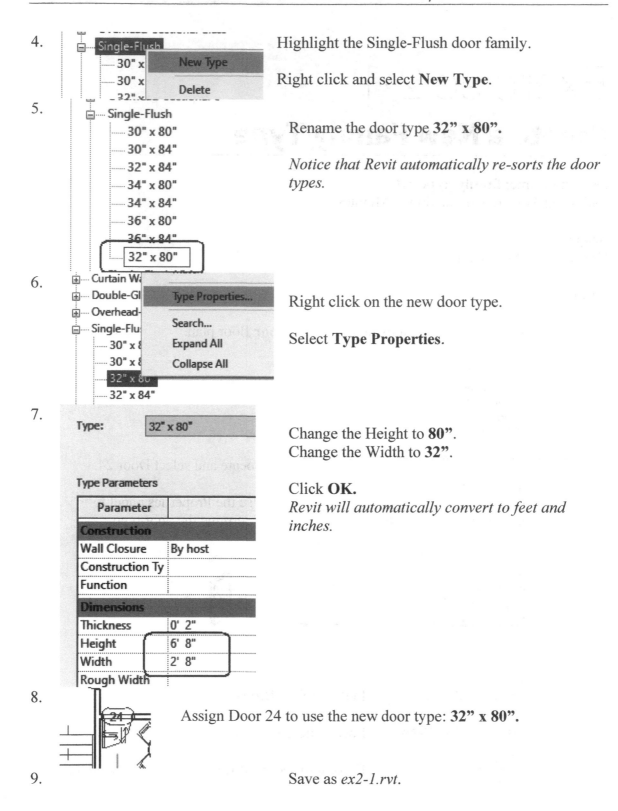 Highlight the Single-Flush door family.

Right click and select **New Type**.

5. Rename the door type **32" x 80"**.

Notice that Revit automatically re-sorts the door types.

6. Right click on the new door type.

Select **Type Properties**.

7. Change the Height to **80"**.
Change the Width to **32"**.

Click **OK**.
Revit will automatically convert to feet and inches.

8. Assign Door 24 to use the new door type: **32" x 80".**

9. Save as *ex2-1.rvt*.

Exercise 2-2

Indentifying a Family

Drawing Name: **i_firestation_elem.rvt**
Estimated Time to Completion: 5 Minutes

Scope
Identify different elements and their families.

Solution

1. Open *i_firestation_elem.rvt*.

2.  Activate the **South** elevation.

3. Select the second window from the left.

4. 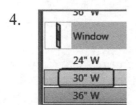 Select the **30″ W** from the window type list in the Properties pane.

5. Use the Measure tool from the Quick Access toolbar to check the distance between the outside edges of the two left windows.

6.

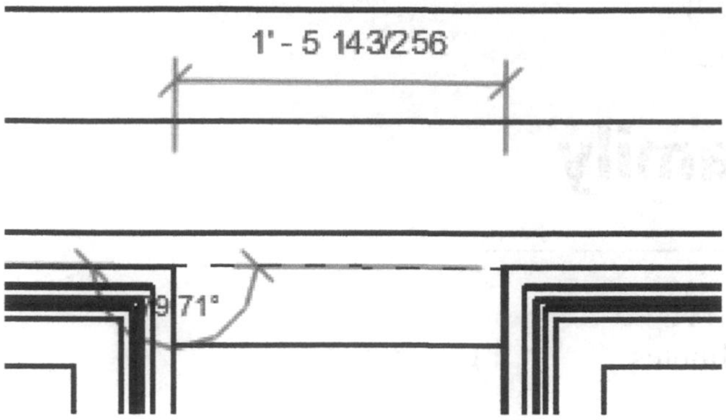

Check to see if you got the same value.

7.   Close without saving.

Transfer Project Standards

System families are only available in the project where they were created and used. However, often you will create a system family and find it useful enough that you will want to use it in another project. Rather than have to redo all the work of recreating the system family, you can use the Transfer Project Standards tool to import the desired system family over to the second project.

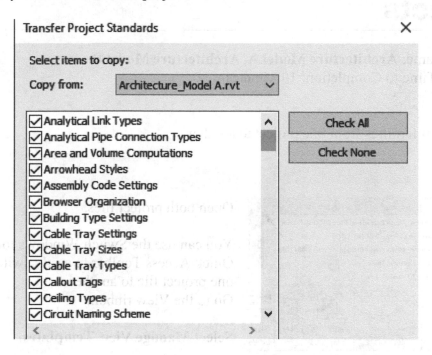

The Transfer Project Standards dialog provides a list of system families available to be imported. You need to have both files open (the project file with the desired system families and the project file missing the desired system families). You make the target file active and then launch the Transfer Project Standards tool. Select the desired system families and click OK. It is *not* recommended to copy all system families from one project to another.

Steps:

1. Open both source and target project files.
2. Make the target project file active.
3. Launch Transfer Project Standards from the Manage ribbon.
4. Select desired system families.
5. Click OK.

Exercise 2-3
Transfer System Families Between Projects

Drawing Name: **Architecture Model A, Architecture Model B**
Estimated Time to Completion: 10 Minutes

Scope
Copy system families from one project to another.

Solution

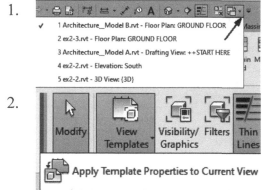

1. Open both project files.

   You can use the Switch Windows tool on the Quick Access Toolbar to easily switch from one project file to another.
   Go to the View ribbon.

2. Select **Manage View Templates**.

3.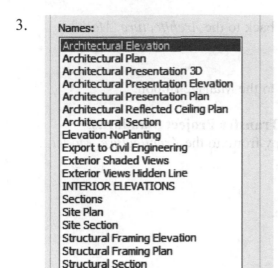

These are the view templates available in the *Model B* project.

Click **OK**.

4.

Switch to the *Architecture_Model A* project.

5.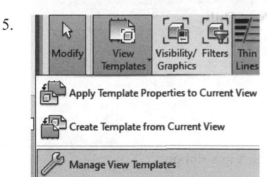

Go to the View ribbon.

Select **Manage View Templates**.

6.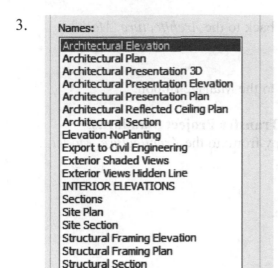

These are the view templates available in the Model A project.

Click **OK**.

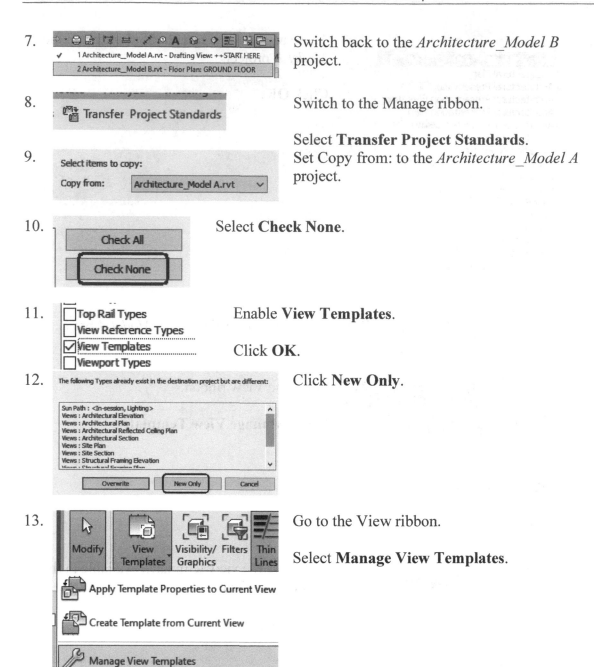

7. Switch back to the *Architecture_Model B* project.

8. Switch to the Manage ribbon.

Select **Transfer Project Standards**.

9. Set Copy from: to the *Architecture_Model A* project.

10. Select **Check None**.

11. Enable **View Templates**.

Click **OK**.

12. Click **New Only**.

13. Go to the View ribbon.

Select **Manage View Templates**.

14.

Names:
Architectural Elevation
Architectural Plan
Architectural Presentation 3D
Architectural Presentation Elevatic
Architectural Presentation Plan
Architectural Reflected Ceiling Plan
Architectural Section
Area - 1/8"
Elevation-NoPlanting
Equipment - 1/8"
Export to Civil Engineering
Exterior Shaded Views
Exterior Views Hidden Line
Floor Plan - 1/8"
Floor Plan - 1/16"
Furniture - 1/8"
INTERIOR ELEVATIONS
Presentation - 1/8"
RCP - 1/8"
Reflected Ceiling Plan - 1/8"
Sections

The View Templates were copied over.

Click **OK**.

15.   Save as *ex2-3.rvt*.

Family Parameters

Family parameters define behaviors or Identity Data that apply across all types in that family. Different categories have different family parameters based on how Revit expects the component to be used. Family parameters can be Type or Instance. A Type parameter does not change regardless of where the element is placed. Type parameters are usually dimensions (such as length or height) or material. These parameters don't change unless a different type is selected. An Instance parameter is unique to each element. For example, the location mark of an element is changed depending on where it is placed. Door hardware can be different depending on the instance. You should be familiar with the concept of type and instance parameters and be able to identify whether or not a parameter is a type or instance parameter.

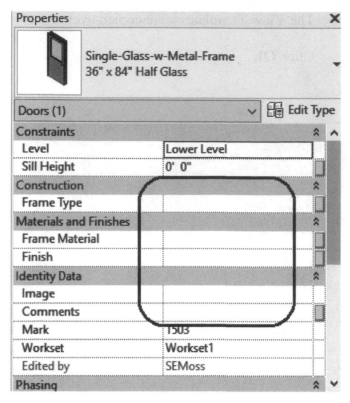

One way to determine whether a parameter is an instance parameter is to select the element and look in the Properties palette.

If you can enter or modify new data next to a parameter, it is an instance parameter.

If you have to select Edit Type to access and modify the parameter, then it is a type parameter.

Exercise 2-4

Create a Casework Family

Drawing Name: **new (using Casework wall-based.rft)**
Estimated Time to Completion: 5 Minutes

Scope
Create a new casework family.

Solution

1. 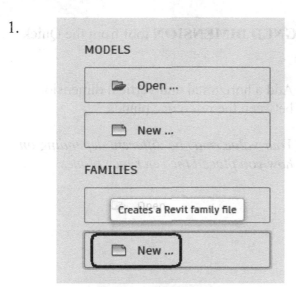 Select **New** under Families.

2. Select *Casework wall-based.rft*.

   Click **Open**.

   By selecting this template, is the element to be hosted or non-hosted?

   If it is hosted, what element will be the host?

3. Switch to the **Right** elevation.

4. 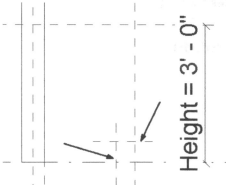 *We want to add reference planes for the toe kick.*

Select the **Reference Plane** tool from the ribbon.

5. 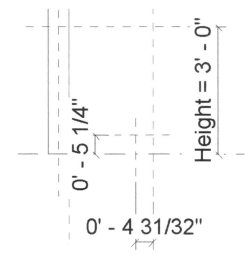 Draw a horizontal plane and a vertical plane.

6. Select the **ALIGNED DIMENSION** tool from the Quick Access toolbar.

7. Add a horizontal and vertical dimension between the reference planes.

Your values may be different depending on how you placed the reference planes.

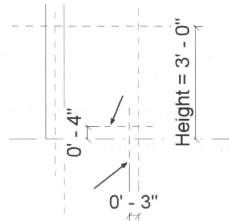

8. Change the horizontal dimension to 3" and the vertical dimension to 4".

Hint: *Select the reference plane and then the dimension in order to modify the dimension.*

9.	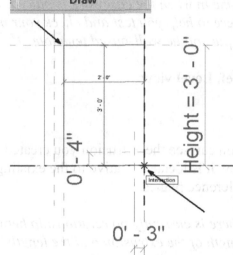	Switch to the Create ribbon.

Select the **Extrusion** tool.

10.	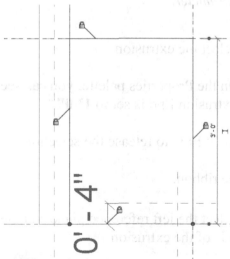	Select **Rectangle** from the Draw panel.

11.	Place a rectangle using the reference plane intersections to define the outer edge of the casework.

12.	Enable the locks on all four sides of the rectangle.

This constrains the rectangle to the reference planes.

13.		**Green check** to complete the extrusion.

14. 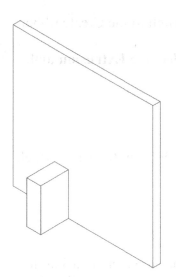 Switch to a 3D view.

Use the Viewcube to re-orient the view.

The casework is not very wide. The larger rectangle represents a wall. This wall does not come in when the casework is placed. It is there to help you test and check your model. It is part of the wall-based template.

15. 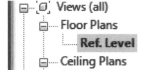 Switch to a **Ref. Level** view.

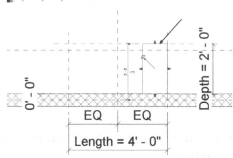

You can see the extrusion you created and how it is located relative to the existing reference planes.

There is currently no relationship between the length of the extrusion and the length parameter.

16. 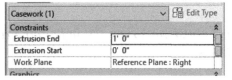 Select the extrusion.

On the Properties palette, you can see the Extrusion End is set to **1'-0"**.

Click ESC to release the selection.

17. Select the **ALIGN** tool from the ribbon.

18. 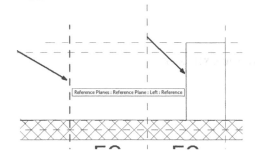 Select the left reference plane and then the left side of the extrusion.

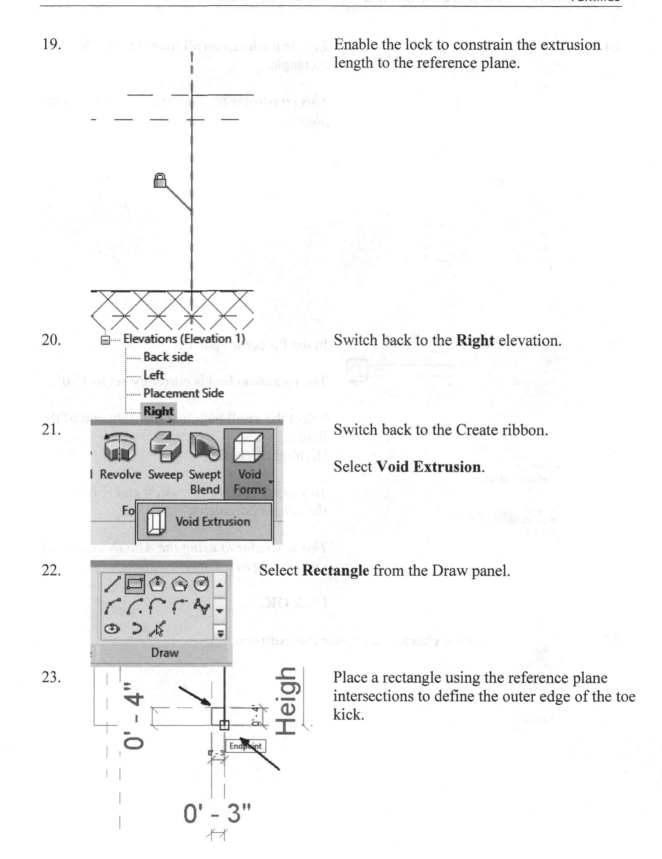

19.   Enable the lock to constrain the extrusion length to the reference plane.

20.   Elevations (Elevation 1)
         Back side
         Left
         Placement Side
         Right

     Switch back to the **Right** elevation.

21.   Revolve   Sweep   Swept Blend   Void Forms

     Void Extrusion

     Switch back to the Create ribbon.

     Select **Void Extrusion**.

22.   Draw

     Select **Rectangle** from the Draw panel.

23.   0' - 4"

     Height

     Endpoint

     0' - 3"

     Place a rectangle using the reference plane intersections to define the outer edge of the toe kick.

24.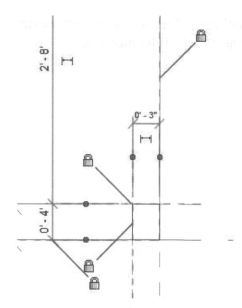

Enable the locks on all four sides of the rectangle.

This constrains the rectangle to the reference planes.

25.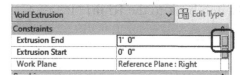

In the Properties palette:

The Extrusion End is currently set to 1'-0".

Select the small button at the right end of the field.

26.

| <none> |
| Default Elevation |
| Depth |
| Height |
| Length |
| Width |

Highlight **Length**.

This associates the extrusion end value with the Length parameter.

This is similar to using the ALIGN command and locking to a reference plane.

Click **OK**.

27.

Green check to complete the extrusion.

28.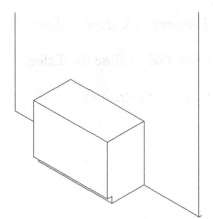

Switch to a 3D view.

Right now, the casework is a single type with one size and no material assigned.

29. Select **Family Types** from the Properties panel.

30.

Parameter	Value	Formula
Constraints		
Default Elevation	0' 0"	=
Construction		
Construction Type		=
Materials and Finishes		
Finish		=
Dimensions		
Length	4' 0"	=
Height	3' 0"	=
Depth	2' 0"	=
Width		=

Select **New** to create a family type.

31. **Name:** White Pine 48" Long

Type **White Pine 48" Long**.

Click **OK**.

32.

Materials and Finishes	
Finish	White Pine

Type **White Pine** in the **Finish** field.

Click **Apply**.

The Finish value is simply text, so it will not actually control the appearance of the extrusion.

33. White Pine 48" Long

Select **New** to create a family type.

34. **Name:** White Pine 36" Long

Type **White Pine 36" Long**.

Click **OK**.

35.

Dimensions	
Length	3' 0"
Height	3' 0"
Depth	2' 0"
Width	

Change the Length to **3'-0"**

Click **Apply.**

36.

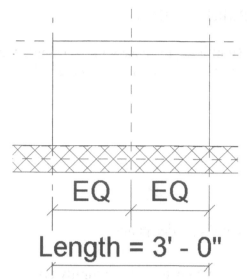

EQ | EQ

Length = 3' - 0"

The casework should adjust.

Click **OK** to close the dialog.

Select the extrusion.

37.

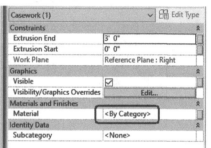

Click in the **Material** field to assign a material.

This will control the appearance of the casework.

38.

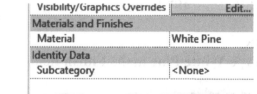

Locate the **White Pine** material and add it to your document.

Click **ESC** to release the selection.

39.

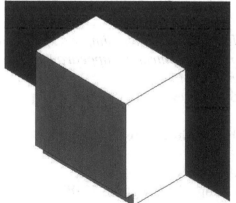

Switch to a 3D view.

Change the display to **Realistic**.

Save as *casework_2.rfa.*

Room Calculation Point

In most instances, families placed within a room are associated with the room in a schedule. There are a few circumstances in which a family instance does not report its room correctly. When families such as furniture, doors, windows, casework, specialty equipment and generic models are placed in a project, sometimes parts of their geometry are located outside a room, space, or within another family, which results in no calculable values being reported.

Room Calculation Points are used to make loadable families "room aware." Once you enable the room calculation point, the family will display the correct room in the project schedule.

To enable and modify the Room Calculation Point to reorient room aware families:

1. Select the family instance in the drawing area.

2. Click Modify | <Element> tab ➤ Mode panel ➤ 📝 Edit Family.

3. In the Family Editor, open a floor plan view of the family.

4. On the Properties palette, select the Room Calculation Point parameter in the Other section. The point is now visible in the drawing area as a green dot.

5. Select the Room Calculation Point and move it to a location that will not be obscured by geometry when placed in the project.

6. Click Modify tab ➤ Family Editor panel ➤ 📤 Load into Project.

Exercise 2-5
Room Calculation Point

Drawing Name: **i_room_calculation_point, casework_2.rfa**
Estimated Time to Completion: 20 Minutes

Scope
Modify a room calculation point on a family so it associates with the correct room in a schedule.

Solution

1.  Open the **Furniture Schedule**.

   There is one piece of furniture not associated with a room.

2.  Highlight the chair with no room shown.

3. 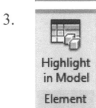 Select **Highlight in Model** on the tab.

4.  Click **OK**.

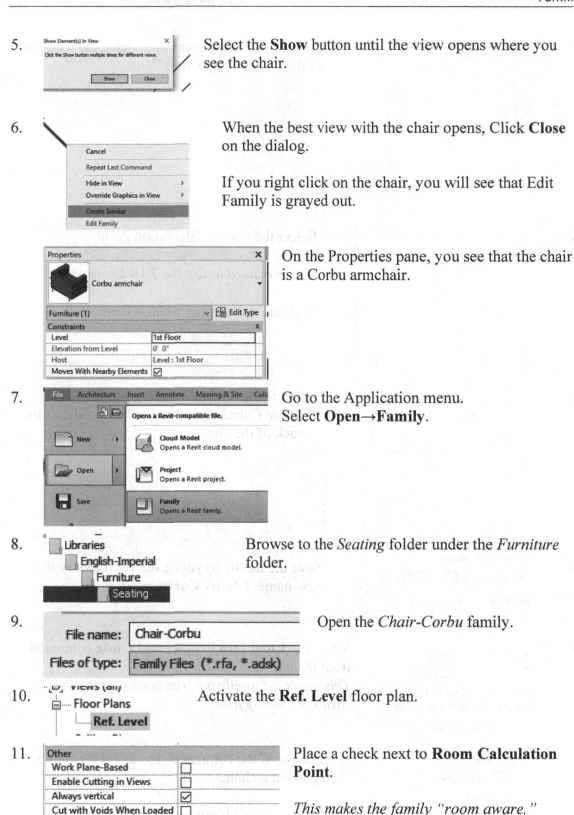

5. Select the **Show** button until the view opens where you see the chair.

6. When the best view with the chair opens, Click **Close** on the dialog.

   If you right click on the chair, you will see that Edit Family is grayed out.

   On the Properties pane, you see that the chair is a Corbu armchair.

7. Go to the Application menu.
   Select **Open→Family**.

8. Browse to the *Seating* folder under the *Furniture* folder.

9. Open the *Chair-Corbu* family.

10. Activate the **Ref. Level** floor plan.

11. Place a check next to **Room Calculation Point**.

    This makes the family "room aware."

12.

A green dot will appear to indicate the location of the Room Calculation Point.

13.

Select the Room Calculation Point.

You may need to use the TAB key to select.

A gizmo will appear when it is selected.

14.

Use the green vertical arrow to move the Room Calculation Point to the middle of the back of the chair.

15.

File name: Chair-Corbu w RCP

Files of type: Family Files (*.rfa)

Save the family to your exercise folder with a new name: **Chair- Corbu w RCP**.

16.

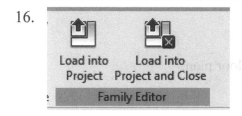

Select the **Load into Project and Close** command from the tab.
This loads the family into the open project and closes the family file.

17.

Floor Plans
  1st Floor
  2nd Floor
  Ground Floor

Cancel out of the command if your cursor shows you placing a chair.

Verify that the 1st Floor view is open and active.

18.

Edit Family
Select Previous
Select All Instances
Delete
Visible in View
In Entire Project

Select the chair. Right click and **Select All Instances→Visible in View**.

19. 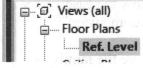 Use the Type Selector to select the **Chair- Corbu w RCP.**

The chair will shift position based on the new RCP.

20. Schedules/Quantities (all) Open the **Furniture Schedule**.
    Furniture Schedule
    Specialty Equipment Schedule

21.

<Furniture Schedule>			
A	**B**	**C**	**D**
Family and Type	Count	Room	Level
Chair-Corbu w RCP: Chair	1	LIVING	1st Floor
Chair-Corbu w RCP: Chair	1	LIVING	1st Floor
Chair-Corbu w RCP: Chair	1	LIVING	1st Floor
Chair-Corbu w RCP: Chair	1	LIVING	1st Floor
Chair-Corbu w RCP: Chair	1	LIVING	1st Floor

Note the chair is now associated with a room.

Close without saving.

22. Open *casework_2.rfa.*

This is the family that was created in the previous exercise.

23.

Family: Casework	⌄	Edit Type
Constraints		☆
Host	Wall	
Identity Data		☆
OmniClass Number	23.40.35.00	
OmniClass Title	Casework	
Other		☆
Always vertical	☑	
Cut with Voids When Loaded	☐	
Shared	☐	
Room Calculation Point	☑	

On the Properties palette:

Enable **Room Calculation Point**.

This allows the casework to report what room it is placed in.

24. Views (all) Switch the view to the **Ref. Level**.
    Floor Plans
    Ref. Level

25.

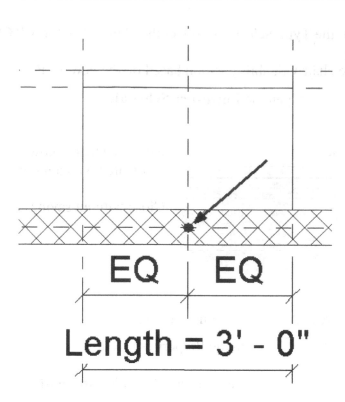

The Room Calculation Point is located inside a wall.

Select the RCP.

26.

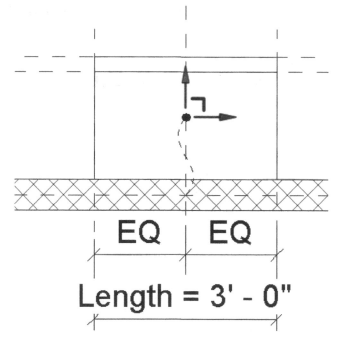

Adjust the position of the point so it is centered in the casework.

Now the casework will be "room aware".

Save the family as *casework_3.rfa*.

Symbolic Lines

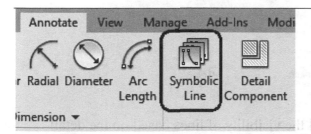

The **Symbolic Line** tool is available when you are creating or editing a family. It is located on the Annotate ribbon→Detail panel. This tool allows you to sketch lines that are meant for symbolic purposes only. Symbolic lines are not part of the actual geometry of the family. Symbolic lines are visible parallel to the view in which you sketch them.

For Example: You might sketch symbolic lines in an elevation view to represent a door swing.

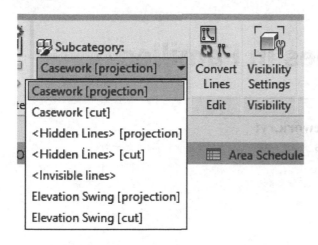

You can control symbolic line visibility on cut instances. Select the symbolic line and click Modify |Lines tab→Visibility panel (Visibility Settings). Select Show only if instance is cut.

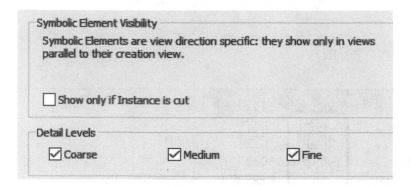

In the displayed dialog, you can also control the visibility of lines based on the detail level of the view. For example, if you select Coarse, that means that when you load the family into a project and place it in a view at the Coarse detail level, the symbolic lines are visible.

Exercise 2-6

Using Symbolic Lines in Families

Drawing Name: **casework_3.rfa, test_casework.rvt**
Estimated Time to Completion: 20 Minutes

Scope
Add symbolic lines to a family.
Control the visibility of family elements.
Load a family into a project.
Change the display of a loadable family.

Solution

1.   Views (all)   Open the **Ref. Level.**
              Floor Plans
                 Ref. Level

2. Select the **Reference Plane** tool from the Create ribbon.

3. Draw a horizontal reference plane through the casework.

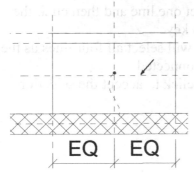

4. Use the ALIGNED DIMENSION to center the reference plane on the casework.

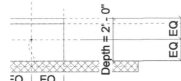

   Hint: *Place a segmented dimension and set the dimensions equal. Make sure you select reference planes. Use the TAB key to cycle through the selections if necessary.*

5. Switch to the Annotate ribbon.

   Select **Symbolic Line**.

6. Select **Elevation Swing [projection]** to assign the Subcategory.

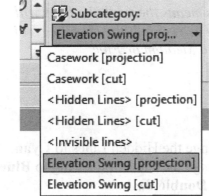

7. Place the four lines as shown using the midpoints.

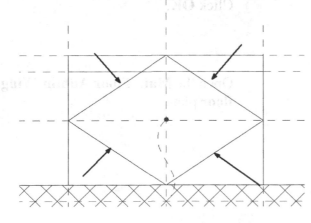

8.

Select one line and then click the TAB key.
This will select all four lines as they are connected.
Left click to accept the selection.

9. Select **Visibility Settings**.

10. Disable **Coarse**.

Click **OK**.

This means the symbolic lines will only be visible if the Detail Level is set to Medium or Fine.

11. Switch to the Manage ribbon.

Click **Object Styles**.

12.

Category	Line Weight		Line Color	Line Pattern
	Projection	Cut		
⊟ Casework	1	3	■ Black	
<Hidden Lines>	1	1	■ Cyan	Dash
Elevation Swing	1	1	■ Blue	Double dash
⊟ Walls	2	2	■ Black	
<Hidden Lines>	2	2	■ Black	Dash
Common Edges	2	2	■ Black	

Change the Hidden Lines to **Cyan**.
Change the Elevation Swing to **Blue** and **Double dash**.

Click **OK**.

13. Save as *casework_4.rfa*.

Open *test_casework.rvt*.

14.

Open the **Main Floor Admin Wing floor** plan.

15. 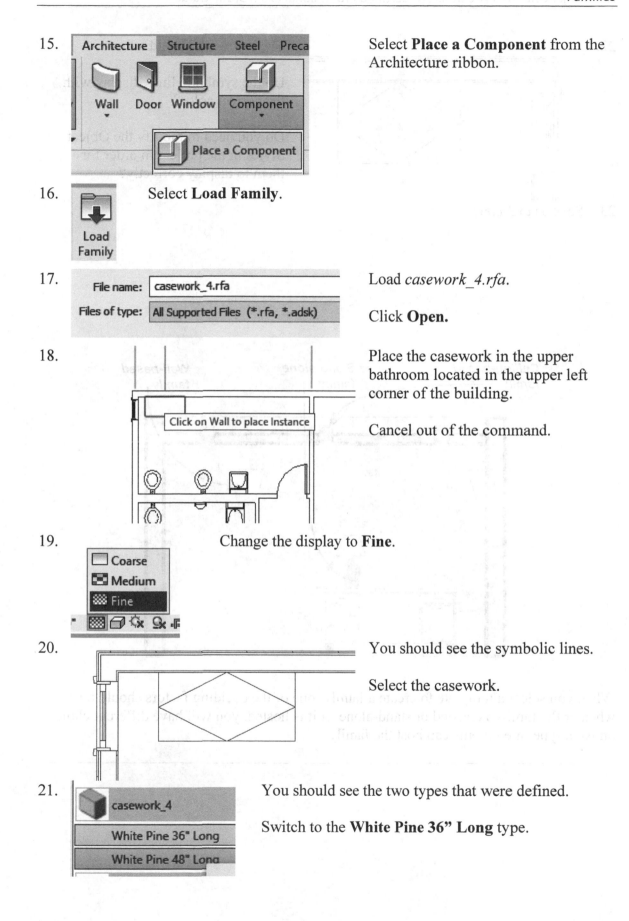 Select **Place a Component** from the Architecture ribbon.

16. Select **Load Family**.

17. Load *casework_4.rfa*.

    Click **Open.**

18. Place the casework in the upper bathroom located in the upper left corner of the building.

    Cancel out of the command.

19. Change the display to **Fine**.

20. You should see the symbolic lines.

    Select the casework.

21. You should see the two types that were defined.

    Switch to the **White Pine 36" Long** type.

22.

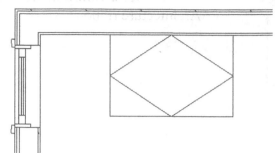

The casework should adjust.

Do the symbolic lines display with the correct color?

Do you need to modify the Object Styles in the project in order for them to display correctly?

23. Save as *ex2-6.rvt*.

Hosted vs. Non-Hosted Families

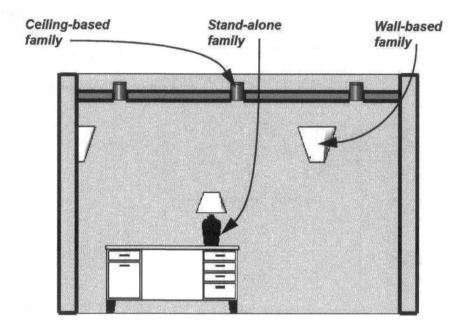

When you select a template to create a family, one of the deciding factors should be whether the family is hosted or stand-alone. If it is hosted, you will have different choices on what type of elements can host the family.

RFT Metric Generic Model ceiling based.rft
RFT Metric Generic Model face based.rft
RFT Metric Generic Model floor based.rft
RFT Metric Generic Model line based.rft
RFT Metric Generic Model Pattern Based.rft
RFT Metric Generic Model roof based.rft
RFT Metric Generic Model two level based.rft
RFT Metric Generic Model wall based.rft
RFT Metric Generic Model.rft

You should be familiar with the different options of how a family can be controlled using a host. For example, doors and windows are wall based. You cannot place a door or window without a wall.

Sub-Categories

When you create a family, the template assigns it to a category that defines the default display of the family (line weight, line color, line pattern, and material assignment of the family geometry) when the family is loaded into a project. To assign different line weights, line colors, line patterns, and material assignments to different geometric components of the family, you need to create subcategories within the category. Later, when you create the family geometry, you assign the appropriate components to the subcategories.

For example, in a window family, you could assign the frame, sash, and mullions to one subcategory and the glass to another. You could then assign different materials (wood and glass) to each subcategory to achieve the following effect.

Detail Components

Detail components are line-based 2D elements that you can add to detail views or drafting views. They are visible only in those views. They scale with the model, rather than the sheet.

Detail components are not associated with the model elements that are part of the building model. Instead, they provide construction details or other information in a specific view.

For example, in the following drafting view, the studs, insulation, and siding are detail components.

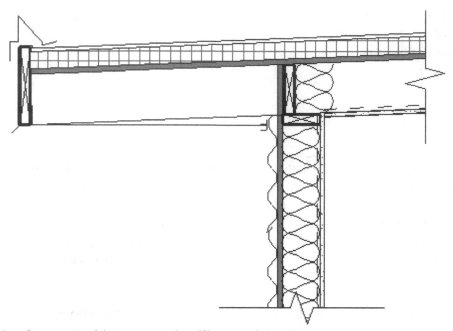

Revit comes with an extensive library of detail components. Many companies have AutoCAD legacy data of detail views for different construction views. For example, threshold and jamb details were drawn in AutoCAD and can be re-used and imported into Revit, so you don't have to re-create those details. These details may be used across different projects if you use the same construction materials in different buildings.

Filled Regions

The Filled Region tool creates a 2-dimensional, view-specific graphic with a boundary line style and fill pattern within the closed boundary. The tool is useful for defining a filled area in a detail view or for adding a filled region to an annotation family.

The filled region is parallel to the view's sketch plane.

A filled region contains a fill pattern. Fill patterns are of 2 types: Drafting or Model. Drafting fill patterns are based on the scale of the view. Model fill patterns are based on the actual dimensions in the building model.

A filled region created for a detail view is part of the Detail Items category. Revit lists the region in the Project Browser under Families ➤ Detail Items ➤ Filled Region. If you create a filled region as part of an annotation family, Revit identifies it as a Filled Region, but does not store it in the Project Browser.

Exercise 2-7

Create a Detail Component Family

Drawing Name: **new using detail item template**
Estimated Time to Completion: 20 Minutes

Scope
Create a detail component family.

Solution

1. 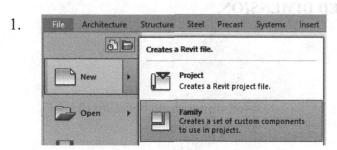 From the Application menu:

   Select **File→New→Family**.

2. 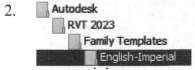 Browse to the *English-Imperial* folder under Family Templates.

3. 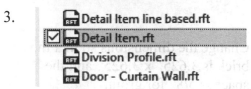 Select *Detail Item.rft*.

4. 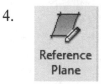 Select the **Reference Plane** tool.

5.

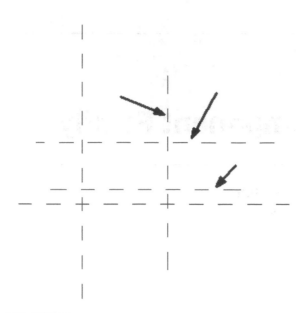

Draw two horizontal reference planes and one vertical reference plane.

6.

Select the **ALIGNED DIMENSION**.

7.

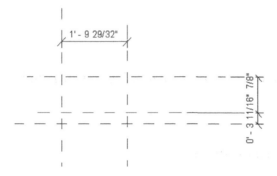

Add dimensions between the reference planes.

8.

Change the dimensions so that the brick is 3.625" x 3.625" and the space is 5/8" for grout.

9.

Select the **Filled Region** tool on the Create ribbon.

10.  Select the **Rectangle** tool from the Draw panel.

11. Verify that the Subcategory is set to **Detail Items**.

12. Use the Type Selector to set the Filled Region to **Brick**.

13. Draw the rectangle in the upper area.

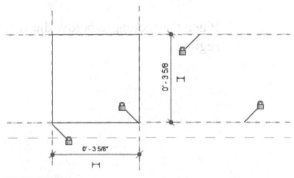

14. Enable all the locks to constrain the filled region to the reference planes.

15.  Green check to complete the filled region.

16. 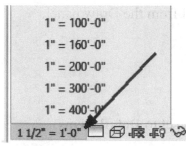 Change the scale to 1 ½" = 1'-0" to see the filled region hatch pattern.

17. Select the **Masking Region** tool from the Create ribbon.

18. 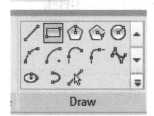 Select the **Rectangle** tool from the Draw panel.

19. 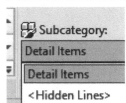 Verify that the Subcategory is set to **Detail Items**.

20. 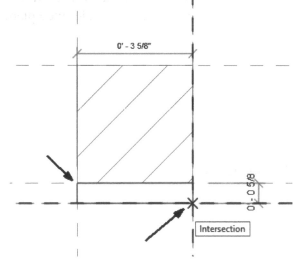 Place the rectangle below the filled region.

21.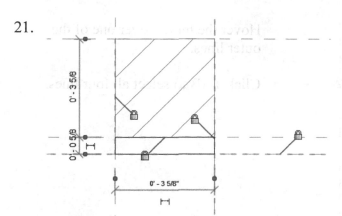
Enable the locks to constrain the masking region to the reference planes.

22.
Green check to complete the masked region.

23.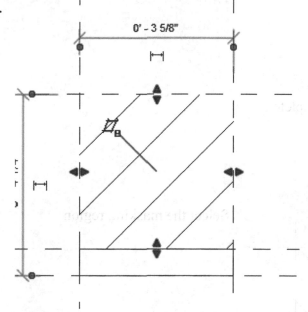
Select the filled region.

24.
Select **Edit Boundary**.

25.

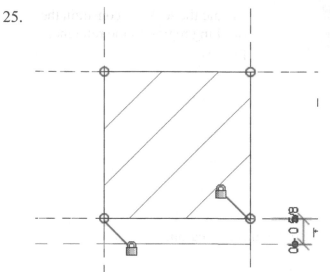

Hover the mouse over one of the outer lines.

Click TAB to select all four lines.

26.

Change the Subcategory to Heavy **Lines**.

27.

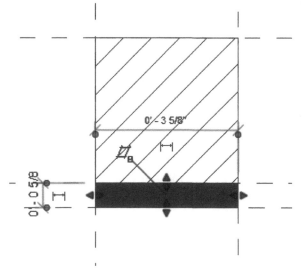

Green check to complete.

28.

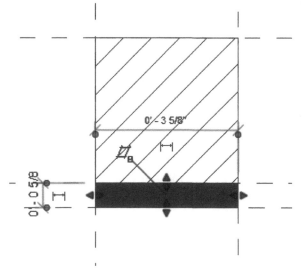

Select the masking region.

29.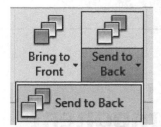

Click **Send to Back** on the ribbon.

This ensures the filled region takes priority over the masking region.

30.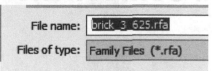

Save as *brick_3_625.rfa*.

Exercise 2-8

Place a Repeating Detail Component Family

Drawing Name: **repeating detail.rvt**
Estimated Time to Completion: 20 Minutes

Scope
Add a repeating detail to a view.

Solution

1. Sections (Detail)
    Callout of Section 1
    Callout of Section 2
    Repeating Detail

    Open the **Repeating Detail** section view.

2. Load Family

    Detail components need to be loaded BEFORE you select the tools to place them.

    Switch to the Insert ribbon.

    Select **Load Family.**

3. File name: brick_3_625.rfa
    Files of type: All Supported Files (*.rfa, *.adsk)

    Select the *brick_3_625.rfa* file.

    Click **Open**.

4.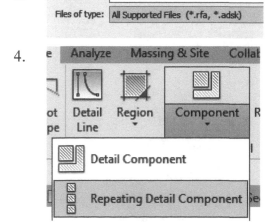

    Switch to the Annotate ribbon.

    Select **Repeating Detail Component**.

5.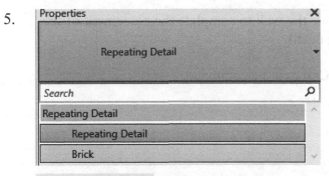

The Repeating Detail has one type called Brick.

Click **Edit Type**.

6. Click **Duplicate**.

7. Type **Repeating Detail – Brick 3 5/8"**.

Click **OK**.

8. Set the Detail to **brick_3_625**.

Set the Layout to **Maximum Spacing**.

Set the Spacing to **4"**.

Click **OK**.

Type Parameters	
Parameter	
Pattern	
Detail	brick_3_625
Layout	Maximum Spacing
Inside	☐
Spacing	0' 4"
Detail Rotation	None

9.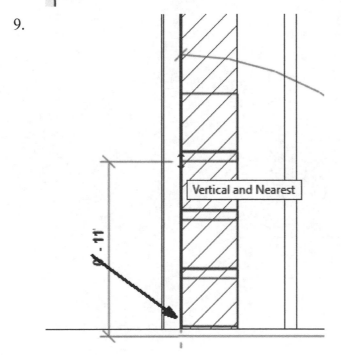

Select the bottom of the wall and move the mouse up the wall to see the brick.

10.

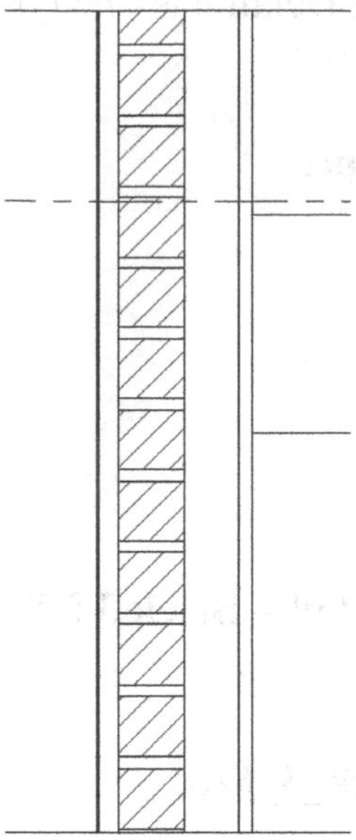

Click above the wall to complete the placement of the repeating detail.

Click ESC to exit the command.

11. Save as *ex2-8.rvt*.

Exercise 2-9

Create a Family using an Array

Drawing Name: **bike rack.rfa**
Estimated Time to Completion: 20 Minutes

Scope
Create a bike rack family using an array formula.
Add a reference plane.
Add an aligned dimension.
Add a family parameter.
Link a dimension to a parameter.
Link an array to a parameter.
Create new family types.

Solution

1. 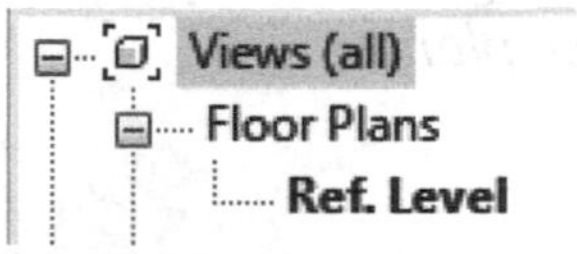 Views (all)
    Floor Plans
    Ref. Level

    Start in the **Ref. Level** view.

2.  Reference Plane

    Select the **Reference Plane** tool from the Create ribbon.

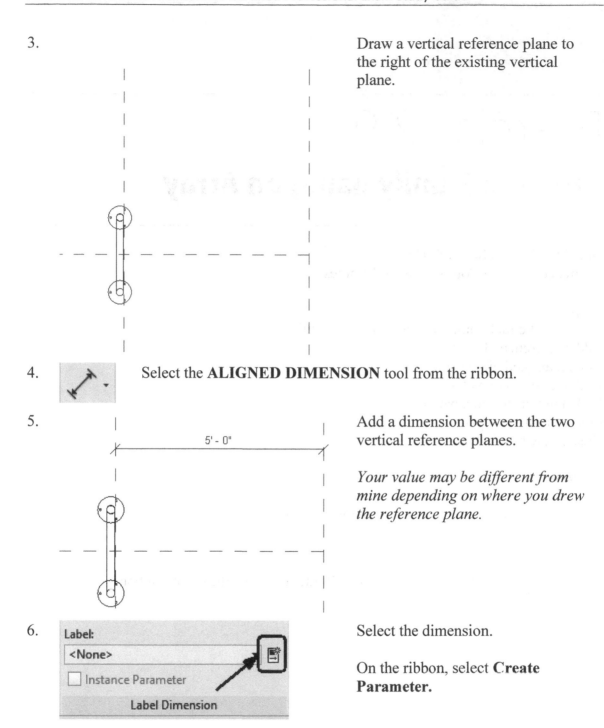

3.   Draw a vertical reference plane to the right of the existing vertical plane.

4.   Select the **ALIGNED DIMENSION** tool from the ribbon.

5.   Add a dimension between the two vertical reference planes.

5' - 0"

Your value may be different from mine depending on where you drew the reference plane.

6.   Select the dimension.

Label:

<None>

☐ Instance Parameter

Label Dimension

On the ribbon, select **Create Parameter.**

7.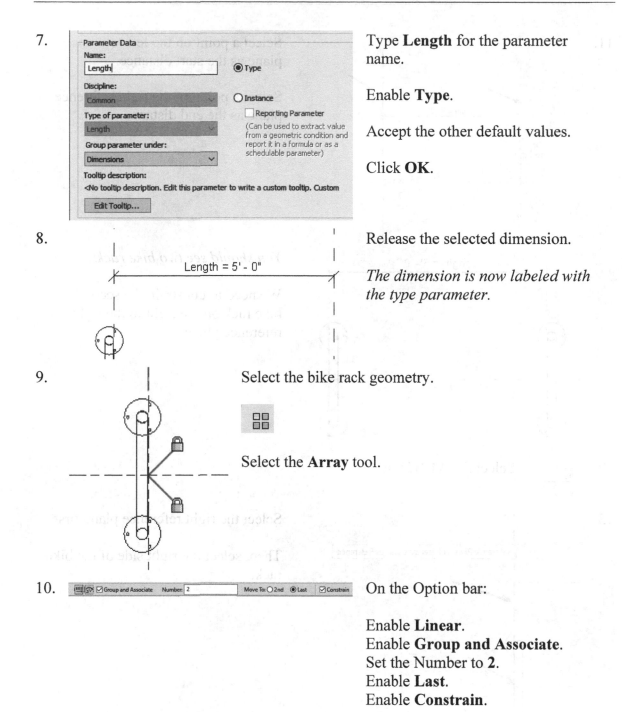

Type **Length** for the parameter name.

Enable **Type**.

Accept the other default values.

Click **OK**.

8.

Release the selected dimension.

The dimension is now labeled with the type parameter.

9.

Select the bike rack geometry.

Select the **Array** tool.

10.

On the Option bar:

Enable **Linear**.
Enable **Group and Associate**.
Set the Number to **2**.
Enable **Last**.
Enable **Constrain**.

11.

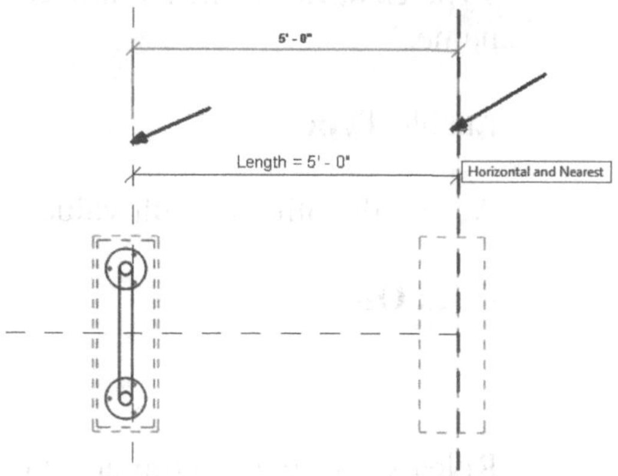

5' - 0"

Length = 5' - 0" | Horizontal and Nearest |

Length = 5' - 0"

Select a point on the left reference plane as the start distance.

Select a point on the right reference plane as the end distance.

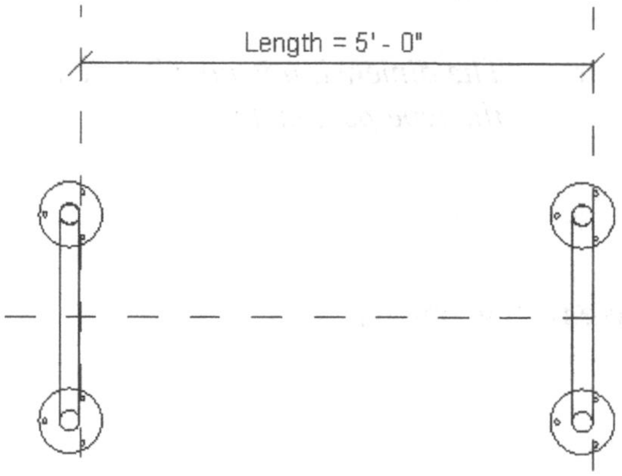

You should see two bike racks.

We need to constrain the second bike rack on the right to the right reference plane.

12. 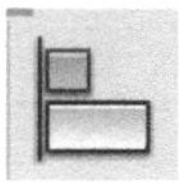   Select the **ALIGN** tool.

13. 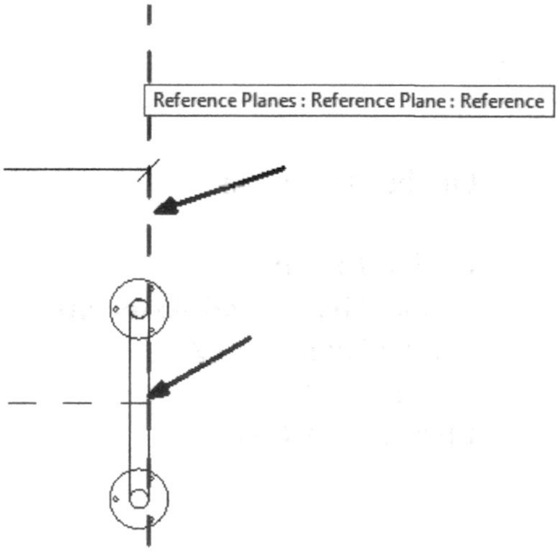 | Reference Planes : Reference Plane : Reference |

Select the right reference plane first.

Then, select the right side of the bike rack.

14. 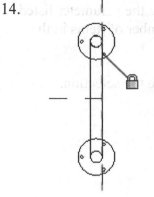 Enable the lock to constrain the bike rack to the reference plane.

15. 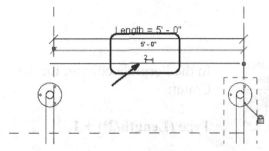 Select the bike rack on the right to "wake up" the array constraint.

If you did not enable Group and Associate when creating the array, you will not have an array constraint applied. You will have to go back to Step 9 and do it over.

16. 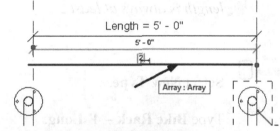 Select the line below the number 2.

17. 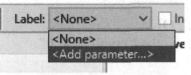 Select **Add parameter** from the Options bar.

18. 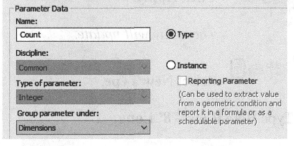 Type **Count** for the Name.

Enable **Type**.

Group parameter under **Dimensions**.

Click **OK**.

19.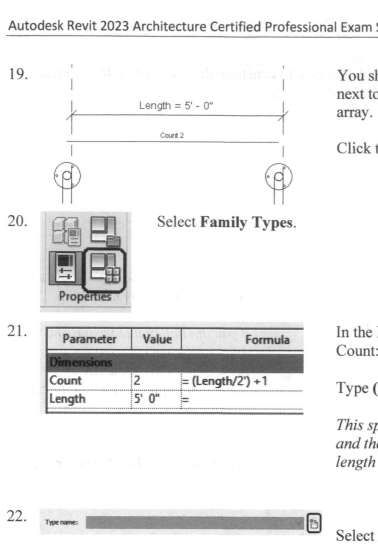

You should see the parameter listed next to the number of items in the array.

Click to release the selection.

20. Select **Family Types**.

21.

Parameter	Value	Formula
Dimensions		
Count	2	= (Length/2') +1
Length	5' 0"	=

In the Formula column for the Count:

Type **(Length/2') + 1**.

This spaces the bike racks 2' apart and the +1 ensures that the total length is always at least 2'.

22. Type name: Select **New Type**.

23. Name: Bike Rack - 4' Long

Type **Bike Rack – 4' Long**.

Click **OK**.

24. Change the Length to **4' 0"**.

Parameter	Value	Formula
Dimensions		
Count	3	= (Length / 2') + 1
Length	4' 0"	=

Click **Apply**.

The array will update.

25. Type name: Select **New Type**.

26. Name: Bike Rack – 8' Long

Type **Bike Rack – 8' Long**.

Click **OK**.

27.

Parameter	Value	Formula
Dimensions		
Count	5	= (Length / 2') + 1
Length	8' 0"	=

Change the Length to **8' 0"**.

Click **Apply**.

The array will update.

28.

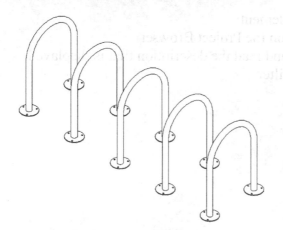

Click **OK to close the dialog**.

Switch to a 3D view, so you can inspect your bike rack.

Save your bike rack file.

Try loading the bike rack into a project using the Component tool and experiment with different lengths.

Practice Exam

1. To identify a family type of an element:
    A. Select the element, look on the Project Browser
    B. Hover over the element and read the description that is displayed
    C. Select the element, use Filter
    D. Use Insert, Load Family

2. True or False

    Furniture families are hosted.

3. True or False

    Window families are hosted.

4. To add a loadable family to a project, use the [fill in the blank] tab:

    A. Create
    B. Insert
    C. Family
    D. Architecture

5. To create a new system family you:

    A. Use a Revit family template
    B. Modify the instance parameters of an existing similar family
    C. Duplicate a similar system family and modify the type parameters
    D. Modify the type parameters of a similar system family

6. When a room schedule does not reflect the correct room for an element, the element family should be modified by:

    A. Moving it into the correct room
    B. Modifying the element's base point
    C. Making the element "room-aware"
    D. Modifying the room schedule

7. Room Calculation Points are used to:
   A. Calculate the square footage of a room
   B. Make loadable families "room aware"
   C. Calculate the volume of a room
   D. Create room schedules

8. To create a new door type of a door family without exiting the active project:
   A. Select the door, right click and select Edit Family
   B. Go to the Properties panel, select Edit Type
   C. Go to the Project Browser, locate the door family, right click and select New Type
   D. Go to the Insert tab, select Load Family

9. A designer is creating a door family. The designer wants to show the door swing in plan views but not 3D views. What should the designer use to create the door swing?
   A. Sweep
   B. Model Line
   C. Symbolic Line
   D. Revolve

10. A designer is creating a door family. The designer creates and extrusion and wants the length and width to adjust for various family types. How can this be accomplished?
   A. Dimension the length and width of the extrusion and label the dimensions with length and width parameters.
   B. Add family parameters for Length and Width.
   C. Associate length and width family parameters to the extrusion's Length and Width parameters.
   D. Constrain the extrusion to reference planes that are driven by length and width parameters.

11. Which of the following best describes the hierarchy of elements in a Revit project?
   A. Roofs > Walls > Floors > Ceilings
   B. Category > Family > Type > Instance
   C. Category > Type > Family > Instance
   D. Family > Category > Instance > Type

12. A designer is modeling doors and windows, but the families are not available in the project. What should the designer do to get additional door and window families in the project?
    A. Click the File tab and select Open > Family and then select the family file.
    B. Doors and windows are system families and are defined within the Revit project.
    C. On the Insert ribbon, in the Load from Library panel, click Load Family.
    D. Find the window or door family in a File Explorer, right-click, and select Import into Revit Project.

13. T F   You can change the category for loadable families, but not system families.

14. Floors, roofs, and walls typically are examples of which kind of family?
    A. In-place
    B. Generic
    C. Loadable
    D. Systems

15. You select an element in a building project. The Edit Family tool appears on the ribbon. This indicates that the selected element is a _____ family.:

    A. In-place
    B. Generic
    C. Loadable
    D. Systems

16. Dimensions are _____ families.

    A. In-place
    B. Generic
    C. Loadable
    D. Systems

Answers:
   1) B; 2) F; 3) T; 4) B; 5) C; 6) C' 7) B; 8) C' 9) C; 10) D; 11) B; 12) C; 13) T; 14) D; 15) C; 16) D

Lesson

03

Documentation

This lesson addresses the following certification exam questions:

- Multi-Segmented Dimensions
- Dimensional Constraints
- Global Parameters
- Project Parameters
- Shared Parameters
- Adding Tags
- Color Schemes
- Color Scheme Legends
- Design Options
- Phases
- Revision Control
- Exporting Sheets

Dimensions

Dimensions are system families. They are in the Annotation category. They have type and instance properties.

Revit has three types of dimensions: listening, temporary and permanent.

A listening dimension is the dimension that is displayed as you are drawing, modifying, or moving an element.

A temporary dimension is displayed when an element is placed or selected. In order to modify a temporary dimension, you must select the element.

A permanent dimension is placed using the Dimension tool. In order to modify a permanent dimension, you must move the element to a new position. The permanent dimension will automatically update. To reposition an element, you can modify the temporary dimension or move the element using listening dimensions.

When you enter dimension values using feet and inches, you do not have to enter the units. You can separate the feet and inches values with a space and Revit will fill in the units. If a single unit is entered, for example '10', Revit assumes that value is 10 feet, not 10 inches.

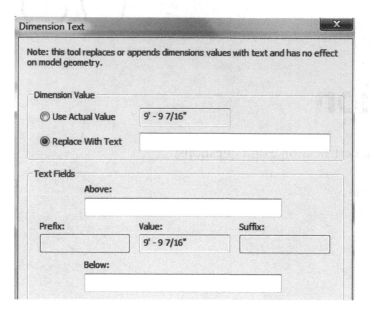

To override a dimension value, select the permanent dimension and enable Replace with Text and put in the desired text.

You can also add additional notes using the Text Fields for Above, Prefix, Suffix, or Below.

Revit has the capability of including symbols in textual notes. It is highly likely to come up as an exam question on the certification exam, so be prepared to demonstrate how to add a symbol to a note.

Exercise 3-1
Create Dimensions

Drawing Name: **i_dimensions.rvt**
Estimated Time to Completion: 20 Minutes

Scope
Match Properties
Placing dimensions

Solution

1. 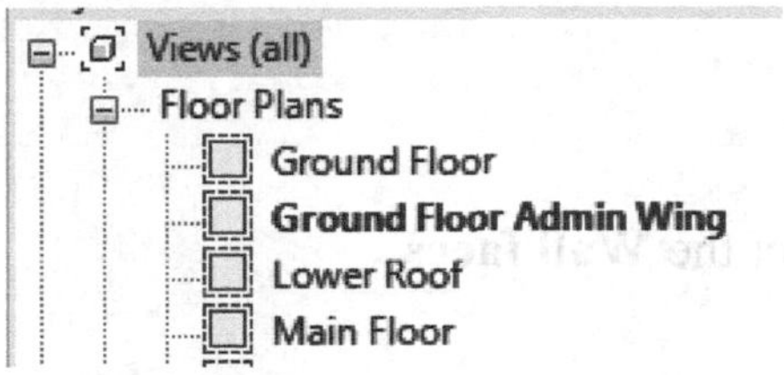 Activate the **Ground Floor Admin Wing** floor plan.

    ⊟ ◻ Views (all)
    　⊟ Floor Plans
    　　　Ground Floor
    　　　Ground Floor Admin Wing
    　　　Lower Roof
    　　　Main Floor

2.  Activate the Modify tab on tab on ribbon.
    Select the **Match Properties** tool from the Clipboard panel.

3. 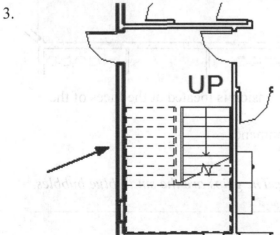 Select the wall indicated as the source object.

4. 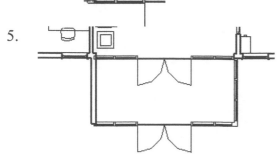 Select the wall indicated as the target object.

5. Note that the curtain wall changed to an Exterior-Siding wall.

   Cancel out of the Match Properties command.

6. 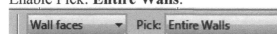 Activate the Annotate ribbon.
   Select the **Aligned Dimension** tool.

   Aligned I

7. On the Options bar, set the dimensions to select the **Wall faces**.
   Enable Pick: **Entire Walls**.

   | Wall faces | ▼ | Pick: Entire Walls | ▼ |

8. Pick the wall indicated. Move the mouse above the selected wall to place the dimension.

   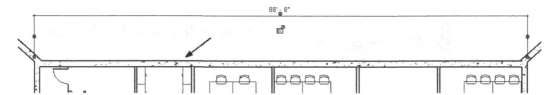

9. Note the entire wall is selected and the dimension is located at the faces of the walls.
   Cancel or escape to end the Dimension command.

10. Select the dimension.
    Note that there are several grips available. The grips are the small blue bubbles.

    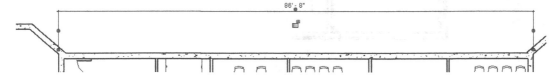

11. 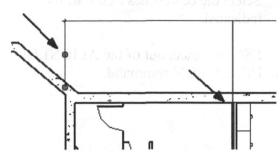 Select the middle grip indicated on the left.
Move the witness line to the wall face indicated by the arrow in the center of the image.

Note that the witness line automatically snaps to the wall face.

12. 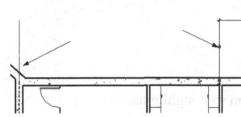 With the dimension selected,
on the Options bar, change the Prefer to **Wall centerlines**.

13. 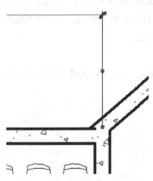 Select the dimension and activate the grip indicated.
Drag the witness line to the center of the left wall.

14. Left click once on the grip on the right side of the dimension to shift the witness line to the wall centerline.

15. *Note how the dimension value updates.*

16. 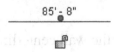 Select the **Aligned** tool.

17. Set **Pick to Individual References** on the options bar.

18.

Select the centerlines of the walls indicated.

ESC or Cancel out of the ALIGNED DIMENSION command.

19.

Select the dimension so it highlights.

Select the two locks on the top dimensions to switch the permanent dimensions to *locked* dimensions. This means these distances will not be changed.

You should see two of the padlocks as closed and one as open.

20.

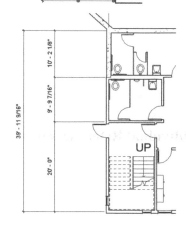

Place an overall dimension using the wall centerlines.

ESC or Cancel out of the ALIGNED DIMENSION command.

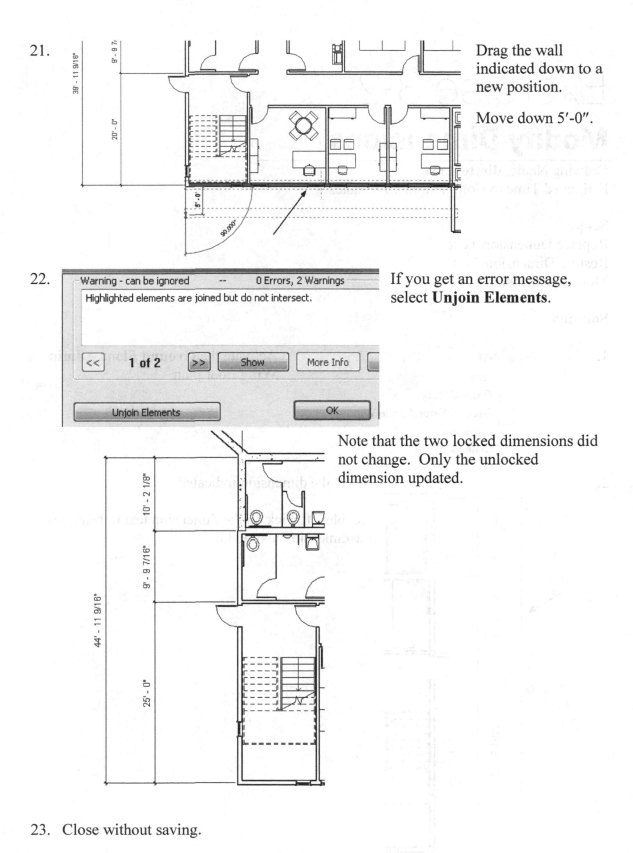

21. Drag the wall indicated down to a new position.

    Move down 5'-0".

22. If you get an error message, select **Unjoin Elements**.

    Warning - can be ignored -- 0 Errors, 2 Warnings

    Highlighted elements are joined but do not intersect.

    << 1 of 2 >> Show More Info

    Unjoin Elements OK

    Note that the two locked dimensions did not change. Only the unlocked dimension updated.

23. Close without saving.

Exercise 3-2
Modify Dimensions

Drawing Name: **dimtext.rvt**
Estimated Time to Completion: 15 Minutes

Scope
Replace Dimension Text
Restore Dimension Text
Modify Dimension Text

Solution

1.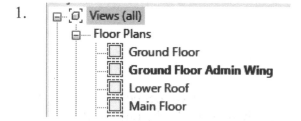
Activate the **Ground Floor Admin Wing** floor plan.

2.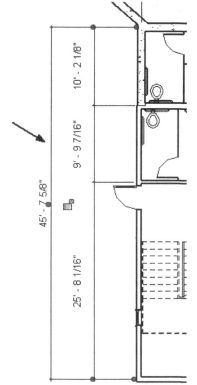
Select the dimension indicated.

Double left click on the dimension text to bring up the dimension text dialog.

3.

Dimension Value

○ Use Actual Value   45' - 7 5/8"

● Replace With Text   45' 8"

Enable **Replace with Text**.
Enter **45′ 8″**.
Click **OK**.

4.

Invalid Dimension Value   ×

Specify descriptive text for a dimension
segment instead of a numeric value.

To change the dimension value for a length or angle of a
segment, select the element the dimension refers to, and click
the value to edit it.

You will get an error message stating that
you cannot change the numeric value
without moving the element to correspond
with that value.

Click **Close**.

5.

Dimension Value

○ Use Actual Value   45' - 7 5/8"

● Replace With Text   OVERALL

Enable **Replace with Text**.
Enter **OVERALL**.
Click **OK**.

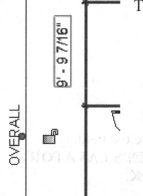

The dimension text updates.

6.

Project
Units

Activate the **Manage** ribbon.
Select **Project Units** on the Settings panel.

7.

Discipline: Common

Units	Format
Angle	12.35°
Area	1235 SF
Cost per Area	[$/ft²] 1235
Distance	1235 [']
Length	1' - 5 11/32"
Mass Density	1234.57 lb/ft
Rotation Angle	12.35°

Select the **Length** button under the Format
column.

8.

Units:

Rounding:

To the nearest 1/2"

Set the Rounding **To the nearest ½″**.
Click **OK** until all dialogs are closed.

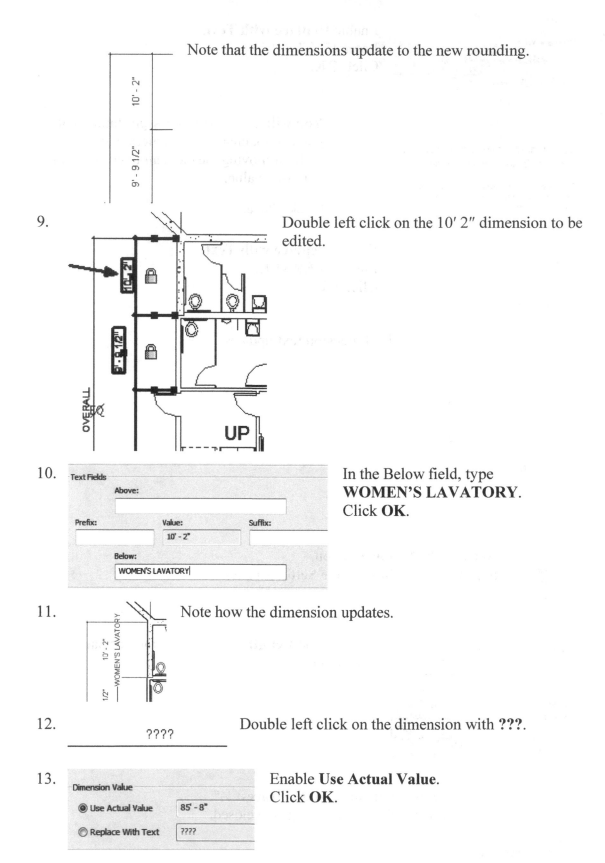

Note that the dimensions update to the new rounding.

9. Double left click on the 10′ 2″ dimension to be edited.

10. In the Below field, type **WOMEN'S LAVATORY**. Click **OK**.

11. Note how the dimension updates.

12. Double left click on the dimension with **???**.

13. Enable **Use Actual Value**. Click **OK**.

14. Close without saving.

Exercise 3-3

Converting Temporary Dimensions to Permanent Dimensions

Drawing Name: **i_dimensions.rvt**
Estimated Time to Completion: 10 Minutes

Scope
Place dimensions using different options
Convert temporary dimension to permanent

Solution

1.

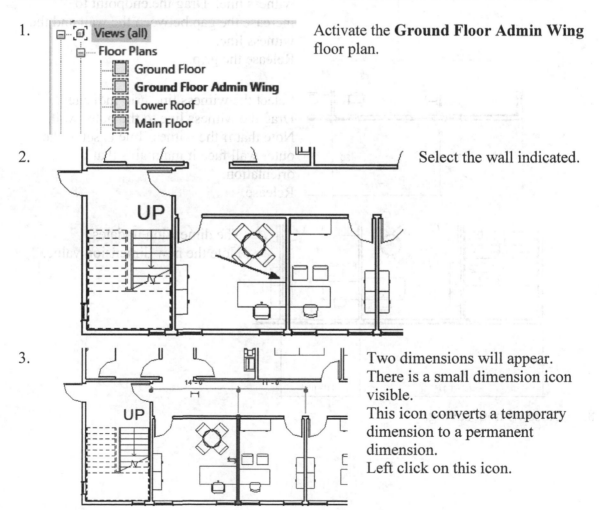

Activate the **Ground Floor Admin Wing** floor plan.

2.

Select the wall indicated.

3.

Two dimensions will appear.
There is a small dimension icon visible.
This icon converts a temporary dimension to a permanent dimension.
Left click on this icon.

4.

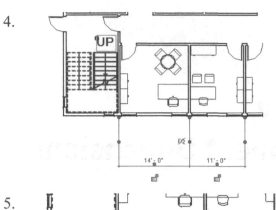

The dimensions are now converted to permanent dimensions.
Drag the dimensions below the view.

5.

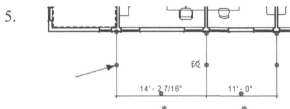

Click on the dimensions so they highlight.
Click on the witness line grip indicated.

Notice how each time the grip is clicked on, the witness line shifts from the wall centerline to the face of the wall.

6.

Select the grip at the endpoint of the witness line. Drag the endpoint to increase the gap between the wall and the witness line.
Release the grip.

7.

Select the witness line grip indicated.
Drag the witness line to the outer wall.
Note that if the witness line is set to the outer wall face it maintains that orientation.
Release.

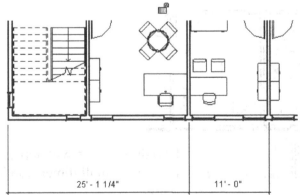

The dimension updates.
Note the new dimension value.

8.   Close without saving.

Exercise 3-4
Multi-Segmented Dimensions

Drawing Name: **multisegment.rvt**
Estimated Time to Completion: 10 Minutes

Scope
Add and delete witness lines to a multi-segment dimension

Solution

1. 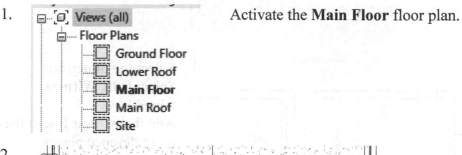 Activate the **Main Floor** floor plan.

2. 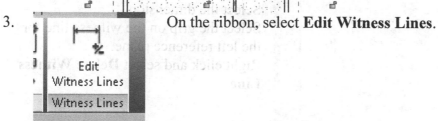 Select the lower dimension so it highlights.

3. 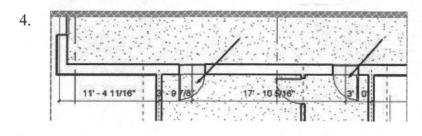 On the ribbon, select **Edit Witness Lines**.

4. Select the center of each door indicated. Then, left pick below the building to accept the

additions.

5.

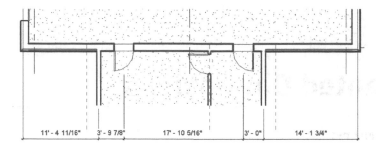

Drag the dimensions down so you can see them easily.

6.

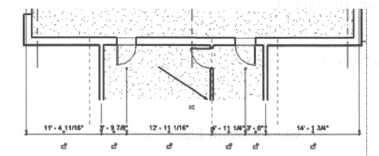

Select the dimension.
Select **Edit Witness Lines**.
Add the witness line at the wall indicated.

7.

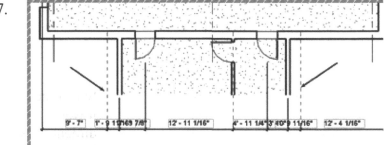

Select the dimension.
Select **Edit Witness Lines**.
Add the witness line at the two reference planes indicated.

Left pick below the building to accept the additions.

8.

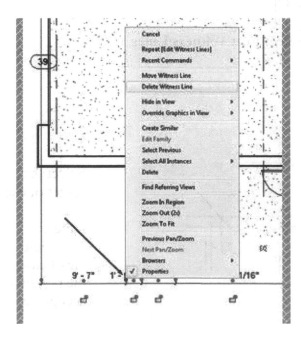

Select the dimension.
Select the grip on the witness line for the left reference plane.
Right click and select **Delete Witness Line**.

9. 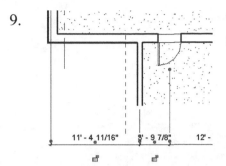 The dimension should update with the witness line removed.

10. 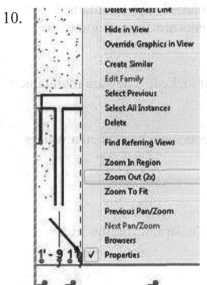 Repeat to delete the witness line on the right reference plane.

11. 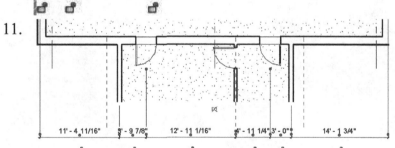 The dimension should update.

12. Close without saving.

Constraints

Revit uses three types of constraints:

- Explicit
- Loose
- Implied

An explicit constraint determines that defined relationships are always maintained. An example of an explicit constraint would be when two elements are aligned and then locked or when a dimension is locked.

A loose constraint is a dimension or alignment which is not locked. They are maintained unless a conflict occurs.

Implied constraints are also maintained unless a conflict occurs, such as when a wall is attached to a roof or two walls are joined at a corner.

An Equality constraint is considered an explicit constraint.

You can display constraints in a view by selecting Reveal Constraints on the View Control bar.

Exercise 3-5
Applying Constraints

Drawing Name: **i_constraints.rvt**
Estimated Time to Completion: 20 Minutes

Scope
Using the Align tool to constrain elements

Solution

1.   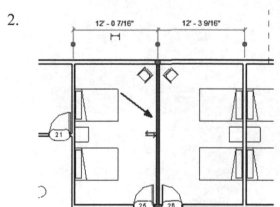   Activate the **Main Floor** floor plan.

2.   Select the interior wall located between Door
     25 and Door 26 as indicated.

3.

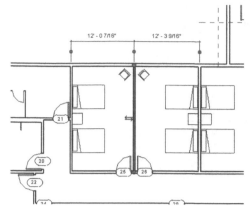

The associated temporary dimensions will become visible.
Left click on the permanent dimension toggle to convert the dimensions to permanent dimensions.

Left click anywhere in the display window to release the selection.

4.   In the Properties pane scroll down to the **Underlay** parameter.
Set the Underlay
Range: Base Level to **Ground Floor**.

Underlay	
Range: Base Level	Ground Floor
Range: Top Level	Main Floor
Underlay Orientation	Look down

This will make the ground floor visible in the graphics window. Elements on the ground floor will be displayed in a lighter shade.

5.

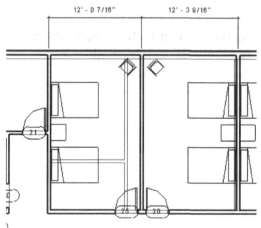

Note in the upper left corner the interior walls are not aligned.

6.   Activate the **Modify** tab on tab on ribbon.
Select the **Align** tool from the Modify panel.

7.

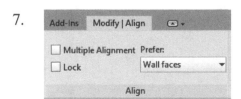

Set Prefer to **Wall faces** on the ribbon.

8.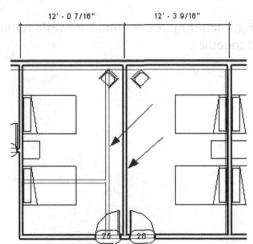

Select the right side of the ground floor interior wall as the source for the alignment.

This is the wall that is grayed out on the left.

9.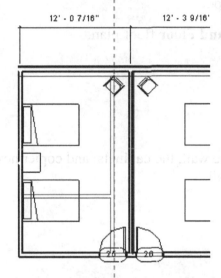

Select the right side of the main floor interior wall as the target for the alignment (this is the wall that will be shifted).

10.

Warning

Insert conflicts with joined Wall.

You will get an error that the door is now interfering with the wall; you can ignore the error.

11. 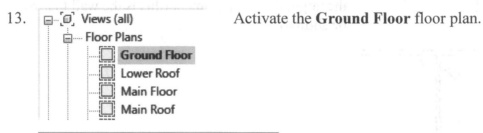 Lock the alignment in place so that the walls remain constrained together.

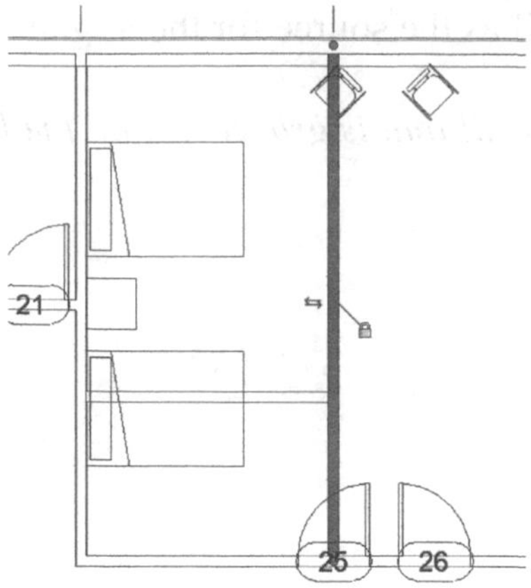

12. Right click and select Cancel to exit the Align mode.

13. Activate the **Ground Floor** floor plan.

Views (all)
  Floor Plans
    Ground Floor
    Lower Roof
    Main Floor
    Main Roof

14. Select the wall, the cabinets, and copier next to the wall.

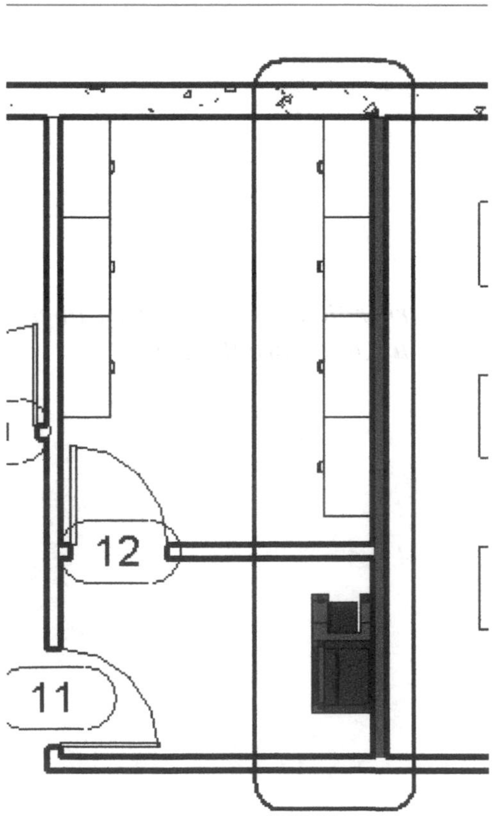

15. Select the **Move** tool on the Modify tab on tab on ribbon.

16. 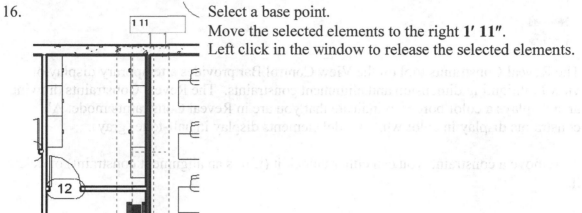 Select a base point.
Move the selected elements to the right **1' 11"**.
Left click in the window to release the selected elements.

17. 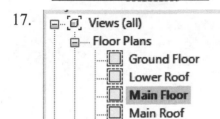 Activate the **Main Floor** floor plan.

18. 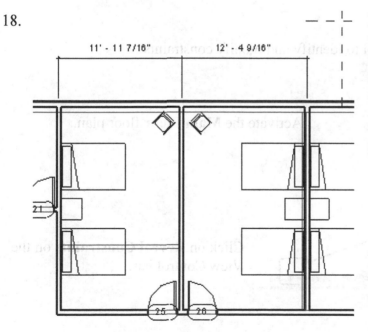 Note that the main floor interior wall has also shifted (the walls remained aligned because they are locked together) and the dimension has updated.

The door is no longer interfering with the wall.

19. Close without saving.

Reveal Constraints

The Reveal Constraints tool on the View Control Bar provides a temporary display of view highlighting dimension and alignment constraints. The Reveal Constraints drawing area displays a color border to indicate that you are in Reveal Constraints mode. All constraints display in color while model elements display in half-tone (gray).

To remove a constraint, you can either unlock it (if it is an alignment constraint) or delete it.

Exercise 3-6

Reveal Constraints

Drawing Name: **reveal_constraints.rvt**
Estimated Time to Completion: 10 Minutes

Scope
Using the Reveal Constraints tool to identify and modify constraints

Solution

1.

Activate the **Main Floor** floor plan.

2.

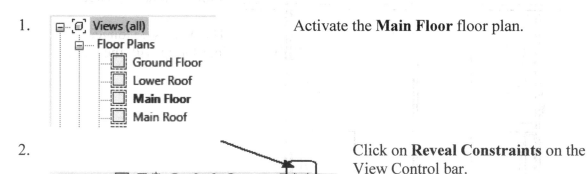

Click on **Reveal Constraints** on the View Control bar.

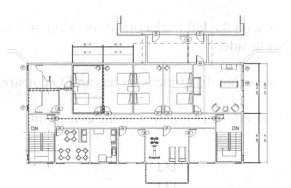

Notice that there are dashed lines as well as dimensions highlighted.

The dashed lines indicate alignment constraints.

3.

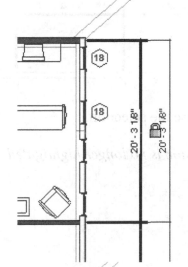

Click on the top vertical dimension on the east side.

Notice this dimension has a lock on it.

Click on the lock to toggle the lock OFF.

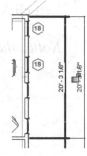

4.

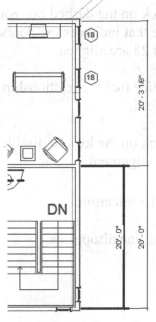

Left click to release the selection.

Notice the unlocked dimension is no longer highlighted.

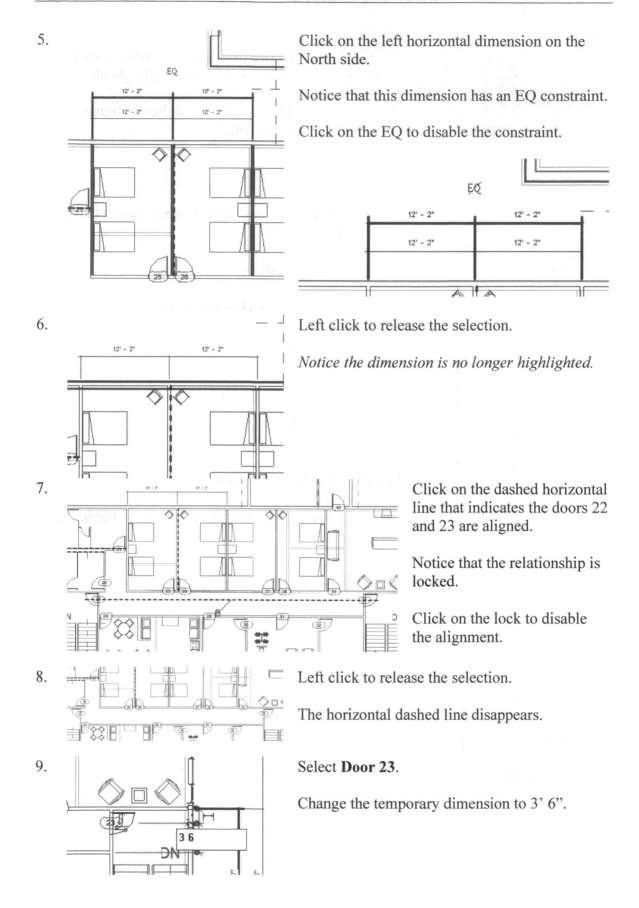

5. Click on the left horizontal dimension on the North side.

Notice that this dimension has an EQ constraint.

Click on the EQ to disable the constraint.

6. Left click to release the selection.

Notice the dimension is no longer highlighted.

7. Click on the dashed horizontal line that indicates the doors 22 and 23 are aligned.

Notice that the relationship is locked.

Click on the lock to disable the alignment.

8. Left click to release the selection.

The horizontal dashed line disappears.

9. Select **Door 23**.

Change the temporary dimension to 3' 6".

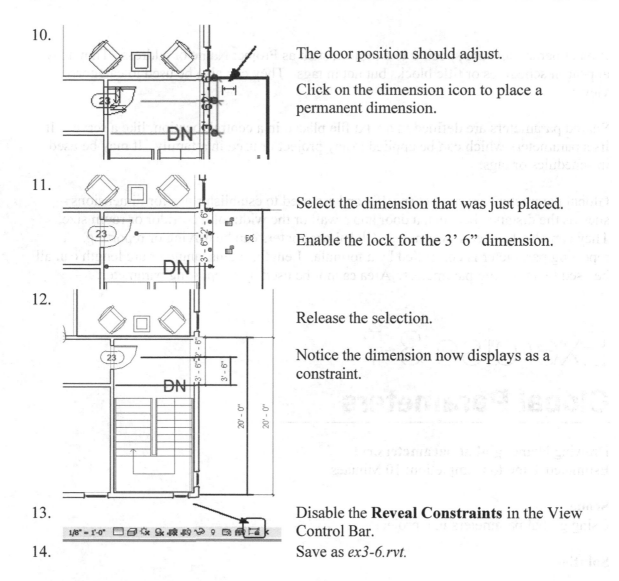

10.   The door position should adjust.

Click on the dimension icon to place a permanent dimension.

11.   Select the dimension that was just placed.

Enable the lock for the 3' 6" dimension.

12.   Release the selection.

Notice the dimension now displays as a constraint.

13.   Disable the **Reveal Constraints** in the View Control Bar.

14.   Save as *ex3-6.rvt*.

Parameters

Revit uses four different types of parameters:

- Family
- Project
- Shared
- Global

Users should be familiar with all the different types of parameters and how they are applied.

Family Parameters are used in loadable families, such as doors and windows. They normally don't appear in schedules or tags. They are used to control size and materials. They can be linked in nested families.

Project parameters are specific to a project, such as Project Name or address. They may appear in schedules or title blocks but not in tags. They can also be used to categorize views.

Shared parameters are defined in a *.txt file placed in a central location, like a server. It lists parameters which can be applied to any project or used in a family. It may be used in schedules or tags.

Global parameters are project specific and are used to establish rules for dimensions – such as the distance between a door and a wall or the width of a corridor or room size. They can also be used in formulas. Global parameters can be driving or reporting. A reporting parameter is controlled by a formula. Length, radius, angle or arc length can all be used by reporting parameters. Area cannot be used by a reporting parameter.

Exercise 3-7

Global Parameters

Drawing Name: **global parameters.rvt**
Estimated Time to Completion: 10 Minutes

Scope
Using global parameters in a project

Solution

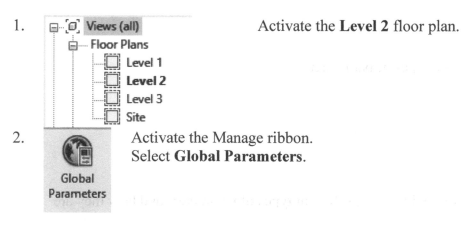

1.   Activate the **Level 2** floor plan.

2.   Activate the Manage ribbon.
Select **Global Parameters**.

3.   Select **New Global Parameter**.

4.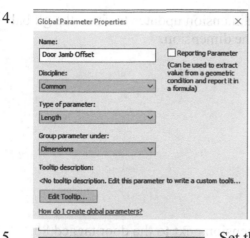

Type **Door Jamb Offset**.
Click **OK**.

5.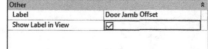

Set the value to **1' 0"**.
Click **OK**.

6.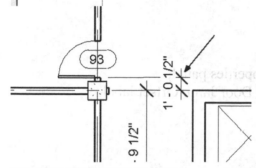

Select the dimension indicating the door jamb offset for the door with the tag labeled 93.

7.

On the ribbon, select **Door Jamb Offset** under Label.

8.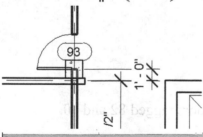

Note that the dimension updates.
Select the dimension.

9.

In the Properties panel, enable **Show Label in View**. Click **Apply**.

10.

Note that the dimension updates to show the global parameter in the dimension.

11.

Select the dimension next to the door tagged 86.

12.

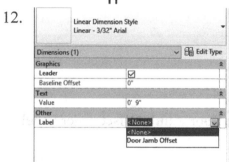

In the Properties panel,
apply the Door Jamb Offset label.

13.

Note that the dimension updates.

14.

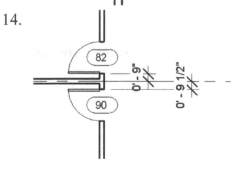

Zoom into the doors tagged 82 and 90.

15. 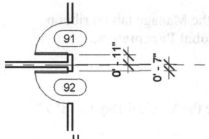 Modify the dimensions using the global parameter called Door Jamb Offset.

You can use the CTL key to select both dimensions and apply the global parameter to both dimensions simultaneously.

16. Zoom into the doors tagged 91 and 92.

17. Modify the dimensions using the global parameter called Door Jamb Offset.

18. Select the dimension between Grids 4 and 5.

19. 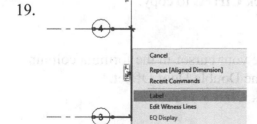 Right click and select **Label**.

Cancel
Repeat [Aligned Dimension]
Recent Commands
Label
Edit Witness Lines
EQ Display
Flip Dimension Direction

20. 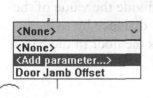 Select **Add parameter**.

<None>
<None>
<Add parameter...>
Door Jamb Offset

21. Type **Vertical Bay** for the name.
Click **OK**.

22. Activate the Manage tab on ribbon.
Select **Global Parameters**.

23.

Parameter	Value
Dimensions	
Door Jamb Offset	1' 0"
Vertical Bay	18' 2"

Set the value for the Vertical Bay to **18' 2"**.

24. Select **New Global Parameter**.

25. Type **Double Door Offset**.
Click **OK**.

26.

Dimensions		
Door Jamb Offset	1' 0"	=
Vertical Bay	18' 2"	=
Double Door Offset	0' 0"	=

Highlight the **Vertical Bay** text.
Click **Ctrl+C** to copy.

27.

Parameter	Value	Formula
Dimensions		
Door Jamb Offset	1' 0"	=
Vertical Bay	18' 2"	=
Double Door Offset	0' 0"	=

Place your cursor in the formula column for the Double Door Offset.
Click **Ctrl+V** to paste.

28.

Parameter	Value	Formula
Dimensions		
Door Jamb Offset	1' 0"	=
Vertical Bay	18' 2"	=
Double Door Offset	0' 0"	= Vertical Bay/2

Type **/2** to divide the value of the vertical bay distance by two.
This centers the door in the bay.

29. 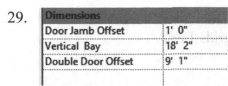 Note that the dimension for the Double Door Offset updates.
Click **OK**.

30. 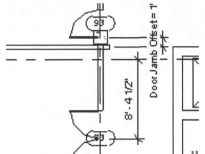 Locate the double door tagged 83.

31. Select the dimension locating the door.
Right click and select **Label.**
Select the **Double Door Offset** parameter.

32. Note that the door position updates.

33. Assign the Vertical Bay parameter to the dimensions between grids 2 & 3 and grids 1 & 2.

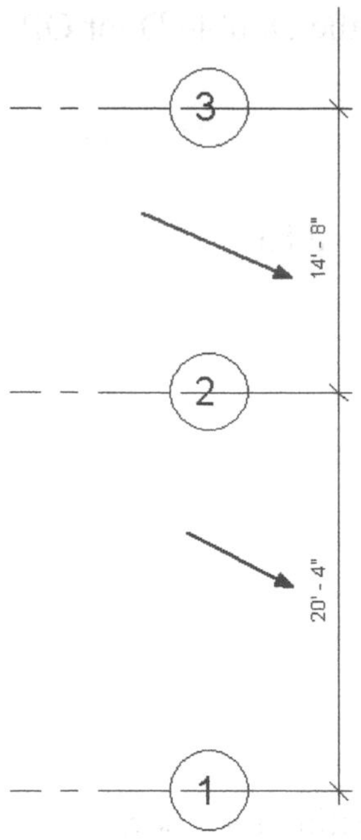

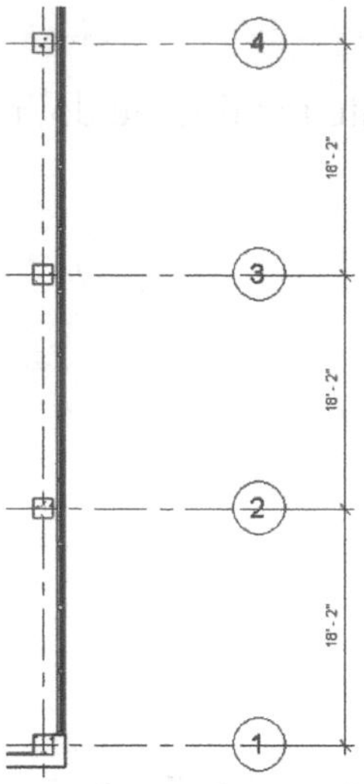

34. Apply the double door offset parameter to the doors labeled 84 and 85.

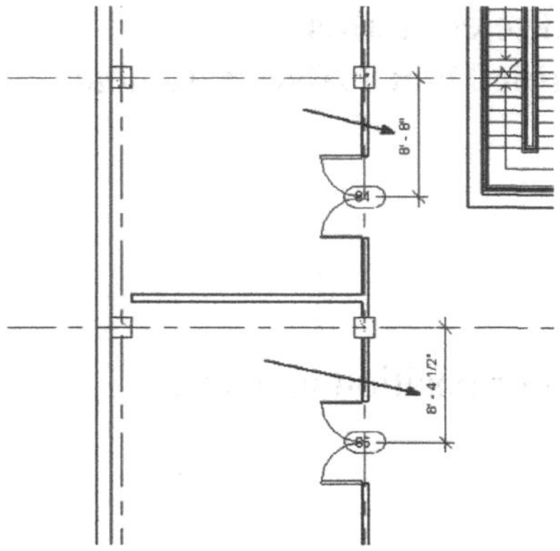

35. Use the **ALIGN** tool to move the wall above the door labeled 85 to center it on Grid 2.

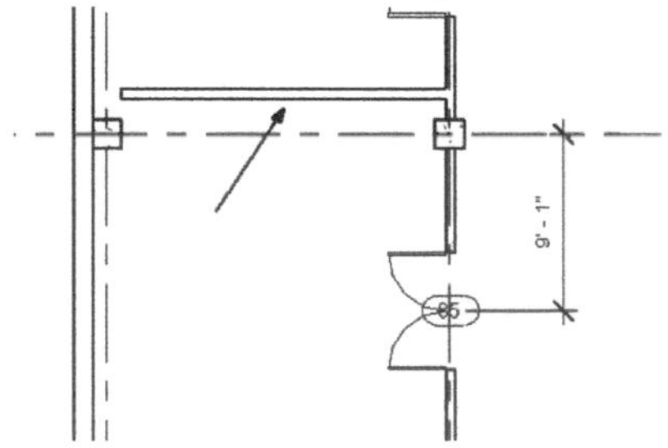

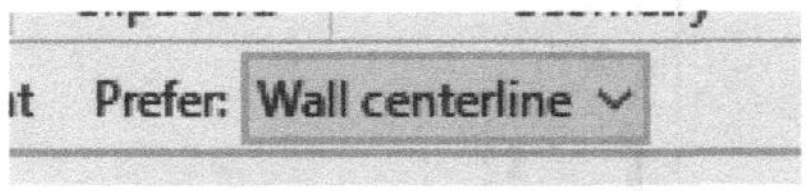

Set the Option to **Prefer Wall Centerline**.

36. Lock the wall to the grid.

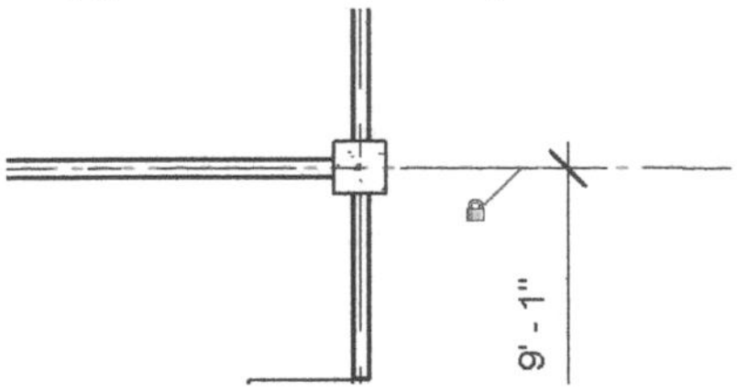

37.

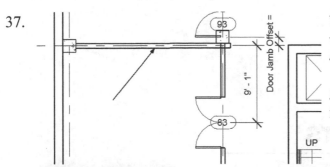

Use the ALIGN tool to position the wall above the door labeled 83 to Grid 4 and lock into position.

Zoom out so you can see the entire floor plan.

38.

📇 Project Parameters
📑 Shared Parameters
📇 Global Parameters

Activate the Manage ribbon.
Select **Global Parameters**.

39.

Parameter	Value
Dimensions	
Door Jamb Offset	1' 0"
Vertical Bay	18' 6"
Double Door Offset	9' 1"

Change the value of the Vertical Bay to **18' 6"**.
Click **OK**.

40. Note that the dimensions update and the double doors remain centered in the bays. Close without saving.

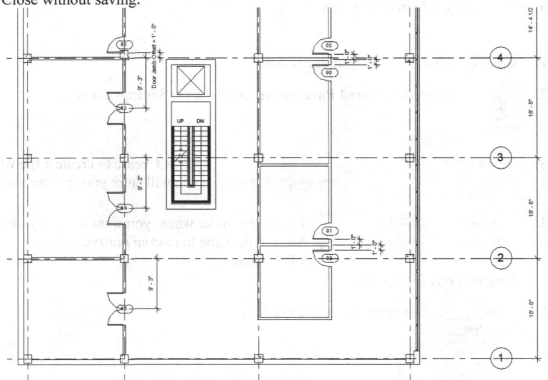

Exercise 3-8
Shared Parameters

Drawing Name: shared_parameters.rvt
Estimated Time: 30 minutes

This exercise reinforces the following skills:

- ❑ Shared Parameters
- ❑ Schedules
- ❑ Family Properties

Many architectural firms have a specific format for schedules. The parameters for these schedules may not be included in the pre-defined parameters in Revit. If you are going to be working in multiple projects, you can define a single file to store parameters to be used in any schedule in any project.

1. **[Manage]** Activate the **Manage** ribbon.

2. *[Shared Parameters icon]* Select the **Shared Parameters** tool from the Settings panel.

3.  Click **Create** to create a file where you will store your parameters.

4. File name: **custom parameters**
    Files of type: Shared Parameter Files (*.txt)

    Locate the folder where you want to store your file. Set the file name to *custom parameters.txt*.
    Click **Save**.

    Note that this is a txt file.

5. Groups
    New...
    Rename...
    Delete

    Under Groups, select **New**.

6. 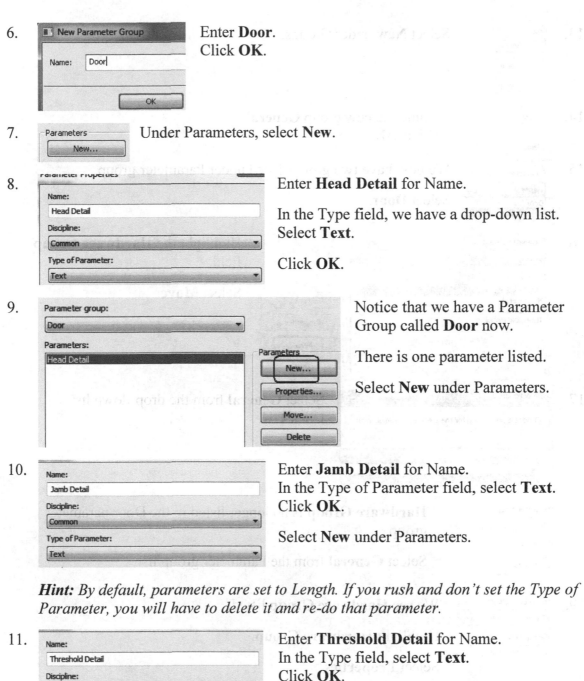 Enter **Door**.
Click **OK**.

7. Under Parameters, select **New**.

8. Enter **Head Detail** for Name.

In the Type field, we have a drop-down list.
Select **Text**.

Click **OK**.

9. Notice that we have a Parameter Group called **Door** now.

There is one parameter listed.

Select **New** under Parameters.

10. Enter **Jamb Detail** for Name.
In the Type of Parameter field, select **Text**.
Click **OK**.

Select **New** under Parameters.

Hint: By default, parameters are set to Length. If you rush and don't set the Type of Parameter, you will have to delete it and re-do that parameter.

11. Enter **Threshold Detail** for Name.
In the Type field, select **Text**.
Click **OK**.

Select **New** under Parameters.

12. 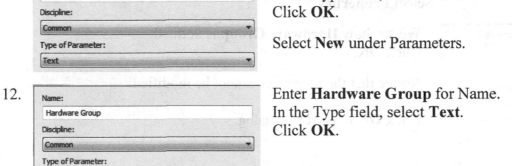 Enter **Hardware Group** for Name.
In the Type field, select **Text**.
Click **OK**.

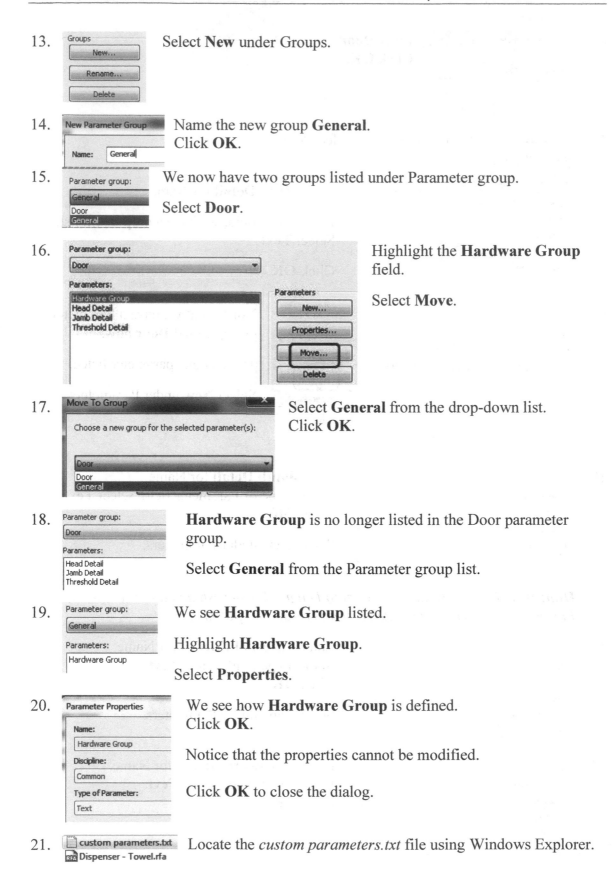

13. Select **New** under Groups.

14. Name the new group **General**.
    Click **OK**.

15. We now have two groups listed under Parameter group.

    Select **Door**.

16. Highlight the **Hardware Group** field.

    Select **Move**.

17. Select **General** from the drop-down list.
    Click **OK**.

18. **Hardware Group** is no longer listed in the Door parameter group.

    Select **General** from the Parameter group list.

19. We see **Hardware Group** listed.

    Highlight **Hardware Group**.

    Select **Properties**.

20. We see how **Hardware Group** is defined.
    Click **OK**.

    Notice that the properties cannot be modified.

    Click **OK** to close the dialog.

21. Locate the *custom parameters.txt* file using Windows Explorer.

22. 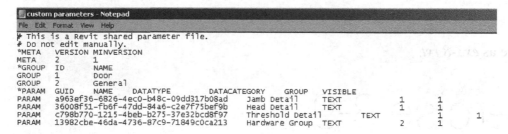 Right click and select **Open**.

This should open the file using Notepad and assumes that you have associated txt files with Notepad.

```
custom parameters - Notepad
File  Edit  Format  View  Help
# This is a Revit shared parameter file.
# Do not edit manually.
*META    VERSION  MINVERSION
META     2        1
*GROUP   ID       NAME
GROUP    1        Door
GROUP    2        General
*PARAM   GUID            NAME          DATATYPE         DATACATEGORY    GROUP    VISIBLE
PARAM    a963ef36-6826-4ec0-b48c-09dd317b08ad    Jamb Detail      TEXT          1       1
PARAM    36008f51-fb6f-47dd-84a6-c2e7f75bef9b    Head Detail      TEXT          1       1
PARAM    c798b770-1215-4beb-b275-37e32bcd8f97    Threshold Detail       TEXT     1       1
PARAM    13982cbe-46da-4736-87c9-71849c0ca213    Hardware Group  TEXT          2       1
```

We see the format of the parameter file.

Note that we are advised not to edit manually.

However, currently this is the only place you can modify the parameter type from Text to Integer, etc.

23. Change the data type to **INTEGER** for Hardware Group.

```
# This is a Revit shared parameter file.
# Do not edit manually.
*META    VERSION  MINVERSION
META     2        1
*GROUP   ID       NAME
GROUP    1        Door
GROUP    2        General
*PARAM   GUID            NAME          DATATYPE         DATACATEGORY    GROUP    VISIBLE  DESCRIPTION    USERMODIFIABLE  HIDEWHENNOVALUE
PARAM    e090884a-02a1-4b7f-b4fc-fcefa8fe2567    Head Detail      TEXT          1       1              1              0
PARAM    ddaea0a3-692f-4958-b6be-3eec38b6e911    Hardware Group  INTEGER       2       1              1              0
PARAM    fea2dbd8-b20e-40be-a1cb-11bd37cfd4f8    Jam Detail       TEXT          1       1              1              0
PARAM    a53f6bfb-0ef7-4298-9467-0d27cb5a2213    Threshold Detail       TEXT     1       1              1              0
```

Do not delete any of the spaces!

24. Save the text file and close.

25. Select the **Shared Parameters** tool on the Manage tab on ribbon.

Shared
Parameters

If you get an error message when you attempt to open the file, it means that you made an error when you edited the file. Re-open the file and check it.

26. 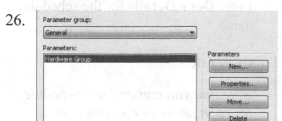 Set the parameter group to **General**.

Highlight **Hardware Group**.
Select **Properties**.

27. 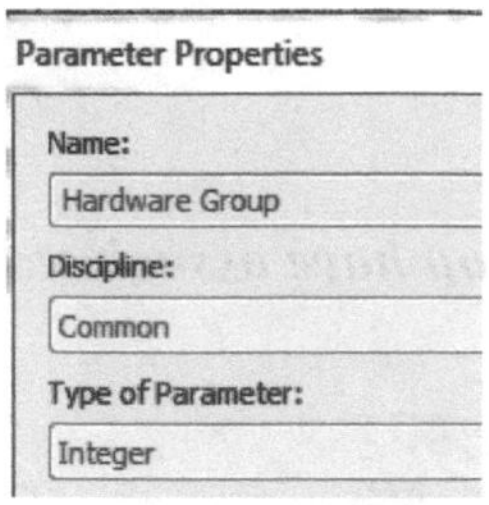 Note that Hardware Group is now defined as an Integer.

Click **OK** twice to exit the Shared Parameters dialog.

Parameter Properties

Name:
Hardware Group

Discipline:
Common

Type of Parameter:
Integer

28. Save as *ex3-8.rvt*.

Exercise 3-9
Using Shared Parameters

Drawing Name: shared_parameters.rvt
Estimated Time: 30 minutes

This exercise reinforces the following skills:

- Shared Parameters
- Family Properties
- Schedules

1.  Activate the **View** tab on ribbon.

2. 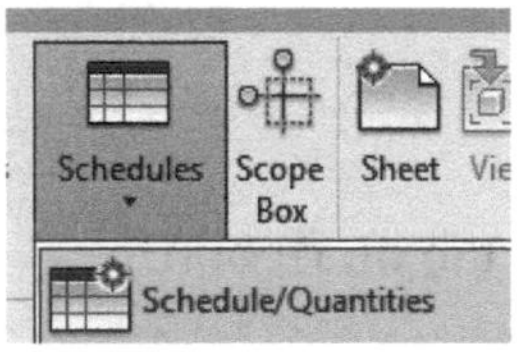 Select **Create→Schedule/ Quantities**.

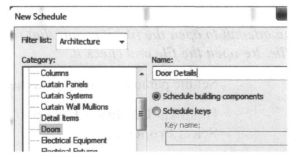

3. Highlight **Doors**.

    Enter **Door Details** for the schedule name.

    Click **OK**.

    Note that you can create a schedule for each phase of construction.

4. Add **Mark**, **Type**, **Width**, **Height**, and **Thickness**.

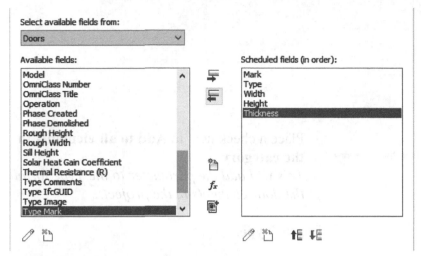

The order of the fields is important. The first field is the first column, etc. You can use the Move Up and Move Down buttons to sort the columns.

5. Select the **Add Parameter** button.

6. 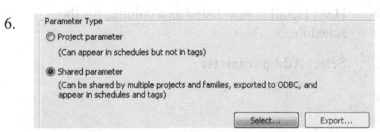 Enable **Shared parameter**.

Click **Select**.

*If you don't see any shared parameters, you need to browse for the custom parameters.txt file to load it. Go to **Shared Parameters** on the Manage tab on ribbon and browse for the file.*

7. Select **Head Detail**.

Click **OK**.

8.

Parameter Data
Name:
Head Detail
Discipline:
Common
Type of Parameter:
Text
Group parameter under:
Construction

Group it under **Construction**.

9.

☑ Add to all elements in the category

Place a check next to **Add to all elements in the category.**
This will add the parameter to the Type for all the door elements in the project.

10.

⦿ Type
◯ Instance

⦿ Values are aligned per group type
◯ Values can vary by group instance

Enable **Type**.

Click **OK**.

11.

Scheduled fields (in order):

Mark
Type
Width
Height
Thickness
Head Detail

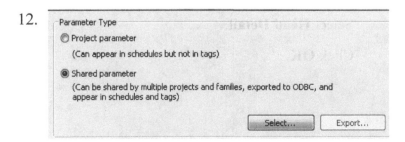

Head Detail is now listed as a column for the schedule.

Select **Add parameter**.

12.

Parameter Type

◉ Project parameter
    (Can appear in schedules but not in tags)

⦿ Shared parameter
    (Can be shared by multiple projects and families, exported to ODBC, and appear in schedules and tags)

Select...   Export...

Enable **Shared parameter**.

Click **Select**.

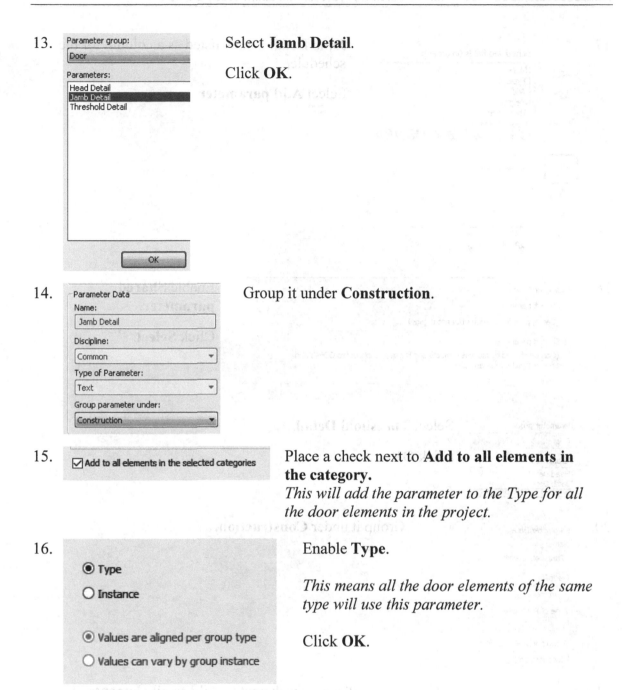

13. Select **Jamb Detail**.

Click **OK**.

14. Group it under **Construction**.

15. Place a check next to **Add to all elements in the category.**

This will add the parameter to the Type for all the door elements in the project.

16. Enable **Type**.

This means all the door elements of the same type will use this parameter.

Click **OK**.

17.

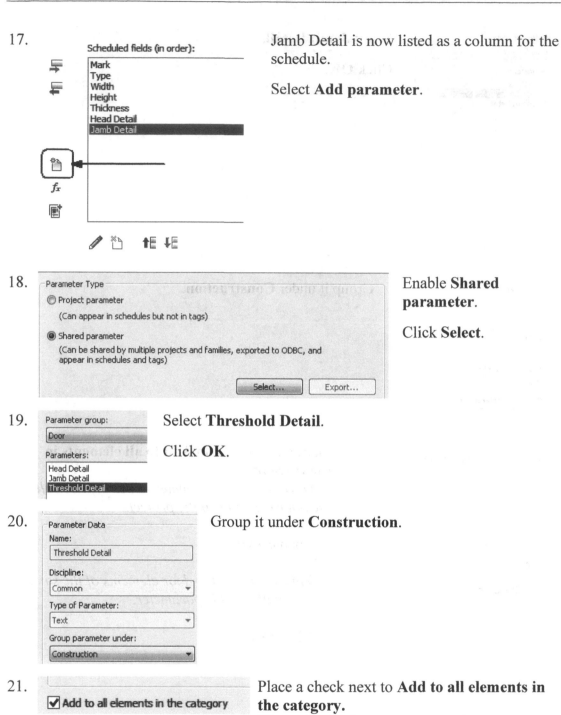

Jamb Detail is now listed as a column for the schedule.

Select **Add parameter**.

18.

Enable **Shared parameter**.

Click **Select**.

19.

Select **Threshold Detail**.

Click **OK**.

20.

Group it under **Construction**.

21.

☑ Add to all elements in the category

Place a check next to **Add to all elements in the category.**
This will add the parameter to the Type Properties for all the door elements in the project.

22.

○ Type
○ Instance

◉ Values are aligned per group type
○ Values can vary by group instance

Enable **Type**.
This means all the door elements of the same type will use this parameter.

Click **OK**.

Scheduled fields (in order):
Mark
Type
Width
Height
Thickness
Head Detail
Jamb Detail
Threshold Detail

The detail columns have now been added to the schedule.

23.

Heading:
Door No

Heading orientation:
Horizontal

Alignment:
Center

Select the **Formatting** tab.

Highlight **Mark**.
Change the Heading to **Door No**.
Change the Alignment to **Center**.

24.

Heading:
Door Type

Heading orientation:
Horizontal

Alignment:
Left

Change the Heading for Type to **Door Type**.

25.

Heading:
W

Heading orientation:
Horizontal

Alignment:
Center

Change the Heading for Width to **W**.

Change the Alignment to **Center**.

26.

Heading:
H

Heading orientation:
Horizontal

Alignment:
Center

Change the Heading for Height to **H**.

Change the Alignment to **Center**.

27.

Heading:
THK

Heading orientation:
Horizontal

Alignment:
Center

Change the Heading for Thickness to **THK**.

Change the Alignment to **Center**.

Click **OK**.

28. A window will appear with your new schedule.

<Door Details>

A	B	C	D	E	F	G	H
Door No.	Door Type	W	H	THK	Head Detail	Jamb Detail	Threshold Detail
1	36" x 84"	3' - 0"	7' - 0"	0' - 2"			
2	36" x 84"	3' - 0"	7' - 0"	0' - 2"			
3	36" x 84"	3' - 0"	7' - 0"	0' - 2"			
4	36" x 84"	3' - 0"	7' - 0"	0' - 2"			
5	72" x 82"	6' - 0"	6' - 10"	0' - 1 3/4"			
6	72" x 82"	6' - 0"	6' - 10"	0' - 1 3/4"			
7	36" x 84"	3' - 0"	7' - 0"	0' - 2"			
8	36" x 84"	3' - 0"	7' - 0"	0' - 2"			
9	36" x 84"	3' - 0"	7' - 0"	0' - 2"			
10	36" x 84"	3' - 0"	7' - 0"	0' - 2"			
11	12000 x 2290 Ope	2' - 11 1/2"	7' - 6 1/4"	0' - 1 1/4"			
12	12000 x 2290 Ope	2' - 11 1/2"	7' - 6 1/4"	0' - 1 1/4"			
21	Door-Curtain-Wall-	6' - 9"	8' - 0"				

29. Select the W column, then drag your mouse to the right to highlight the H and THK columns.

30. Select **Titles & Headers→Group** on the ribbon.

31. Type '**SIZE**' as the header for the three columns.

32. Select the Head Detail column, then drag your mouse to the right to highlight the Jamb Detail and Threshold Detail columns.

33. Select **Titles & Headers→Group**.

34. Type '**DETAILS**' as the header for the three columns.

35. Our schedule now appears in the desired format.

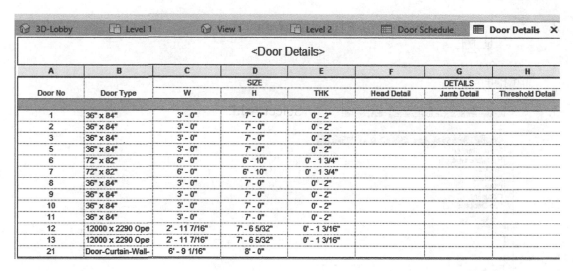

A	B	C	D	E	F	G	H
			SIZE			DETAILS	
Door No	Door Type	W	H	THK	Head Detail	Jamb Detail	Threshold Detail
1	36" x 84"	3' - 0"	7' - 0"	0' - 2"			
2	36" x 84"	3' - 0"	7' - 0"	0' - 2"			
3	36" x 84"	3' - 0"	7' - 0"	0' - 2"			
5	36" x 84"	3' - 0"	7' - 0"	0' - 2"			
6	72" x 82"	6' - 0"	6' - 10"	0' - 1 3/4"			
7	72" x 82"	6' - 0"	6' - 10"	0' - 1 3/4"			
8	36" x 84"	3' - 0"	7' - 0"	0' - 2"			
9	36" x 84"	3' - 0"	7' - 0"	0' - 2"			
10	36" x 84"	3' - 0"	7' - 0"	0' - 2"			
11	36" x 84"	3' - 0"	7' - 0"	0' - 2"			
12	12000 x 2290 Ope	2' - 11 7/16"	7' - 6 5/32"	0' - 1 3/16"			
13	12000 x 2290 Ope	2' - 11 7/16"	7' - 6 5/32"	0' - 1 3/16"			
21	Door-Curtain-Wall-	6' - 9 1/16"	8' - 0"				

36.   Save as *ex3-9.rvt*.

Project Parameters

Project parameters can be used to organize the brower and to standardize information used across sheets.

Exercise 3-10

Project Parameters

Drawing Name: *project_parameters.rvt*
Estimated Time: 20 minutes

This exercise reinforces the following skills:

❑   Project Browser
❑   Project Parameters
❑   Browser Organization

1.   Manage   Select the **Manage** ribbon.

2. 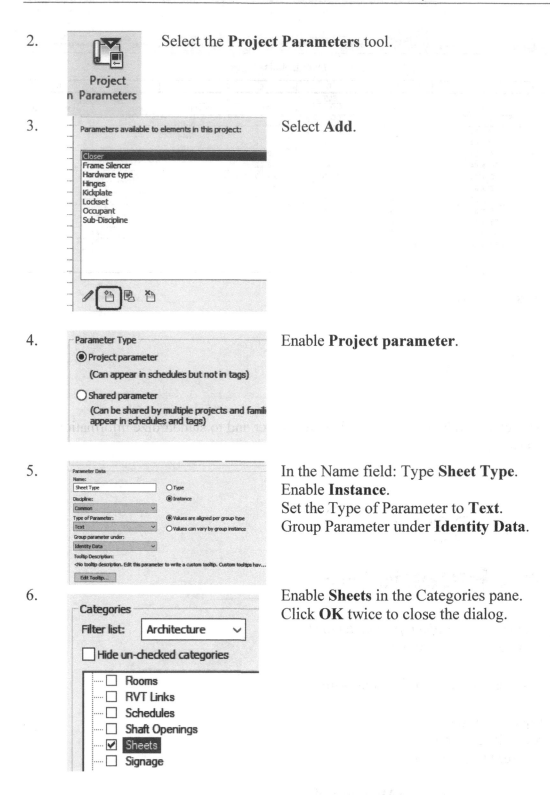 Select the **Project Parameters** tool.

3. Select **Add**.

4. Enable **Project parameter**.

5. In the Name field: Type **Sheet Type**.
Enable **Instance**.
Set the Type of Parameter to **Text**.
Group Parameter under **Identity Data**.

6. Enable **Sheets** in the Categories pane.
Click **OK** twice to close the dialog.

7. 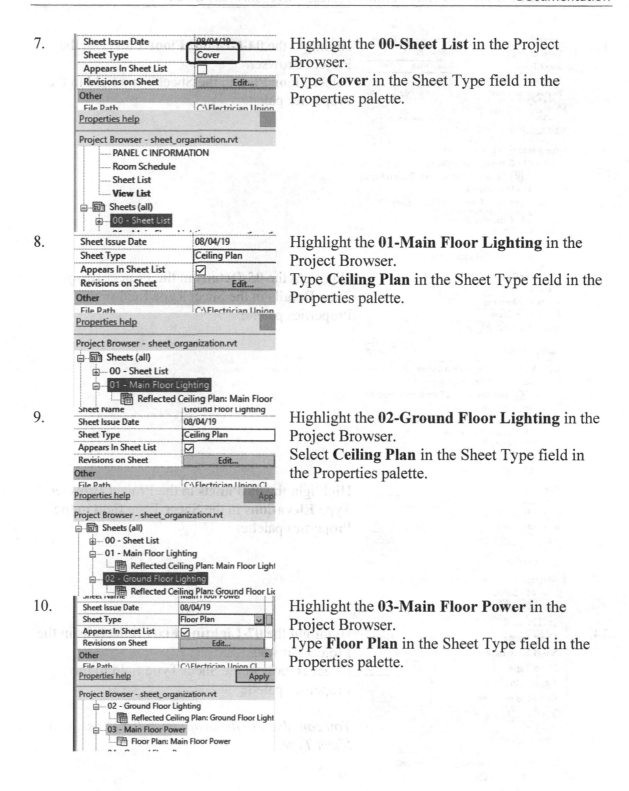 Highlight the **00-Sheet List** in the Project Browser.
Type **Cover** in the Sheet Type field in the Properties palette.

8. Highlight the **01-Main Floor Lighting** in the Project Browser.
Type **Ceiling Plan** in the Sheet Type field in the Properties palette.

9. Highlight the **02-Ground Floor Lighting** in the Project Browser.
Select **Ceiling Plan** in the Sheet Type field in the Properties palette.

10. Highlight the **03-Main Floor Power** in the Project Browser.
Type **Floor Plan** in the Sheet Type field in the Properties palette.

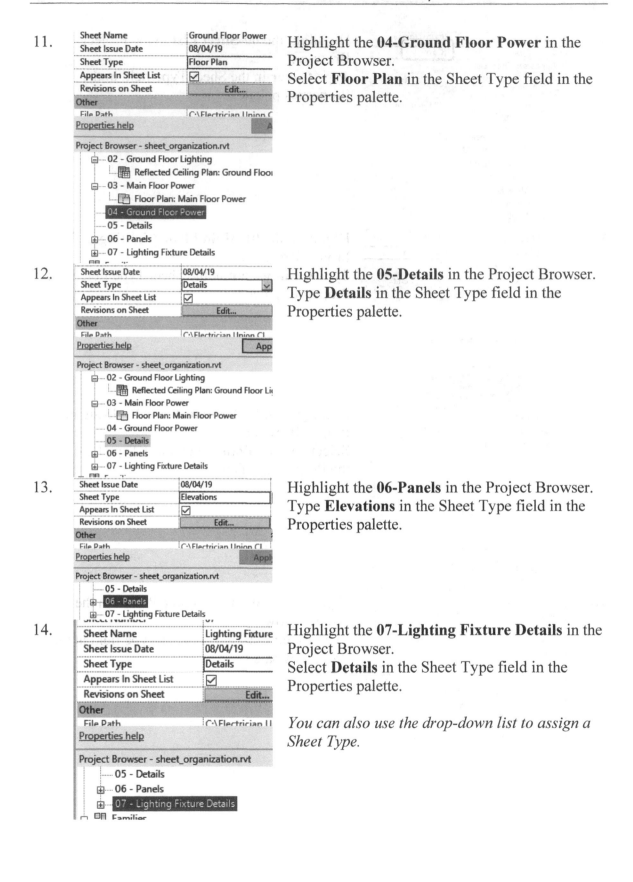

11. Highlight the **04-Ground Floor Power** in the Project Browser.
Select **Floor Plan** in the Sheet Type field in the Properties palette.

12. Highlight the **05-Details** in the Project Browser.
Type **Details** in the Sheet Type field in the Properties palette.

13. Highlight the **06-Panels** in the Project Browser.
Type **Elevations** in the Sheet Type field in the Properties palette.

14. Highlight the **07-Lighting Fixture Details** in the Project Browser.
Select **Details** in the Sheet Type field in the Properties palette.

You can also use the drop-down list to assign a Sheet Type.

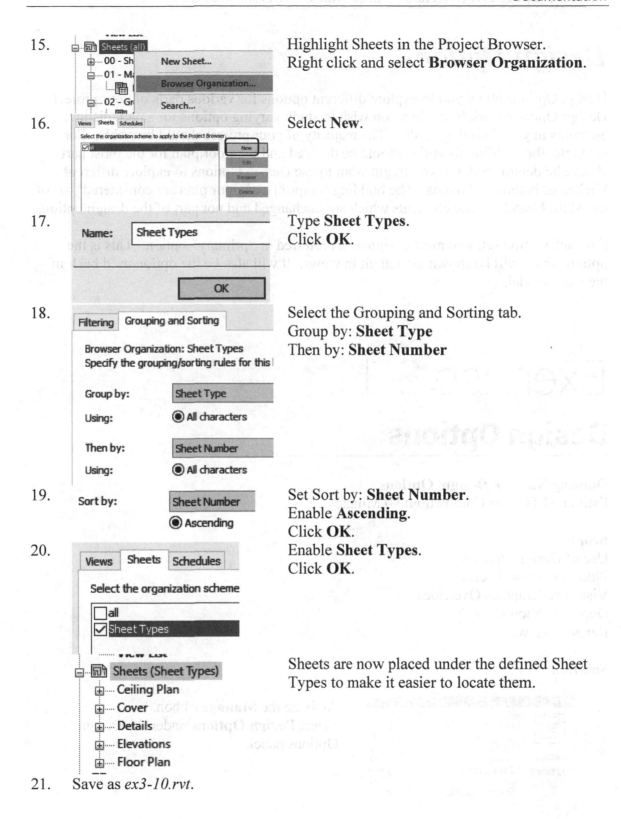

15. Highlight Sheets in the Project Browser.
Right click and select **Browser Organization**.

16. Select **New**.

17. Type **Sheet Types**.
Click **OK**.

18. Select the Grouping and Sorting tab.
Group by: **Sheet Type**
Then by: **Sheet Number**

19. Set Sort by: **Sheet Number**.
Enable **Ascending**.
Click **OK**.

20. Enable **Sheet Types**.
Click **OK**.

Sheets are now placed under the defined Sheet Types to make it easier to locate them.

21. Save as *ex3-10.rvt*.

Design Options

Design Options allow you to explore different options for various parts of your project. Design Options work best when you wish to study varying options for small, distinct elements in your building model. The majority of your project should be stable. For example, the building footprint should be decided and the floor plan for the most part should be determined, but you might want to use Design Options to explore different kitchen or bathroom layouts. The building footprint and floor plan are considered part of the Main Model – those elements which are unchanged and not part of the design options.

For each Option set, you must designate a preferred or "primary" option. This is the option which will be shown by default in views. It will also be the option used back in the main model.

Exercise 3-11

Design Options

Drawing Name: **i_Design_Options**
Estimated Time to Completion: 90 Minutes

Scope
Use of Design Options
Place Views on Sheets
Visibility/Graphics Overrides
Duplicate Views
Rename Views

Solution

1.

Activate the **Manage** ribbon.
Select **Design Options** under the Design Options panel.

2.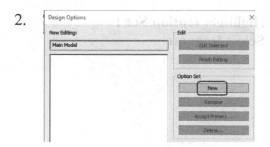

Select **New** under Option Set.
Select **New** a second time.

3.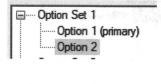

There should be two Option Sets displayed in the left panel.
Each Option set represents a design choice group. The Option set can have as many options as needed. The more options, the larger your file size will become.

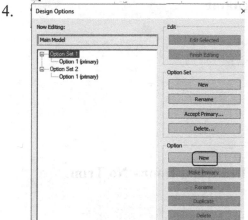

4.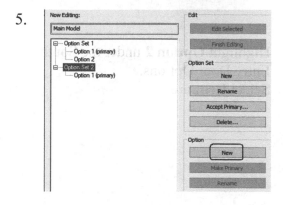

Highlight the **Option Set 1**.
Select the **New** button under Option.
Note that Option Set 1 now has two sub-options.

5.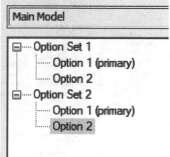

Highlight the **Option Set 2**.
Select the **New** button under Option.
Note that Option Set 2 now has two sub-options.

6.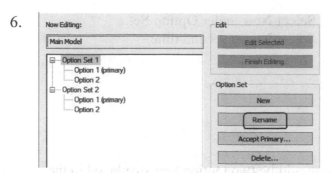
Highlight **Option Set 1**.
Select **Rename**.

7.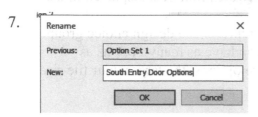
Rename Option Set 1 **South Entry Door Options**.
Click **OK**.

8.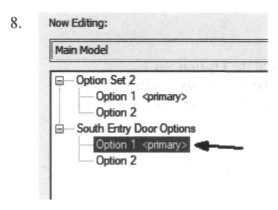
Highlight **Option 1 (primary)** under the South Entry Door Options.
Select **Rename**.

9.
Rename to **Dbl Glass Door - No Trim**.
Click **OK**.

10.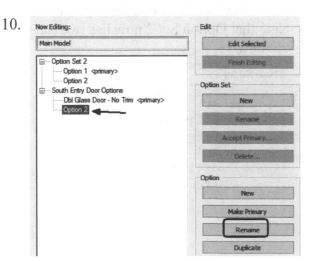
Highlight **Option 2** under the South Entry Door Options.
Select **Rename**.

11. 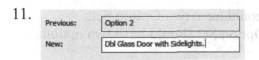 Rename to **Dbl Glass Door with Sidelights**.
Click **OK**.

12. 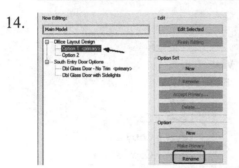 Highlight **Option Set 2**.
Select **Rename**.

13. 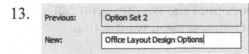 Rename Option Set 2 **Office Layout Design Options**.
Click **OK**.

14. 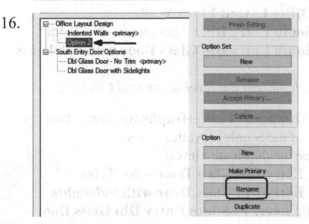 Highlight **Option 1 (primary)** under the **Office Layout Design Options**.
Select **Rename**.

15. 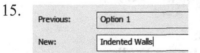 Rename to **Indented Walls**.
Click **OK**.

16. Highlight **Option 2**.
Select **Rename**.

17. 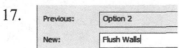 Rename Option 2 **Flush Walls**.
Click **OK**.

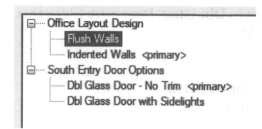

You should have two Option Sets.
Each Option Set should have two options.

Notice that Revit automatically sorted the Options and the sets alphabetically.

An Option Set can have as many options as you like, but the more option sets and options, the larger your file size and the more difficult it becomes to manage.

18. Close the Design Options dialog.

19.

Note in the bottom of the window, you can select which Option set you want active.

20. Using **Duplicate View→Duplicate**, create four copies of the Level 1 view.

Views (all)
   Floor Plans
      Level 1
      Level 1 Copy 1
      Level 1 Copy 2
      Level 1 Copy 3
      Level 1 Copy 4
      Level 2

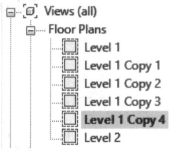

Rename the duplicate views:
Level 1 - Office Layout Indented Walls
Level 1 - Office Layout Flush Walls
Level 1 - South Entry Dbl Glass Door – No Trim
Level 1 - South Entry Dbl Glass Door with Sidelights

To rename, highlight the level name and Click F2.

21. Elevations (Elevation 1)
   East
   North
   South
   South Entry Dbl Glass Door with Sidelights
   South Entry Dbl Glass Door – No Trim
   West

Using **Duplicate View→Duplicate**, create two copies of the South Elevation view.
Rename the duplicate views:
South Entry Dbl Glass Door – No Trim
South Entry Dbl Glass Door with Sidelights

22. Views (all)
   Floor Plans
      Level 1
      Level 1 - Office Layout Flush Walls
      Level 1 - Office Layout Indented Walls
      Level 1 - South Entry Dbl Glass Door with Sidelights
      Level 1 - South Entry Dbl Glass Door – No Trim
      Level 2

Activate **Level 1 - South Entry Dbl Glass Door – No Trim**.

23.

Parts Visibility	Show Original
Visibility/Graphics Overrides	Edit...
Graphic Display Options	Edit...
Underlay	None

In the Properties pane:
Select **Edit** Visibilities/Graphics Overrides.

24. Activate the **Design Options** tab.

Set **Dbl Glass Door – No Trim** on South Entry Door Options.
Click **Apply** and **OK**.

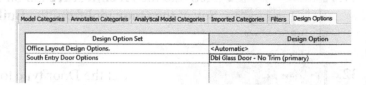

Model Categories	Annotation Categories	Analytical Model Categories	Imported Categories	Filters	Design Options
Design Option Set				Design Option	
Office Layout Design Options.				<Automatic>	
South Entry Door Options				Dbl Glass Door - No Trim (primary)	

25. Select the **Design Options** button.

26. Highlight **Dbl Glass Door with Sidelights**.

Click **Edit Selected**.

Click **Close**.

27. Set the Design Option to **Dbl Glass Door - No Trim (primary)**.

28. Uncheck **Active Only**.

29. Select the south horizontal wall.

30. Activate the **Manage** ribbon.
Under Design Options, select **Add to Set**.
*The selected wall is added to the **Dbl Glass Door - No Trim (primary)** set.*
We need to add the wall to the set so we can place a door. Remember doors are wall-hosted.

31. 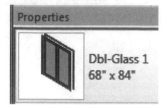 Activate the **Architecture** tab on tab on ribbon.
Select the **Door** tool from the Build panel.

32. 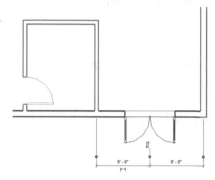 Set the Door type to **Dbl-Glass 1: 68″ x 84″**.

33. 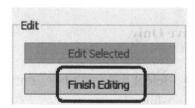 Place the door as shown.

34. 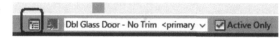 Select the **Design Options** button.

35.  Click **Finish Editing**.

Click **Close**.

36. 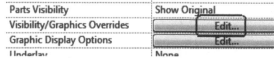 Activate **Level 1 - South Entry Dbl Glass Door with Sidelights**.

37. | Parts Visibility | Show Original |
    | Visibility/Graphics Overrides | Edit... |
    | Graphic Display Options | Edit... |
    | Underlay | None |

In the Properties pane:
Select **Edit** Visibilities/Graphics Overrides.

38. Activate the **Design Options** tab.

Design Option Set	Design Option
South Entry Door Options	Dbl Glass Door with Sidelights
Office Layout Design Options	<Automatic>

Set **Dbl Glass Door with Sidelights** on South Entry Door Options.
Click **OK**.

39.  Select the **Design Options** button.

40. 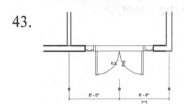 Highlight **Dbl Glass Door with Sidelights**.

Click **Edit Selected**.

Close the dialog.

41.  Verify the Active Design Option is **Dbl Glass Door with Sidelights**.

42. Activate the **Architecture** tab on tab on ribbon.
Select the **Door** tool from the Build panel.

43. Place a **Double-Raised Panel with Sidelights: 68″ x 80″** door as shown.

Double-Raised Panel with Sidelights
68" x 80"

44. Activate the **South Entry Dbl Glass Door – No Trim** elevation.

Elevations (Elevation 1)
- East
- North
- South
- South Entry Dbl Glass Door with Sidelights
- **South Entry Dbl Glass Door – No Trim**
- West

45. In the Properties pane:
Select **Edit** Visibilities/Graphics Overrides.

Parts Visibility	Show Original
Visibility/Graphics Overrides	Edit...
Graphic Display Options	Edit...
Underlay	None

46. Activate the **Design Options** tab.

Set **Dbl Glass Door - No Trim** on South Entry Door Options.
Click **OK**.

Design Option Set	Design Option
South Entry Door Options	Dbl Glass Door - No Trim (primary)
Office Layout Design Options	<Automatic>

47. Activate the **South Entry Dbl Glass Door with Sidelights** elevation.

Elevations (Elevation 1)
- East
- North
- South
- **South Entry Dbl Glass Door with Sidelights**
- South Entry Dbl Glass Door – No Trim
- West

48. In the Properties pane:
Select **Edit** Visibilities/Graphics Overrides.

Parts Visibility	Show Original
Visibility/Graphics Overrides	Edit...
Graphic Display Options	Edit...
Underlay	None

49.

Design Option Set	Design Option
South Entry Door Options	Dbl Glass Door with Sidelights
Office Layout Design Options	<Automatic>

Activate the **Design Options** tab.

Set **Dbl Glass Door with Sidelights** on South Entry Door Options.

Click **OK**.

50.
- Sheets (all)
  - A101 - South Entry Door Option
  - A102 - Office Layout Options

Activate the Sheet named **South Entry Door Options**.

51.

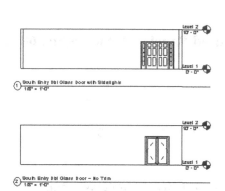

Drag and drop the two South Entry Option elevation views on the sheet.

52.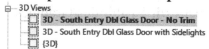

Switch to 3D view.

53. Use **Duplicate View→Duplicate** to create two new 3D views.

- 3D Views
  - 3D - South Entry Dbl Glass Door - No Trim
  - 3D - South Entry Dbl Glass Door with Sidelights
  - {3D}

Rename the views:
3D - South Entry Dbl Glass Door - No Trim
3D - South Entry Dbl Glass Door with Sidelights

54.
- 3D Views
  - 3D - South Entry Dbl Glass Door - No Trim
  - 3D - South Entry Dbl Glass Door with Sidelights
  - {3D}

Activate **3D - South Entry Dbl Glass Door - No Trim**.

55.

Detail Level	Coarse
Visibility/Graphics Overrides	Edit...
Visual Style	Hidden Line
Graphic Display Options	Edit...

In the Properties pane:
Select **Edit** Visibilities/Graphics Overrides.

56.

Design Option Set	
South Entry Door Options	Dbl Glass Door - No Trim (primary)
Office Layout Design Options	<Automatic>

Activate the **Design Options** tab.

Set **Dbl Glass Door - No Trim** on South Entry Door Options.
Click **OK**.

57. Disable the visibility of **LEVELS** on the Annotation Categories tab.

58. Activate **3D - South Entry Dbl Glass Door with Sidelights**.

59. In the Properties pane:
Select **Edit** Visibilities/Graphics Overrides.

60. Activate the **Design Options** tab.

Set **Dbl Glass Door with Sidelights** on South Entry Door Options.
Click **OK**.

If you forget which design option you are working in, check the title at the top of the dialog.

61. Disable the visibility of **LEVELS** on the Annotation Categories tab.

62. 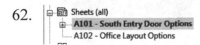 Activate the Sheet named **South Entry Door Options**.

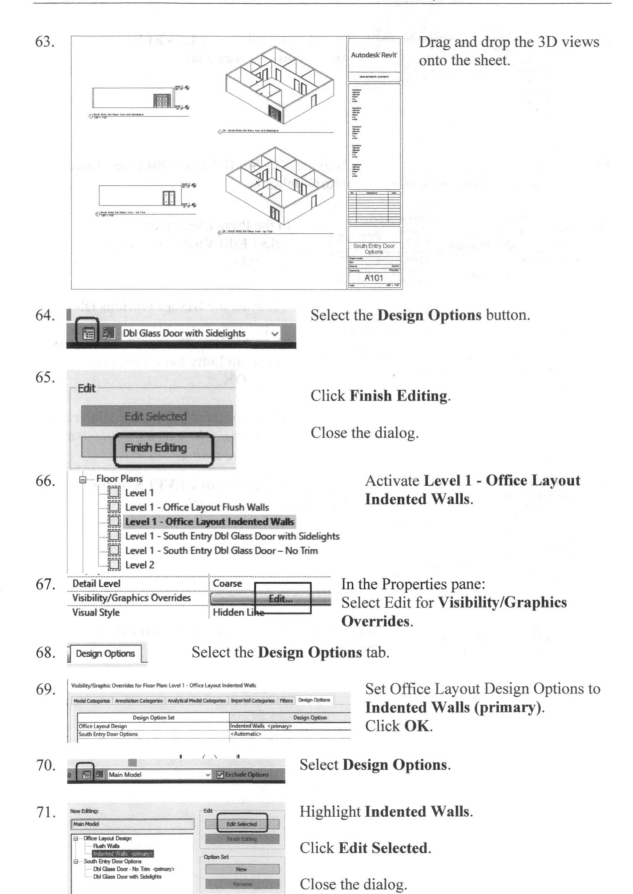

63. Drag and drop the 3D views onto the sheet.

64. Select the **Design Options** button.

65. Click **Finish Editing**.

Close the dialog.

66. Activate **Level 1 - Office Layout Indented Walls**.

67. In the Properties pane:
Select Edit for **Visibility/Graphics Overrides**.

68. Select the **Design Options** tab.

69. Set Office Layout Design Options to **Indented Walls (primary)**.
Click **OK**.

70. Select **Design Options**.

71. Highlight **Indented Walls**.

Click **Edit Selected**.

Close the dialog.

72. 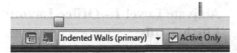 Verify the Design Option is **Indented Walls (primary)**.

73.  Select the **Wall** tool from the Architecture tab on tab on ribbon.

Select the **Basic Wall: Interior - 5" Partition (2-hr)**.

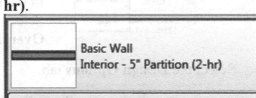

74.  Enable **Chain** on the Options bar.

75. Place the two walls indicated.

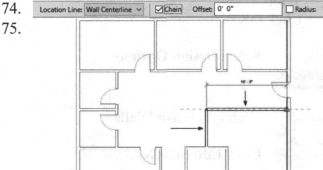 The vertical wall is placed at the midpoint of the small horizontal wall to the south.

The horizontal wall is aligned with the wall indicated by the dashed line.

76. 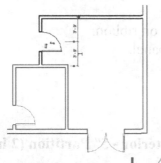 Activate the **Architecture** tab on tab on ribbon.

Select the **Door** tool from the Build panel.

77. Place a **Sgl Flush 36″ x 80″** door as shown.

78.  Select **Design Options**.

79. 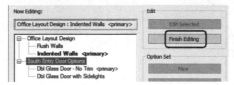 Highlight **Indented Walls**.

Click **Finish Editing**.

Close the dialog.

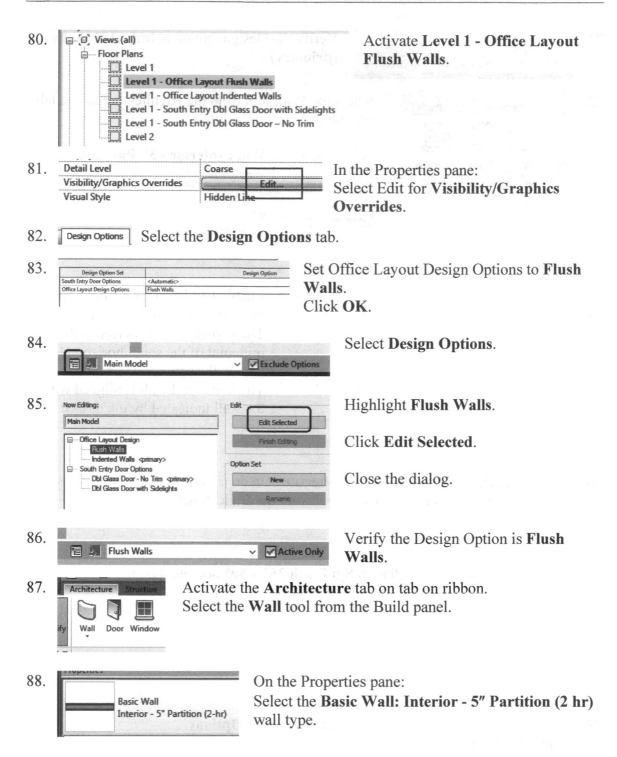

80. Activate **Level 1 - Office Layout Flush Walls**.

81. In the Properties pane:
Select Edit for **Visibility/Graphics Overrides**.

82. Select the **Design Options** tab.

83. Set Office Layout Design Options to **Flush Walls**.
Click **OK**.

84. Select **Design Options**.

85. Highlight **Flush Walls**.

Click **Edit Selected**.

Close the dialog.

86. Verify the Design Option is **Flush Walls**.

87. Activate the **Architecture** tab on tab on ribbon.
Select the **Wall** tool from the Build panel.

88. On the Properties pane:
Select the **Basic Wall: Interior - 5″ Partition (2 hr)** wall type.

89. Add the wall shown.

Note that the walls and door added for the Indented Walls option are not displayed.

90. Activate the **Architecture** tab on tab on ribbon.
Select the **Door** tool from the Build panel.

91. Place a **Sgl Flush 36″ x 84″** door as shown.

92. Activate the **Manage** tab on tab on ribbon.
Select **Design Options** on the Design Options panel.

93. Select **Finish Editing**.
Close the dialog.

94. Verify the Design Option is now set to **Main Model**.

95. Note that if you hover your mouse over the element, it will display which Option set it belongs to.

This only works if Active Only or Exclude Options is disabled.

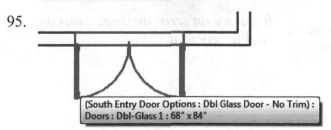

(South Entry Door Options : Dbl Glass Door - No Trim) :
Doors : Dbl-Glass 1 : 68″ x 84″

96. Sheets (all)
    A101 - South Entry Door Options
    A102 - Office Layout Options

    Activate the Sheet named **Office Layout Options**.

97. Drag and drop the two Office Layout options onto the sheet.

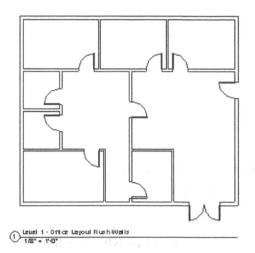 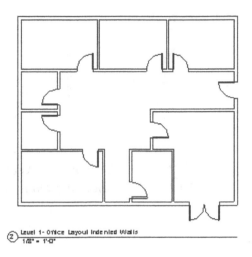

① Level 1 - Office Layout Flush Walls
  1/8" = 1'-0"

② Level 1- Office Layout Indented Walls
  1/8" = 1'-0"

98. Design Options | Add to Set | Pick to Edit | Main Model
    Design Options

    Activate the Manage tab on tab on ribbon.
    Select **Design Options**.

99. Now Editing:

    Main Model

    South Entry Door Options
      Dbl Glass Door - No Trim (primary)
      Dbl Glass Door with Sidelights
    Office Layout Design Options
      Indented Walls (primary)
      Flush Walls

    Let's assume that the client decided they prefer the flush walls option.

    Highlight the **Flush Walls** option.

100. Make Primary

     Office Layout Design Options
       Indented Walls
       Flush Walls (primary)

     Warning: 1 out of 2
     A relationship between Elements in Main Model and Option "Office Layout Design Options : Indented Walls" has been deleted. There is no corresponding relationship in the Primary Option "Office Layout Design Options : Flush Walls".

     Select **Make Primary**.

     Note that (primary) is now next to Flush Walls.

     If you see an error message, you can click to ignore it.

101.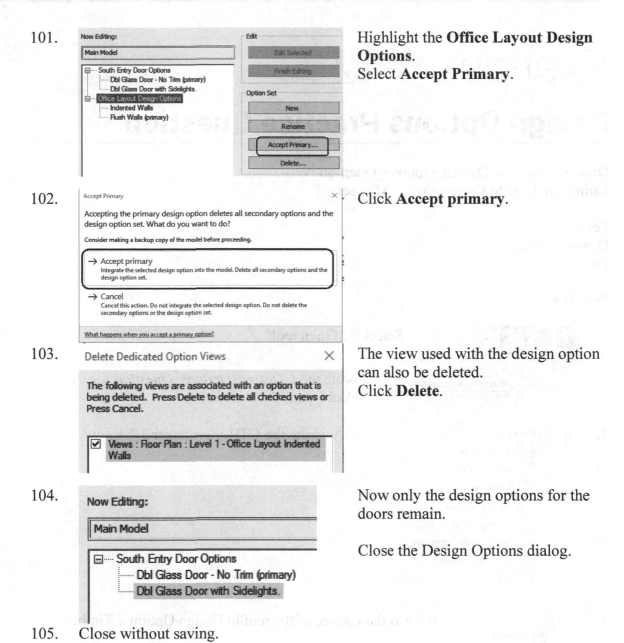
Highlight the **Office Layout Design Options**.
Select **Accept Primary**.

102. Click **Accept primary**.

103. The view used with the design option can also be deleted.
Click **Delete**.

104. Now only the design options for the doors remain.

Close the Design Options dialog.

105. Close without saving.

Exercise 3-12

Design Options Practice Question

Drawing Name: i_Design_Options_Question.rvt
Estimated Time to Completion: 5 Minutes

Scope
Design Options
Properties

Solution

1. Select the **Open** tool.

2. File name: i_Design_Options_Question.rvt   Locate the *i_Design_Options_Question* file.
Select Open.

3. 3D Views   Activate the {3D} view under 3D Views.
   - Approach
   - From Yard
   - Kitchen
   - Living Room
   - Living Room - ISO
   - Section Perspective
   - Solar Analysis
   - {3D}

4. Roof   What is the volume of the roof in Design Option – Timber?
   - **Metal (primary)**
   - Timber
   - Tile

5. Timber ☑ Active Only   Select the **Timber** design option.
Enable **Active Only**.

6.
Select the roof.
Go to the Properties panel.

Scroll down.

What is the volume of the roof?

Properties
Basic Roof
Warm Roof – Timber

Roofs (1)
Construction	
Fascia Depth	0' 0"
Rafter Cut	Plumb Cut
Dimensions	
Slope	
Thickness	1' 1 79/256"
Volume	2066.67 CF
Area	1863.73 SF

Identity Data

7.

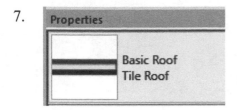

If you switch design options, does the volume of the roof change?

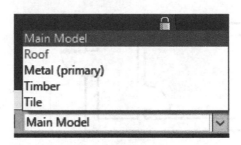

Phases

Phases are distinct, separate time periods within the life of a building. Phases can represent either the time periods or how the building appeared during that time period. By default, every Revit project has two phases or time periods already pre-defined. They are named Existing and New Construction.

The most common use of phases is to keep track of the "Before" and "After" scenarios. If you are a recovering AutoCAD user, you probably created a copy of your project and did a "Save As" to re-work the existing building for the proposed remodel. This has a number of disadvantages – not the least of which is the use of external references, the possibility of missing something, and the duplication of data.

If you are working on a completely new building project on a "clean" site, you still might want to use Phases as a way to control when to schedule special equipment on the site as well as crew. You might have a phase for foundation work, a phase for framing, a phase for electrical and so on. You can create schedules based on phases, so you will know exactly what inventory you might need on hand based on the phase.

You determine what elements are displayed in a view by assigning a phase to the view. You can even control colors and linetypes of different phases so you get a visual cue on which elements were created or placed in which phase.

EXISTING PLAN

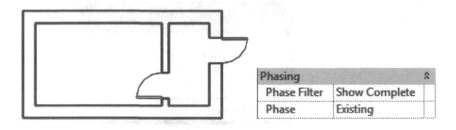

Phasing	
Phase Filter	Show Complete
Phase	Existing

DEMOLITION PLAN

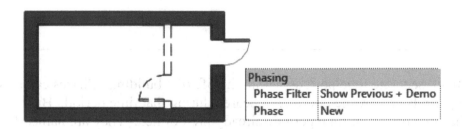

Phasing	
Phase Filter	Show Previous + Demo
Phase	New

NEW FLOOR PLAN

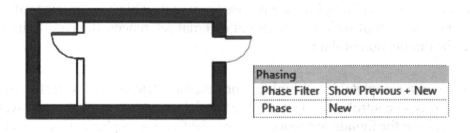

Phasing	
Phase Filter	Show Previous + New
Phase	New

A common error is to create a phase for demolition. This is unnecessary. Instead you can duplicate the view and set one view as the Demolition Plan and one view at the new floor plan.

Exercise 3-13

Phases

Drawing Name: c_phasing.rvt
Estimated Time to Completion: 75 Minutes

Scope
Properties
Filter
Phases
Rename View
Copy View
Graphic Settings for Phases

Solution

1. 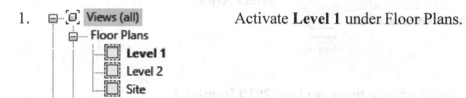   Activate **Level 1** under Floor Plans.

2.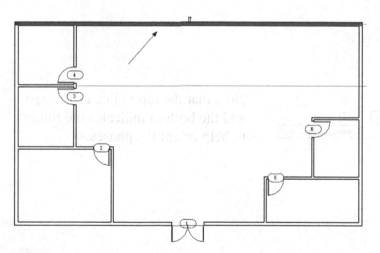

    Select the wall indicated.
    It should highlight.

3.

    Scroll down to the Phasing category in the
    Properties panel on the upper left.

    This wall was created in the New Construction Phase.
    Note that it is not set to be demolished.

4. Right click and Click **Cancel** to deselect the wall.   Cancel

5. 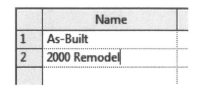 Go to the **Manage** ribbon.
Select **Phases**.

6.
	Name
1	As-Built
2	New Construction

Rename Existing to **As-Built**.

7.
	Name
1	As-Built
2	2000 Remodel

Rename New Construction to **2000 Remodel**.

8.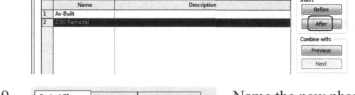

Highlight the **2000 Remodel**.
Select **After**.

9.

	Name
1	As-Built
2	2000 Remodel
3	2010 Remodel

Name the new phase **2010 Remodel**.

Project Phases	Phase Filters	Graphic Overrides	
	PAST		
	Name		Description
1	As-Built		
2	2000 Remodel		
3	2010 Remodel		
	FUTURE		

Note that the top indicates the past and the bottom indicates the future to help orient the phases.

10. Select the **Graphic Overrides** tab.

Project Phases	Phase Filters	Graphic Overrides				

Phase Status	Projection/Surface		Cut		Halftone	Material
	Lines	Patterns	Lines	Patterns		
Existing	————————		————————	Hidden	☐	Phase-Exist
Demolished	- - - - - - - -		- - - - - - - - -	Hidden	☐	Phase-Demo
New	————————		▬▬▬▬▬▬▬▬	████	☐	Phase-New
Temporary	················		··················	/////	☐	Phase-Temp

11. Note that in the Lines column for the Existing Phase, the line color is set to gray.

12.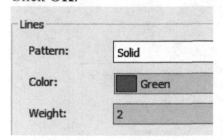

Project Phases	Phase Filters	Graphic Overrides		
Phase Status	Projection/Surface		Cut	
	Lines	Patterns	Lines	
Existing	————————	Override...	————————	
Demolished	- - - - - - -		- - - - - - - -	
New	————————		▬▬▬▬▬▬	
Temporary	···············		··················	

Highlight **Existing**. Click in the **Lines** column and the Line Graphics dialog will display.

Projection/Surface is what is displayed in the floor plan views.
Cut is the display for elevation or section views.
Override indicates you have changed the display from the default settings.

13. Set the Color to **Green** for the Existing phase by selecting the color button. Click **OK**.

Lines	
Pattern:	Solid
Color:	Green
Weight:	2

Set the Color to **Blue** for the Demolished phase.
Set the Color to **Magenta** for the New phase.
Change the colors for both Projection/Surface and Cut.

Phase Status	Projection/Surface		Cut	
	Lines	Patterns	Lines	Patterns
Existing	————————		————————	Hidden
Demolished	- - - - - - - -		- - - - - - - -	Hidden
New	————————	Override...	▬▬▬▬▬▬	████
Temporary	···············		··················	// //

14.

Project Phases	Phase Filters	Graphic Overrides

Select the **Phase Filters** tab.

15.

	Filter Name	New	Existing	Demolished	Temporary
1	Show All	By Category	Overridden	Overridden	Overridden
2	Show Demo + New	By Category	Not Displayed	Overridden	Overridden
3	Show Previous + Dem	Not Displayed	Overridden	Overridden	Not Displayed
4	Show Previous + New	By Category	Overridden	Not Displayed	Not Displayed
5	Show Previous Phase	Not Displayed	Overridden	Not Displayed	Not Displayed

Note that there are already phase filters pre-defined that will control what is displayed in a view.

16. **New**   Click the **New** button on the bottom of the dialog.

17.

	Filter Name
1	Show All
2	Show Demo + New
3	Show Previous + Dem
4	Show Previous + New
5	Show Previous Phase
6	Show Existing

Change the name for the new phase filter to **Show Existing**.
Show Previous + Demo will display existing plus demo elements, but not new.
Show Previous + New will display existing plus new elements, but not demolished elements.

18. In the New column, select **Overridden**.
In the Existing column, select **Overridden**.
This means that the default display settings will use the new color assigned.
In the Demolished column, select **Not Displayed**.

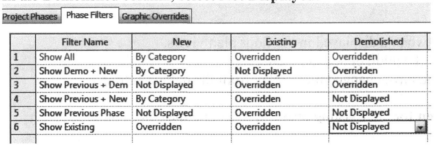

	Filter Name	New	Existing	Demolished
1	Show All	By Category	Overridden	Overridden
2	Show Demo + New	By Category	Not Displayed	Overridden
3	Show Previous + Dem	Not Displayed	Overridden	Overridden
4	Show Previous + New	By Category	Overridden	Not Displayed
5	Show Previous Phase	Not Displayed	Overridden	Not Displayed
6	Show Existing	Overridden	Overridden	Not Displayed

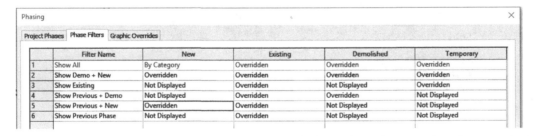

	Filter Name	New	Existing	Demolished	Temporary
1	Show All	By Category	Overridden	Overridden	Overridden
2	Show Demo + New	Overridden	Overridden	Overridden	Overridden
3	Show Existing	Not Displayed	Overridden	Not Displayed	Overridden
4	Show Previous + Demo	Not Displayed	Overridden	Overridden	Not Displayed
5	Show Previous + New	Overridden	Overridden	Not Displayed	Not Displayed
6	Show Previous Phase	Not Displayed	Overridden	Not Displayed	Not Displayed

19. Use Overridden to display the colors you assigned to the different phases.
Verify that in the Show Previous + Demo phase New elements are not displayed.
Verify that in the Show Previous + New phase Demolished elements are not displayed.
Click **Apply** and **OK** to close the Phases dialog.

20. Window around the entire floor plan.
Select the **Filter** button.

Filter

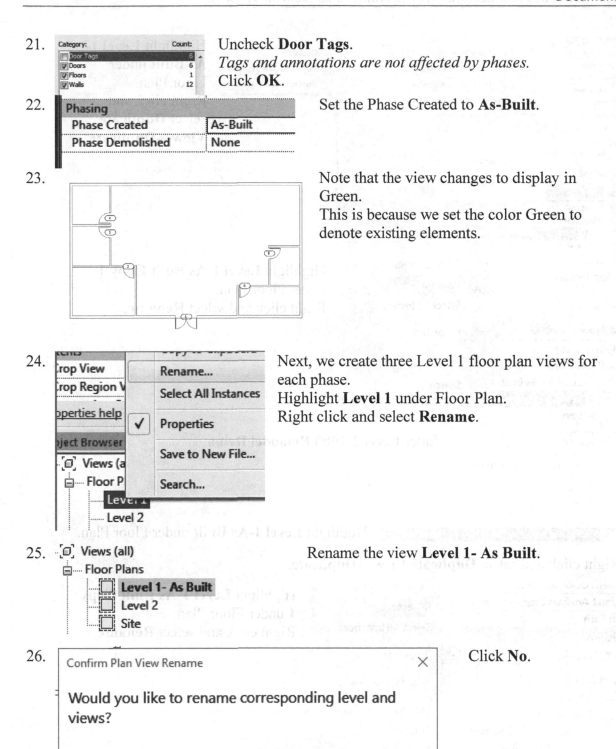

21. Uncheck **Door Tags**.
    Tags and annotations are not affected by phases.
    Click **OK**.

22. Set the Phase Created to **As-Built**.

23. Note that the view changes to display in Green.
    This is because we set the color Green to denote existing elements.

24. Next, we create three Level 1 floor plan views for each phase.
    Highlight **Level 1** under Floor Plan.
    Right click and select **Rename**.

25. Rename the view **Level 1- As Built**.

26. Click **No**.

27.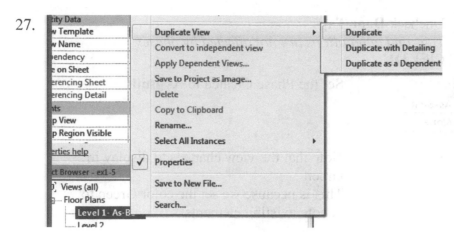

    Highlight **Level 1-As Built** under Floor Plan.
    Right click and select **Duplicate View→Duplicate**.

28.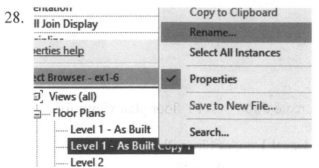

    Highlight **Level 1-As Built Copy 1** under Floor Plan.
    Right click and select **Rename**.

29.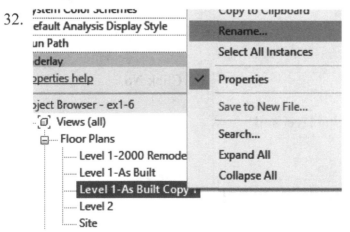

    Enter **Level 1-2000 Remodel Demo**.

30.

    Highlight **Level 1-As Built** under Floor Plan.

31. Right click and select **Duplicate View→Duplicate**.

32. Highlight **Level 1-As Built Copy 1** under Floor Plan.
    Right click and select **Rename**.

33. 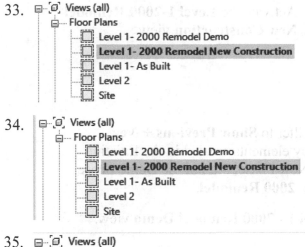 Enter **Level 1-2000 Remodel New Construction**.

34. You should have three Level 1 floor plan views listed:

- As Built
- 2000 Remodel Demo
- 2000 Remodel New Construction.

35. 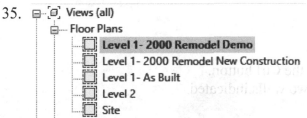 Activate the **Level 1-2000 Remodel Demo** view.

36. In the Properties dialog:

Set the Phase Filter to **Show Previous + Demo**.

The previous phase to demo is As-Built. This means the view will display elements created in the existing and demolished phase.

Depth Clipping	No clip
Phasing	
Phase Filter	Show Previous + Demo
Phase	2000 Remodel

Set the Phase to **2000 Remodel**.

37. 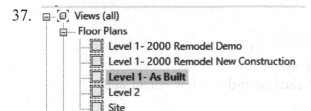 Activate the **Level 1-As Built** view.

38. In the Properties dialog:

Set the Phase Filter to **Show All**.
Set the Phase to **As-Built**.

Phasing	
Phase Filter	Show All
Phase	As-Built

The display does not show the graphic overrides. By default, Revit only allows you to assign graphic overrides to phases AFTER the initial phase. Because the As-Built view is the first phase in the process, no graphic overrides are allowed. The only work-around is to create an initial phase with no graphic overrides and go from there.

39.

```
☐ [□] Views (all)
   ☐ Floor Plans
      ☐ Level 1- 2000 Remodel Demo
      ☐ Level 1- 2000 Remodel New Construction
      ☐ Level 1- As Built
      ☐ Level 2
      ☐ Site
```

Activate the **Level 1-2000 Remodel New Construction** view.

40. In the Properties dialog:

Phasing	
Phase Filter	Show Previous + New
Phase	2000 Remodel

Set the Phase Filter to **Show Previous + New**. This will display elements created in the Existing Phase and the New Phase, but not the Demo phase. Set the Phase to **2000 Remodel**.

41.

```
☐ Floor Plans
   Level 1- 2000 Remodel Demo
   Level 1- 2000 Remodel New Construction
   Level 1- As-Built
   Level 2
   Site
```

Activate the **Level 1 - 2000 Remodel Demo** view.

42.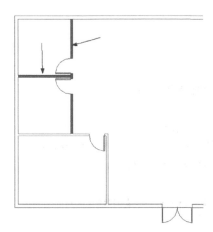

Hold down the Ctrl button.
Select the two walls indicated.

43. In the Properties pane:

Phasing	
Phase Created	As-Built
Phase Demolished	2000 Remodel

Scroll down to the bottom.
In the Phase Demolished drop-down list, select **2000 Remodel**.

44. The demolished walls change appearance based on the graphic overrides.
Release the selected walls using right click→Cancel or by Clicking ESCAPE.

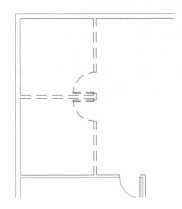

45.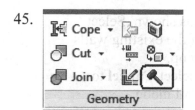

Activate the **Modify** tab on tab on ribbon.
Use the **Demolish** tool on the Geometry panel to demolish the walls indicated.

46.

Note that the doors will automatically be demolished along with the walls. If there were windows placed, these would also be demolished. That is because those elements are considered *wall-hosted*.

Right click and select Cancel to exit the Demolish mode.

47.

This is how the Level 1- 2000 Remodel Demo view should appear.

If it doesn't, check the walls to verify that they are set to Phase Created: As Built, Phase Demolished: 2000 Remodel.

Phasing	
Phase Created	As-Built
Phase Demolished	2000 Remodel

48.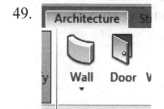

Activate the **Level 1 - 2000 Remodel New Construction** view.

49.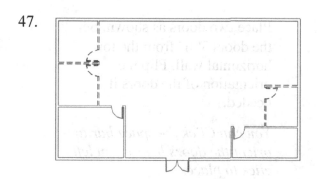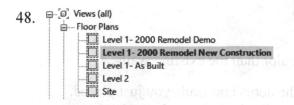

Select the **Wall** tool from the Architecture ribbon.

50. 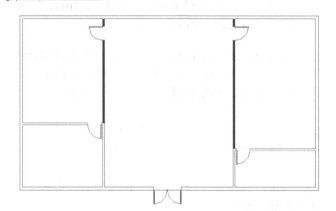 Place two walls as shown. Select the end points of the existing walls and simply draw up.
Right click and select **Cancel** to exit the Draw Wall mode.

51. Select the **Door** tool under the Build panel on the Architecture tab on tab on ribbon.

52. Place two doors as shown. Set the doors 3′ 6″ from the top horizontal wall. Flip the orientation of the doors if needed.

You can Click the space bar to orient the doors before you left click to place.

Note that the new doors and walls are a different color than the existing walls.

53. 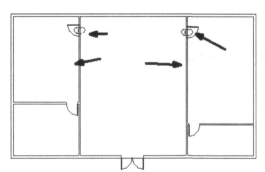 Select the doors and walls you just placed. You can select by holding down the CONTROL key or by windowing around the area.

Note: If Door Tags are selected, you will not be able to access Phases in the Properties dialog.

54. Look in the Properties panel and scroll down to Phasing.

Phasing	
Phase Created	2000 Remodel
Phase Demolished	None

Note that the elements are already set to **2000 Remodel** in the Phase Created field.

55. Switch between the three views to see how they display differently.

56. Remember that Existing should show as Green, Demo as Blue, and New as Magenta.

    If the colors don't display correctly in the Level 1 Remodel Demo or Remodel New Construction, check the Phase Filters again and make sure that the categories to be displayed are Overridden to use the assigned colors.

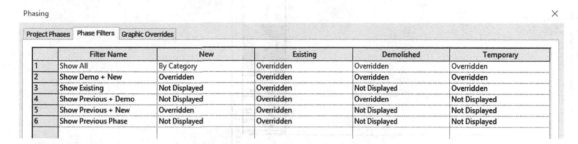

	Filter Name	New	Existing	Demolished	Temporary
1	Show All	By Category	Overridden	Overridden	Overridden
2	Show Demo + New	Overridden	Overridden	Overridden	Overridden
3	Show Existing	Not Displayed	Overridden	Not Displayed	Overridden
4	Show Previous + Demo	Not Displayed	Overridden	Overridden	Not Displayed
5	Show Previous + New	Overridden	Overridden	Not Displayed	Not Displayed
6	Show Previous Phase	Not Displayed	Overridden	Not Displayed	Not Displayed

57. 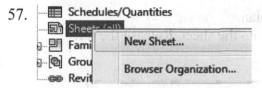   Highlight **Sheets** in the Project Browser. Right click and select **New Sheet**.

58.   Click **OK** to accept the default title block.

59.   A view opens with the new sheet.

60.   Highlight the Level 1 – As-Built Floor plan. Hold down the left mouse button and drag the view onto the sheet. Release the left mouse button to click to place.

61. A preview will appear on your cursor. Left click to place the view on the sheet.

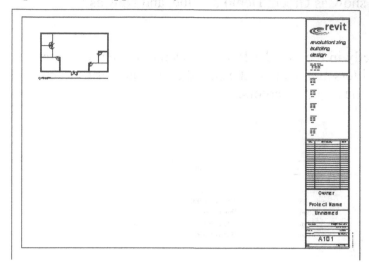

62. Highlight the **Level 1 - 2000 Remodel Demo** Floor plan.
    Hold down the left mouse button and drag the view onto the sheet. Release the left mouse button to click to place.

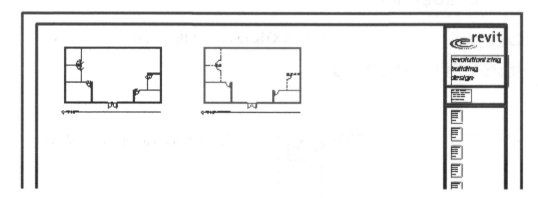

    The two views appear on the sheet.

63. Highlight the Level 1 - 2000 New Construction plan.
    Hold down the left mouse button and drag the view onto the sheet. Release the left mouse button to click to place.

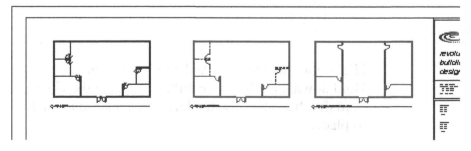

64. Save as *ex3-13.rvt*.

Challenge Exercise:

Create two more views called Level 1 2010 Remodel Demo and Level 1 2010 Remodel New Construction.

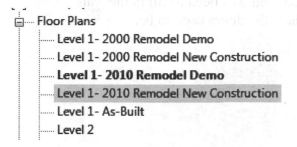

Set the Phases and phase filters to the new views.

The 2010 Remodel Demo view should be set to:

Phasing	
Phase Filter	Show Previous + Demo
Phase	2010 Remodel

The 2010 Remodel New Construction view should be set to:

Phasing	
Phase Filter	Show Previous + New
Phase	2010 Remodel

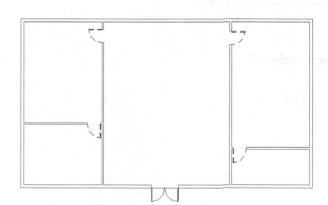

On the 2010 Remodel Demo view:
Demo all the interior doors.

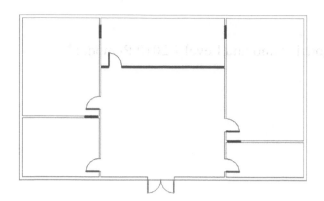

For the 2010 remodel new construction, add the walls and doors as shown.

Note you will need to fill in the walls where the doors used to be.

Add the 2010 views to your sheet.

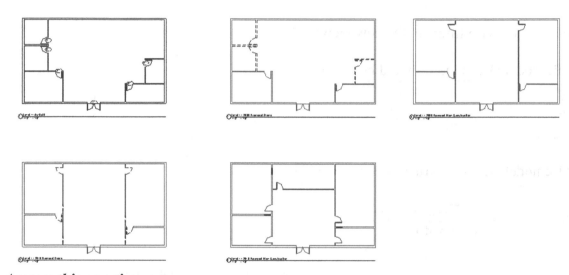

Answer this question:

When should you use phasing as opposed to design options?

Exercise 3-14

Tag Elements by Category

Drawing Name: tag_elements.rvt
Estimated Time: 5 minutes

This exercise reinforces the following skills:

- ❏ Tag All Not Tagged
- ❏ Window Schedules
- ❏ Schedule/Quantities
- ❏ Schedule Properties

1. 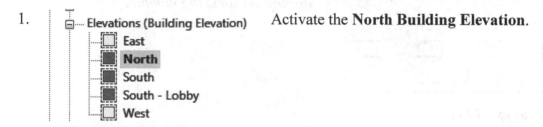 Activate the **North Building Elevation**.

2.  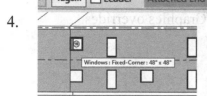 Activate the Annotate ribbon.

   Select the **Tag by Category** tool from the Tag panel.

3. 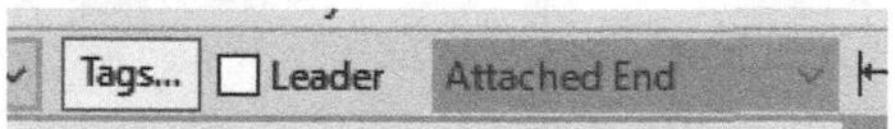 Tags... ☐ Leader Attached End Disable **Leader** on the Options bar.

4. Click on a window in the view to place the window tag. Right click and select Cancel to exit the command.

   Revit automatically knew to place a window tag based on what was selected.

5. Activate the Annotate ribbon.

   Select the **Tag All** tool from the Tag panel.

6. 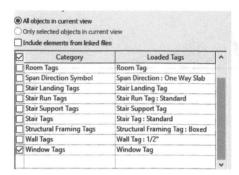 Click **Window Tags**.

   Click **OK**.

   You can select more than one category tag to be placed using Tag All.

7. 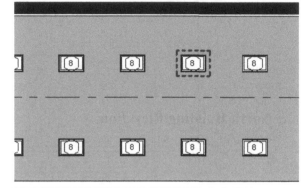 Tags will appear on all the windows.

   The window tag uses the window type as an identifier, not the window mark. So, all windows of the same type will display the same tag number.

8. Save the file as *ex3-14.rvt*.

Matchline

A match line is a line on a design drawing that projects a location or distance from one portion of the drawing to another portion of the drawing. The matchline indicates where a new view splits off.

You can modify the appearance of a matchline by using Visibility/Graphics overrides.

Exercise 3-15
Using a Matchline

Drawing Name: matchline.rvt
Estimated Time: 30 minutes

This exercise reinforces the following skills:

- ❑ Matchline
- ❑ Sheets
- ❑ Views
- ❑ View Reference Annotation

1.  📂 Open *matchline.rvt*.

2. 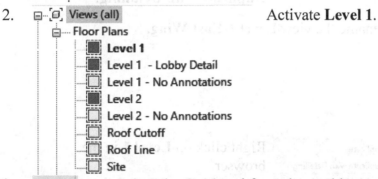   Activate **Level 1**.

3.  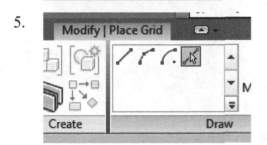   Select the **Grid** tool from the Architecture ribbon.

4. 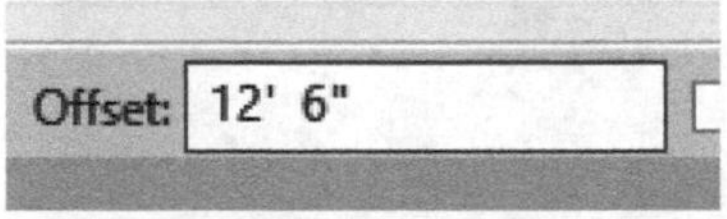 Offset: 12' 6"   On the Options ribbon, set the Offset to **12' 6"**.

5.   Select **Pick Line** mode on the Draw panel on the ribbon.

6. Place a grid line between Grids D & E.

7. Rename the Grid bubble **D.5.**

8. Right click on **Level 1** in the browser.
Select **Duplicate View →**
Duplicate with Detailing.

Duplicate View ▶	Duplicate
Convert to independent view	Duplicate with Detailing
Apply Dependent Views...	Duplicate as a Dependent

9. Rename the view **Level 1-East Wing**.

Views (all)
└ Floor Plans
  ├ Level 1
  ├ Level 1 - Lobby Detail
  ├ **Level 1 - East Wing**
  ├ Level 1 - No Annotations
  └ Level 2

10. Right click on **Level 1** in the browser.
Select **Duplicate View →**
Duplicate with Detailing.

Duplicate View ▶	Duplicate
Convert to independent view	Duplicate with Detailing
Apply Dependent Views...	Duplicate as a Dependent

11. Rename the view **Level 1-West Wing**.

Views (all)
└ Floor Plans
  ├ Level 1
  ├ Level 1 - Lobby Detail
  ├ Level 1 - East Wing
  ├ Level 1 - No Annotations
  ├ **Level 1 - West Wing**
  └ Level 2

12. Activate the view **Level 1-West Wing** floor plan.

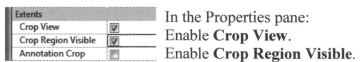

In the Properties pane:
Enable **Crop View**.
Enable **Crop Region Visible**.

13.

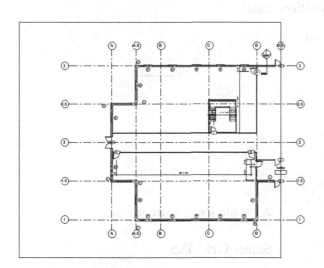

Use the grips on the crop region to only show the west side of the floor plan.

14.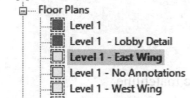

Activate **Level 1 - East Wing**.

15.

Extents	
Crop View	☑
Crop Region Visible	☑
Annotation Crop	☐

In the Properties pane:
Enable **Crop View**.
Enable **Crop Region Visible**.

16.

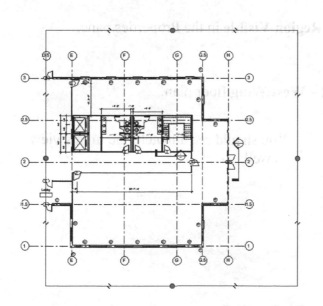

Use the grips on the crop region to only show the east side of the floor plan.

17. View   Activate the **View** ribbon.

18. 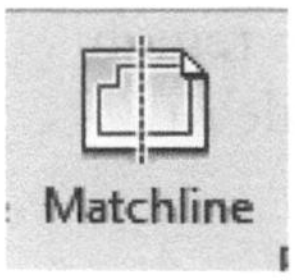 Under the Sheet Composition panel:

Select the **Matchline** tool.

19.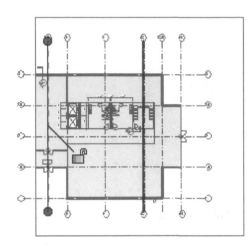

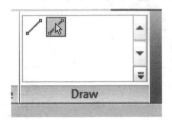

Activate **Pick Line** mode.

Select **Grid D.5**.

20.  Select the **Green Check** under Mode to finish the matchline.

21.  Disable **Crop Region Visible** in the Properties pane.

22. Level 1 - Lobby Detail
Level 1 - West Wing
Level 1 - East Wing

Activate the **Level 1 - West Wing** floor plan.

23. 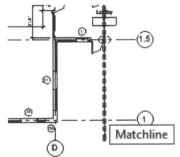 You should see a match line in this view as well.

24. 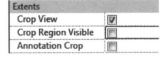 Disable **Crop Region Visible** in the Properties pane.

25. 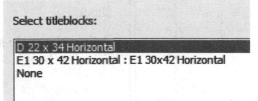 Add a new **Sheet** using the Sheet Composition panel on the View ribbon.

Select titleblocks:

D 22 x 34 Horizontal
E1 30 x 42 Horizontal : E1 30x42 Horizontal
None

Highlight the **D 22 x 34 Horizontal** titleblock.
Click **OK**.

26. 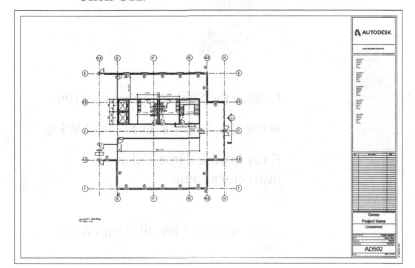 Place the **Level 1-East Wing** floor plan on the sheet.

Adjust the scale so it fills the sheet.

Adjust the grid lines, if needed.

27. 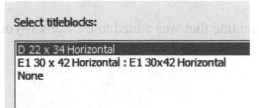 Change the Sheet Number to **A105**.
Name the sheet **Level 1 - East Wing**.

Enter in your name and your instructor's name.

M. Instructor
J. Student
M. Instructor
J. Student
A105
Level 1 - East Wing

28. Add a new **Sheet** using the Sheet Composition panel on the View ribbon.

Select titleblocks:

D 22 x 34 Horizontal
E1 30 x 42 Horizontal : E1 30x42 Horizontal
None

Highlight the **D 22 x 34 Horizontal** titleblock.
Click **OK**.

29.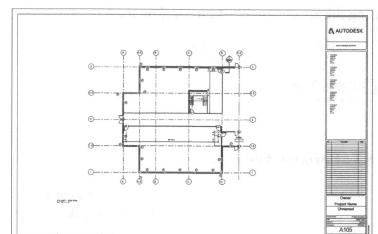

Place the **Level 1- West Wing** floor plan on the sheet.

Adjust the scale so it fills the sheet.

30.

M. Instructor
J. Student
M. Instructor
J. Student
A106
Level 1 - West Wing

Change the Sheet Number to **A106**.

Name the sheet **Level 1 -West Wing**.

Enter in your name and your instructor's name.

31.

Zoom into the view title bar on the sheet.

① Level 1 - West Wing
   1/8" = 1'-0"

32.

⊟ Views (all)
  ▢ Floor Plans
    ■ **Level 1**
    ■ Level 1 - Lobby Detail
    ▢ Level 1 - No Annotations
    ■ Level 1 - West Wing
    ■ Level 1- East Wing

Activate the **Level 1** floor plan.

33.

(D.5)

You can see the matchline that was added to the view if you zoom into the D.5 grid line.

34.

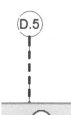

View Reference

Activate the View ribbon.

Select the **View Reference** tool on the Sheet Composition panel.

This adds an annotation to a view.

35.

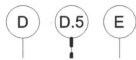

Place the note on the right side of the grid line to correspond with the view associated with the right wing.

Cancel out of the command.

36.

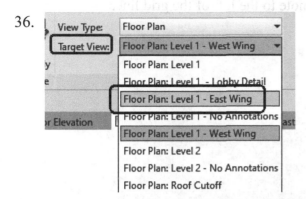

Select the **View Reference** label.

Set the Target View on the ribbon to the **Floor Plan: Level 1 – East Wing.**

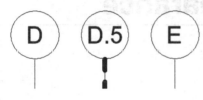

The View Reference label updates to the correct sheet number.

37.

Note that the **A106 – Level 1 – East Wing** sheet view is also labeled 1.

38.

Activate the View ribbon.

Select the **View Reference** tool on the Sheet Composition panel.

39. 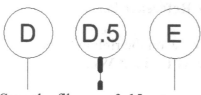 On the ribbon, you can select which view you want to reference.

Select the **Level 1 – West Wing** view.

40.

1 / A106   1 / A105   Place the note to the left of the grid line.

D   D.5   E

41. Save the file as *ex3-15.rvt*.

Exercise 3-16

Modifying a Matchline Appearance

Drawing Name: matchline_appearance.rvt
Estimated Time: 10 minutes

This exercise reinforces the following skills:

- ❏ Matchline
- ❏ Visibility/Graphics Overrides

1. Open *matchline_appearance.rvt*.

2. 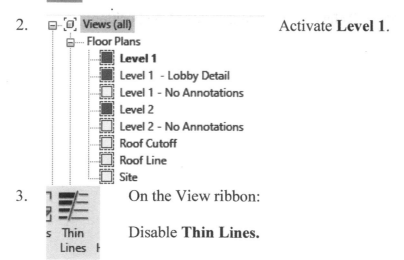 Activate **Level 1**.

3. On the View ribbon:

Disable **Thin Lines.**

This allows you to see line weights of elements.

4. 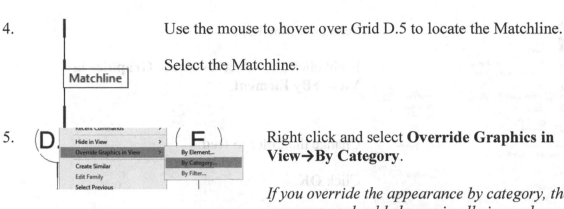 Use the mouse to hover over Grid D.5 to locate the Matchline.

Select the Matchline.

5. Right click and select **Override Graphics in View→By Category**.

If you override the appearance by category, the appearance should change in all views where the element is visible. If you override the appearance by element, then the appearance only changes in the active view.

6. 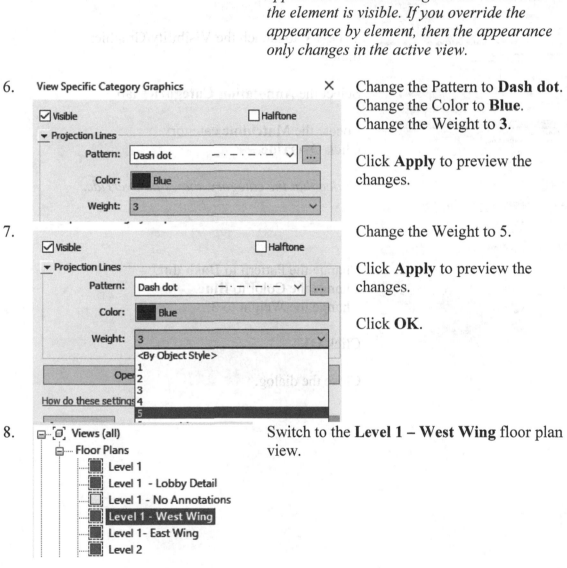 Change the Pattern to **Dash dot**. Change the Color to **Blue**. Change the Weight to **3**.

Click **Apply** to preview the changes.

7. Change the Weight to 5.

Click **Apply** to preview the changes.

Click **OK**.

8. Switch to the **Level 1 – West Wing** floor plan view.

9.

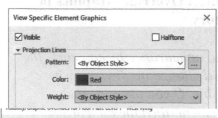

Select the Matchline.

Right click and select **Override Graphics in View→By Element**.

10.

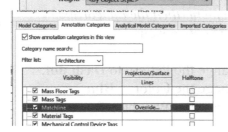

Change the Color to **Red**.

Click **OK**.

11.

Type **VV** to launch the Visibility/Graphics dialog.

Select the **Annotation Categories** tab.

Locate the **Matchline** category.
Click **Override**.

Notice that the category shows no override.

12.

Change the Pattern to **Dash dot**.
Change the Color to **Blue**.
Change the Weight to **3**.

Click **OK**.

Close the dialog.

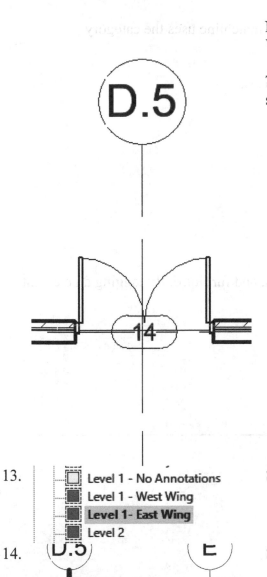

Notice that the matchline in the Level 1 – West Wing view is still red.

That is because we applied the override specifically to the view.

13. Switch to the **Level 1 – East Wing** view.

14. Select the Matchline.

Right click and select **Override Graphics in View→By Element**.

15. Change the Color to **Cyan**.

Click **OK**.

16. Switch to the **Level 1** Floor plan view.

17.

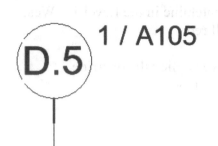

Notice that the matchline uses the category settings.

Save as *ex3-16.rvt*.

Tags

Tags are used to identify elements, such as rooms, doors, and furniture. By tagging an element, you can use that data in schedules.

Exercise 3-17

Create a Tag

Drawing Name: tags.rvt
Estimated Time: 15 minutes

This exercise reinforces the following skills:

- Tag
- Families
- Categories
- Labels

1.  Open *tags.rvt*.

2.

Activate **Livingroom Furniture Plan**.

We want to create a furniture tag for the furniture in the room. This is a rental unit and we need to keep track of the inventory.

3.

Go to **File→New→Annotation Symbol**.

4.

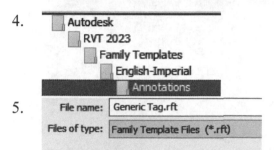

Browse to the *English-Imperial/Annotations* folder under *Family Templates*.

5.

File name: Generic Tag.rft

Files of type: Family Template Files (*.rft)

Select the **Generic Tag** template.

6.

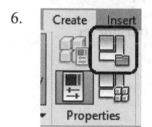

Select **Categories**.

This determines which elements the tag will recognize.

7.

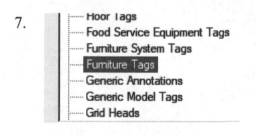

Highlight **Furniture Tags**.

Click **OK**.

8.

Select **Label**.

Labels are associated with parameters.

Click above the intersection of the reference planes.

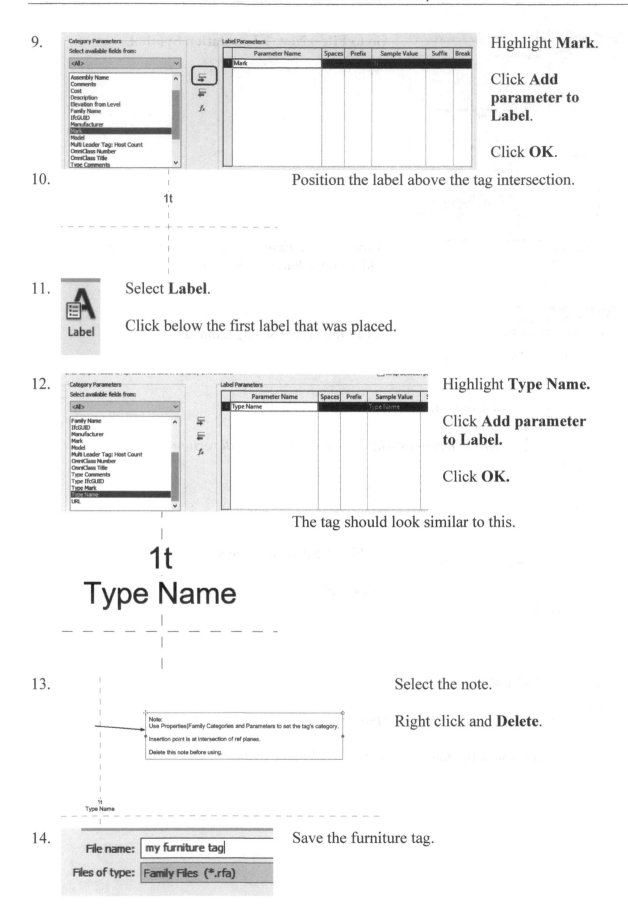

9. Highlight **Mark**.

Click **Add parameter to Label**.

Click **OK**.

10. Position the label above the tag intersection.

11. Select **Label**.

Click below the first label that was placed.

12. Highlight **Type Name**.

Click **Add parameter to Label**.

Click **OK.**

The tag should look similar to this.

13. Select the note.

Right click and **Delete**.

14. Save the furniture tag.

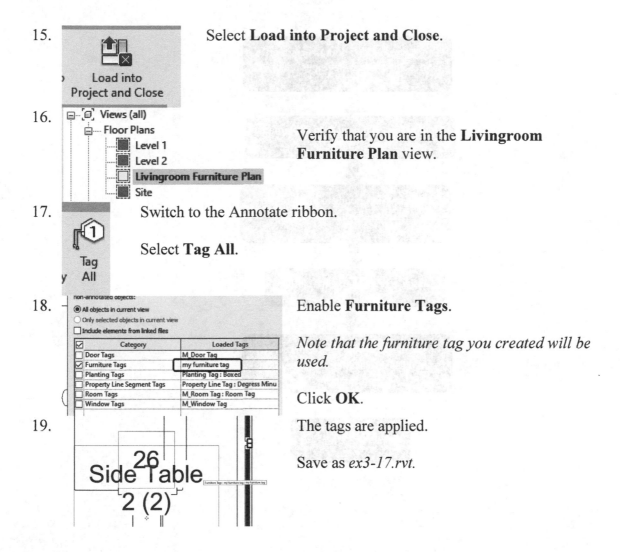

15.   Select **Load into Project and Close**.

16.   Verify that you are in the **Livingroom Furniture Plan** view.

17.   Switch to the Annotate ribbon.

       Select **Tag All**.

18.   Enable **Furniture Tags**.

Note that the furniture tag you created will be used.

Click **OK**.

19.   The tags are applied.

Save as *ex3-17.rvt*.

Revisions

Revision tracking is the process of recording changes made to a building model after sheets have been issued.

When working on building projects, changes are often required to meet client or regulatory requirements.

Revisions need to be tracked for future reference. For example, you may want to check the revision history to identify when, why, and by whom a change was made. Revit provides tools that enable you to track revisions and include revision information on sheets in a construction document set.

Use the workflow below for entering and managing revisions. On the exam, you may be asked to organize this workflow or how revision clouds work.

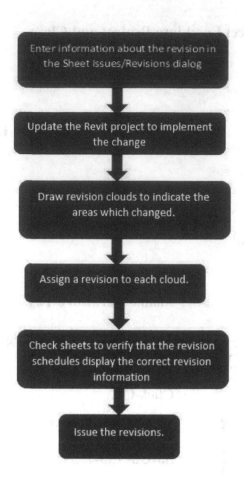

Exercise 3-18

Revision Control

Drawing Name: **i_Revisions.rvt**
Estimated Time to Completion: 15 Minutes

Scope

Add a sheet.
Add a view to a sheet.
Setting up Revision Control in a project.

Solution

1. Activate the **View** ribbon. Select the **New Sheet** tool on the Sheet
    Composition panel.

    Sheet

2.  Select the **Load** button.

3. 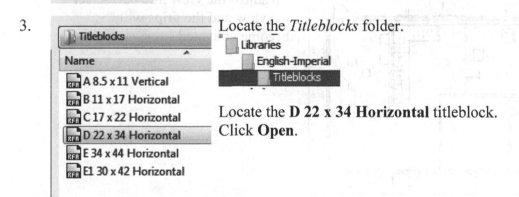 Locate the *Titleblocks* folder.

    Locate the **D 22 x 34 Horizontal** titleblock.
    Click **Open**.

4.  Highlight the **D 22 x 34 Horizontal** titleblock.
    Click **OK**.

5.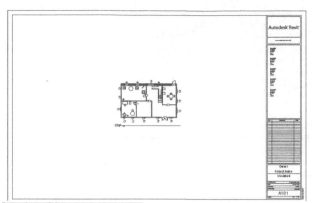

Drag and drop the **Level 1** floor plan onto the sheet.

6.

Graphics	
View Scale	1/4" = 1'-0"
Scale Value 1:	48

Select the view.
In the Properties pane,
set the View Scale to **1/4″ = 1′-0″**.

7.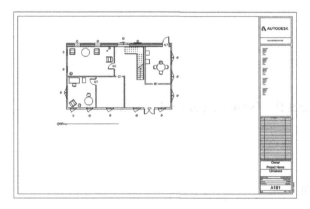

Position the view on the sheet.

8.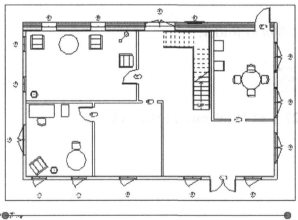

To adjust the view title bar, select the view and the grips will become activated.

9.

    Activate the **View** ribbon.

Revisions (

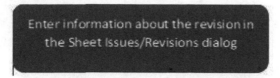

This is the first step in the Revisions workflow.

    Select **Revisions** on the **Sheet Composition** panel.

10. This dialog manages revision control settings and history.

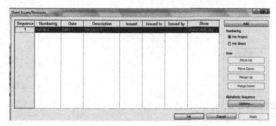

Numbering can be controlled per project or per sheet. The setting used depends on your company's standards.

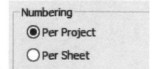

Enable **Per Project**.

One Revision is available by default. Additional revisions are added using the **Add** button.

11. The visibility of revisions can be set to **None**, **Tag** or **Cloud and Tag**. Use the **None** setting for older revisions which are no longer applicable so as not to confuse the contractors.

Set the revision to show **Cloud and Tag**.

12.   Click **Numbering**.

13.

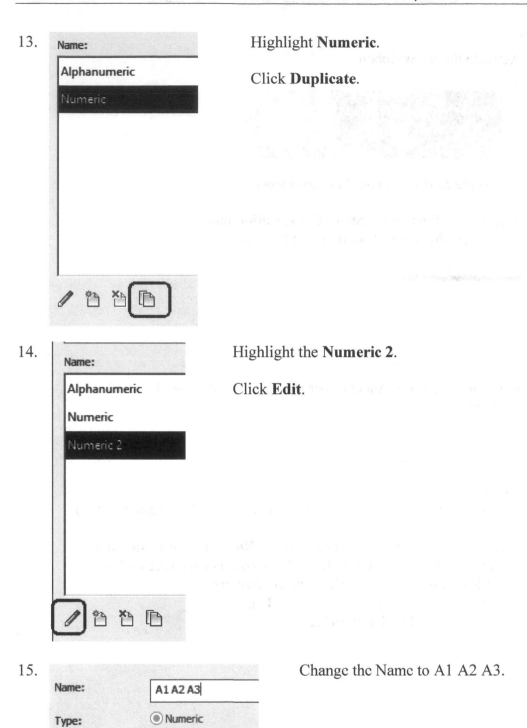

Highlight **Numeric**.

Click **Duplicate**.

14.

Highlight the **Numeric 2**.

Click **Edit**.

15.

Change the Name to A1 A2 A3.

16.

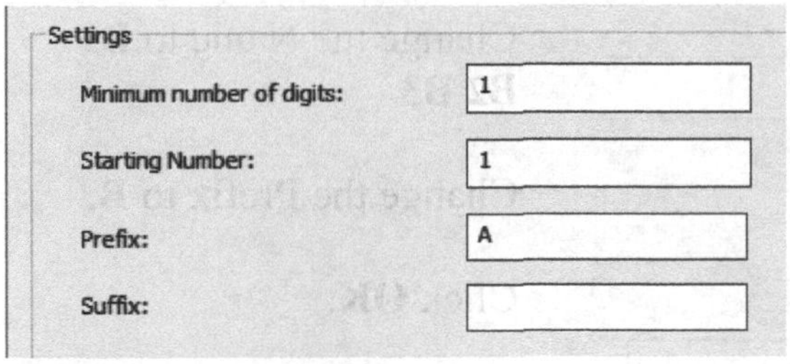

Settings

Minimum number of digits: 1

Starting Number: 1

Prefix: A

Suffix:

Add the Prefix **A**.

Click **OK**.

17.

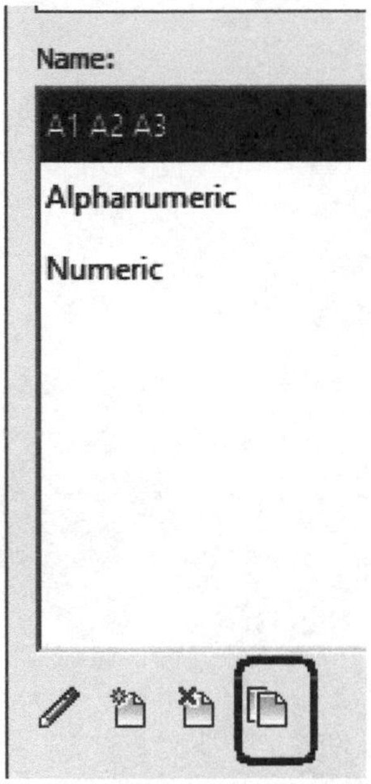

Name:

A1 A2 A3

Alphanumeric

Numeric

Highlight **A1 A2 A3**.

Click **Duplicate**.

18.

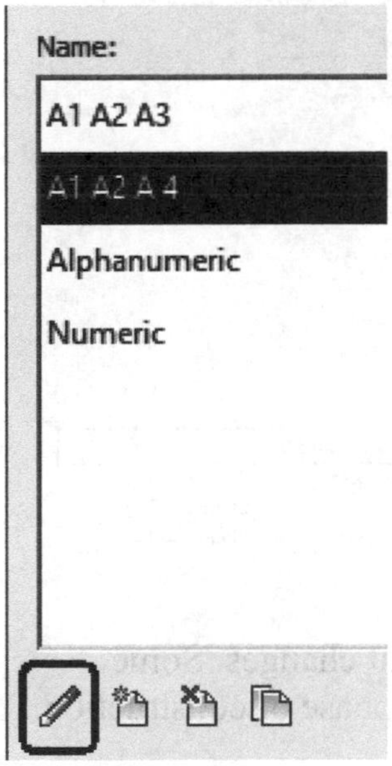

Name:

A1 A2 A3

A1 A2 A 4

Alphanumeric

Numeric

Highlight **A1 A2 A4**.

Click **Edit**.

19.

Name:	B1 B2 B3
Type:	⦿ Numeric
	○ Alphanumeric

Settings

Minimum number of digits:	1
Starting Number:	1
Prefix:	B|
Suffix:	

Change the Name to **B1 B2 B3.**

Change the Prefix to **B.**

Click **OK.**

20.

Name:

A1 A2 A3

Alphanumeric

B1 B2 B3

Numeric

Click **OK.**

21. Enter the three revision changes shown.

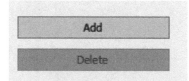

Use the Add button to add the additional lines.

Sequence	Revision Number	Numbering	Date	Description	Issued	Issued to	Issued by	Show
1	A1	A1 A2 A3	08.02	Wall Finish Change	☐	Joe	Sam	Cloud and Tag
2	B1	B1 B2 B3	08.04	Window Style Change	☐	Joe	Sam	Cloud and Tag
3	A	Alphanumeric	08.13	Door Hardware Change	☐	Joe	Sam	Cloud and Tag

Notice you can select different Numbering schemes for different changes. Some companies use different numbering schemes depending on the phase of construction.

22. Click **OK** to close the dialog.
    You can delete revisions if you make a mistake. Just highlight the row and click
    Delete.

Sequence	Numbering	Date	Description	Issued	Issued to	Issued by	Show	
1	Alphabetic	Date 1	Revision 1	☐			Cloud and Tag	Add
2	Alphabetic	Date 2	Revision 2				Cloud and Tag	Delete

23. Save the project as *ex3-18.rvt*.

Exercise 3-19
Modify a Revision Schedule

Drawing Name: **revision_schedule.rvt**
Estimated Time to Completion: 20 Minutes

Scope
Modify a revision schedule in a title block.

Solution

1. ⊟ 🔲 Sheets (all)
        ⊞ A100 - Cover Sheet
        ⊞ **A101 - Level 1 Floorplan**

    Open the **A101 – Level 1 Floorplan** sheet.

2. Zoom in to the **Revision Block** area on the sheet. Note that the title block includes a revision schedule by default.

3. 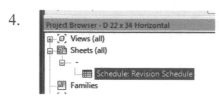 Select the title block.
Right click and select **Edit Family**.

4. 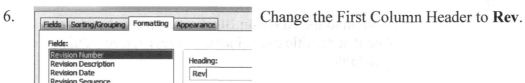 Select the **Revision Schedule** in the Project Browser.

5. 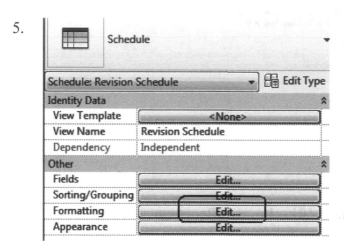 Select **Edit** next to Formatting in the Properties pane.

6. Change the First Column Header to **Rev**.

7.  Select the Fields tab.
Add the **Issued By** field.

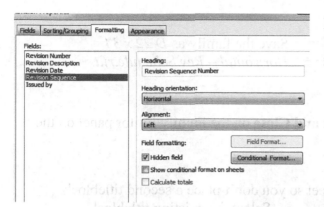

Do **NOT** remove the Revision Sequence field. This is a hidden field.

Note that the Hidden field control is located on the Formatting tab. This is a possible question on the certification exam.

8.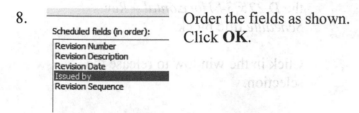

Order the fields as shown.
Click **OK**.

9.
The Revision Schedule updates.

10.

To return to the title block sheet view, double left click on the – below Sheets (all) in the browser.

11.

Adjust the column width of the schedule so it fits properly in the title block.

12.

File name: D 22 x 34 Horizontal - Rev Schedule.rfa

Files of type: Family Files (*.rfa)

Save the family as *D 22 x 34 Horizontal – Rev Schedule.rfa*.

13.

Load into Project and Close

Select **Load into Project and Close** on the Family Editor panel on the ribbon.

14. Click Cancel when you return to the sheet so you don't place a second titleblock.

15.

Properties

D 22 x 34 Horizontal

R

D 22 x 34 Horizontal

D 22 x 34 Horizontal

D 22 x 34 Horizontal - Rev Schedule

D 22 x 34 Horizontal - Rev Schedule

Select the existing titleblock.

Use the Type Selector to switch to the *D 22 x 34 Horizontal – Rev Schedule* title block.

Click in the window to release the selection.

16.

Rev	Description	Date	Issued by

Owner

Note the title block updates with the new revision schedule format.

Save as *ex3-19.rvt*.

Step Two of the workflow

Update the Revit project to implement the change

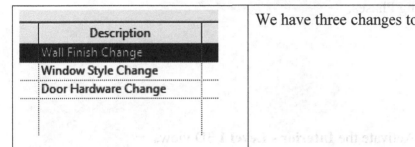

Description
Wall Finish Change
Window Style Change
Door Hardware Change

We have three changes to make in our project.

Exercise 3-20
Update a Project using Phases

Drawing Name: **update_project.rvt**
Estimated Time to Completion: 5 Minutes

Scope
Make changes to a Revit project using Phases.
Change a wall finish.
Change a window style.
Change a door style.

Solution

1. 3D Views
    Interior - Level 1
    Southeast
    {3D}

Activate the **Interior - Level 1** 3D view.

2.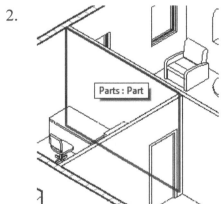

Parts : Part

If you hover over an interior wall, you will see that the wall has been divided into parts.

This allows you to replace the finish on the walls with a different finish.

Select the wall part.

3. Uncheck Material By Original.

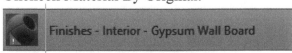

Finishes - Interior - Gypsum Wall Board

Change the Material to **Finishes – Interior – Gypsum Wall Board**.
Click **OK**.
Uncheck **Phase Demolished By Original**.

4. 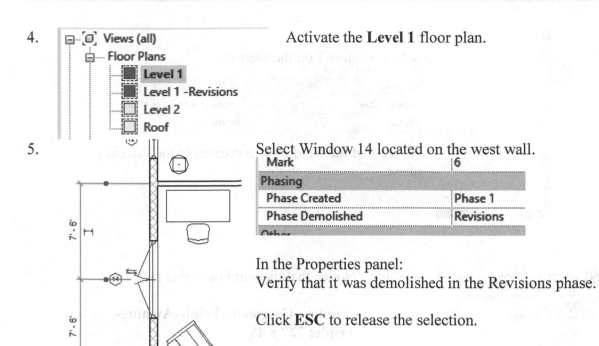 Activate the **Level 1** floor plan.

5. Select Window 14 located on the west wall.

Mark	6
Phasing	
Phase Created	Phase 1
Phase Demolished	Revisions
Other	

In the Properties panel:
Verify that it was demolished in the Revisions phase.

Click **ESC** to release the selection.

6. 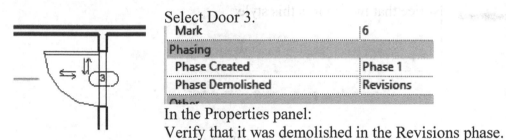 Select Door 3.

Mark	6
Phasing	
Phase Created	Phase 1
Phase Demolished	Revisions
Other	

In the Properties panel:
Verify that it was demolished in the Revisions phase.

Click **ESC** to release the selection.

7. 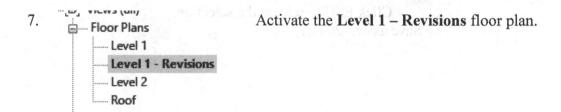 Activate the **Level 1 – Revisions** floor plan.

8.

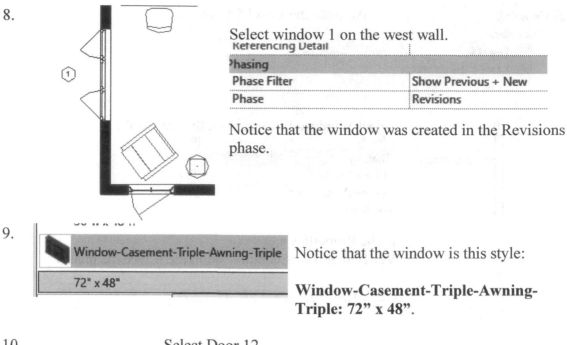

Select window 1 on the west wall.

Referencing Detail	
Phasing	
Phase Filter	Show Previous + New
Phase	Revisions

Notice that the window was created in the Revisions phase.

9.

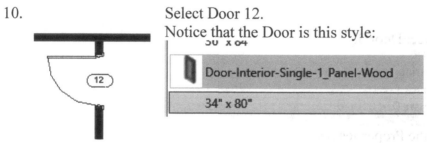

Notice that the window is this style:

Window-Casement-Triple-Awning-Triple: 72" x 48".

10.

Select Door 12.
Notice that the Door is this style:

Door-Interior-Single-1_Panel-Wood	
34" x 80"	

11.

Phasing	
Phase Created	Revisions
Phase Demolished	None

With the door still selected:
Set the Phase Created to **Revisions**.

Click ESC to release the selection.

12.
Save as *ex3-20.rvt*.

Step Three of the workflow

> Draw revision clouds to indicate the
> areas which changed.

Revision Clouds

Revision clouds are used to designate areas in the drawing where changes have been made. Revision clouds are annotation objects and are specific to a view. You can assign the same revision to multiple clouds.

Exercise 3-21

Add Revision Clouds

Drawing Name: **revision clouds.rvt**
Estimated Time to Completion: 10 Minutes

Scope
Add revision clouds to a view.

Solution

1.  Sheets (all)
    A100 - Cover Sheet
    A101 - Level 1 Floorplan
    Activate the **A101 - Level 1** floor plan sheet.

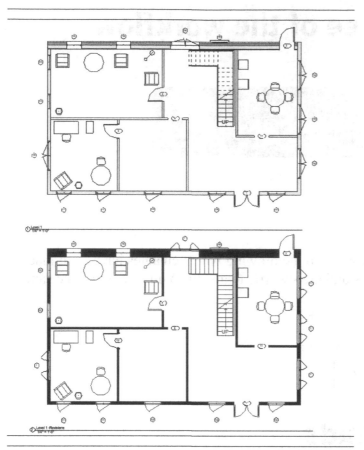

The top view shows the existing floor plan prior to the revisions.
The bottom view shows the revisions that were made.
We will be adding revision clouds to the bottom view.

2.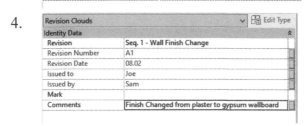
Activate the **Annotate** ribbon.
Select **Revision Cloud** from the Detail panel.

3.

Revision Clouds	
Identity Data	
Revision	Seq. 1 - Wall Finish Change
Revision Number	A1
Revision Date	08.02
Issued to	Joe
Issued by	Sam
Mark	
Comments	

On the Properties pane,
select the Revision that is tied to the
revision cloud – **Wall Finish Change**.

4.

Revision Clouds		Edit Type
Identity Data		
Revision	Seq. 1 - Wall Finish Change	
Revision Number	A1	
Revision Date	08.02	
Issued to	Joe	
Issued by	Sam	
Mark		
Comments	Finish Changed from plaster to gypsum wallboard	

Under Comments, enter **Finish Changed from plaster to gypsum wallboard**.

5.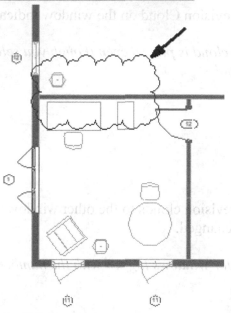

    You can use any of the available Draw tools to create your revision cloud. Select the **Rectangle** tool.

6.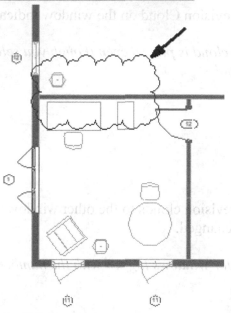

    Draw the Revision Cloud on the wall indicated.

    Note that a cloud is placed even though you selected the rectangle tool.

7.

    Select the **Green Check** on the Mode panel to **Finish Cloud**.

8.

    Activate the **Annotate** ribbon.
    Select **Revision Cloud**.

9.

Revision Clouds	
Identity Data	
Revision	Seq. 2 - Window Style Change
Revision Number	B1
Revision Date	08.04
Issued to	Joe
Issued by	Sam
Mark	
Comments	Use Assy Code 301

    Select the Revision that is tied to the revision cloud – **Window Style Change**.

    Under Comments, enter **Use Assy Code 301**.

10. 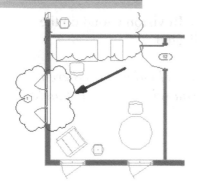 Select the **Circle** tool.

11. 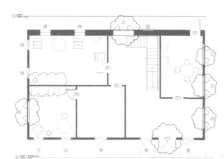 Draw the Revision Cloud on the window indicated.

Note that a cloud is placed even though you selected the circle tool.

12.  Add revision clouds to the other windows that were changed.

Look for windows tagged with the number 1.

13. Select the **Green Check** on the Mode panel to **Finish Cloud**.

14. 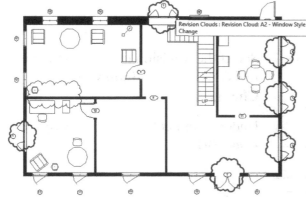 If you hover over the revision cloud, you will see a tooltip to indicate what the revision is.

Notice all the revision clouds highlight.

15. Activate the **Annotate** ribbon.
Select **Revision Cloud** from the Detail panel.

16.

Revision Clouds	
Identity Data	
Revision	Seq. 3 - Door Hardware Change
Revision Number	A
Revision Date	08.13
Issued to	Joe
Issued by	Sam
Mark	
Comments	Use Assy Code 405

On the Properties pane, select the Revision that is tied to the revision cloud – **Door Hardware Change**.
Under Comments, enter **Use Assy Code 405**.

By using different revision schema, you can assign and track revisions for different subcontractors.

17.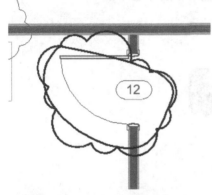

Select the **Spline** tool.

Draw the cloud in the clockwise direction.

18. Draw the Revision Cloud on the door indicated.

19.

Error - cannot be ignored
Self intersecting splines are not allowed in sketches.
Show

You might see this error if the spline has overlapping segments. If you get this error, delete the spline and try again. You do not have to close the spline in order to create the revision cloud.

20. 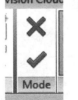 Select the **Green Check** on the Mode panel to **Finish Cloud**.

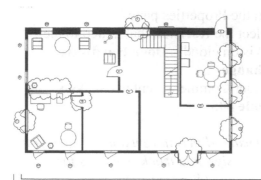

The revision clouds have been placed in the view.

21.

Rev.	Description	Date	Issued by
A1	Wall Finish Change	08.02	Sam
B1	Window Style Change	08.04	Sam
A	Door Hardware Change	08.13	Sam

Zoom into the title block and note that the revision block has updated with the revisions that have been issued.

22.

Save as *ex3-21.rvt*.

Step Four of the workflow

Assign a revision to each cloud.

Exercise 3-22

Tag Revision Clouds

Drawing Name: **tag clouds.rvt**
Estimated Time to Completion: 5 Minutes

Scope
Tag revision clouds in a view.

Solution

1.

Sheets (all)
A100 - Cover Sheet
A101 - Level 1 Floorplan

Activate the **A101- Level 1 floor plan** sheet.

2. **Tag by Category**

   Select the **Tag by Category** tool from the Tag panel on the Annotate tab on the ribbon.

3. Pick one of the revision clouds to identify the category to be tagged.

   Revision Clouds : Revision Cloud: A2 - Window Style Change

4. There is no tag loaded for Revision Clouds. Do you want to load one now?

   If you see this message, click **Yes**.

5. North Arrow 2
   Part Tag
   Revision Tag
   Room Tag
   Section Head - 1 po

   Libraries
   English-Imperial
   Annotations

   Select the *Annotations* folder.
   Locate the **Revision Tag**.
   Click **Open**.

6. Horizontal | Tags... | ☑ Leader | Attached End | ⊢ 1/2"

   Enable **Leader** on the Option bar.

7. Select the revision cloud to add the tag. The tag is placed.

8. Repeat to add tags to the other revision clouds.

9. Save as *ex3-22.rvt*.

Step Five & Six of the workflow

> Check sheets to verify that the revision schedules display the correct revision information

> Issue the revisions.

Exercise 3-23

Issue Revisions

Drawing Name: **issue revisions.rvt**
Estimated Time to Completion: 5 Minutes

Scope
Issue revisions.

Solution

1.
```
Sheets (all)
    A100 - Cover Sheet
    A101 - Level 1 Floorplan
```
Activate the **A101-Level 1 floor plan** sheet.

2.

Rev.	Description	Date	Issued by
A1	Wall Finish Change	08.02	Sam
B1	Window Style Change	08.04	Sam
A	Door Hardware Change	08.13	Sam

Zoom into the title block and verify that the revisions are listed.

3.

Switch to the View tab on the ribbon.

Select Revisions.

4.

Sequence	Revision Number	Numbering	Date
1	A1	A1 A2 A3	08.02
2	B1	B1 B2 B3	08.04
3	A	Alphanumeric	08.13

Highlight the A revision on Sequence 3.

Use the **MOVE UP** button to assign it to Sequence 1.

5.

Sequence	Revision Number	Numbering	Date
1	A	Alphanumeric	08.13
2	A1	A1 A2 A3	08.02
3	B1	B1 B2 B3	08.04

A revision should now be located at Sequence 1.

Click **Apply**.

6.

Rev.	Description	Date	Issued by
A	Door Hardware Change	08.13	Sam
A1	Wall Finish Change	08.02	Sam
B1	Window Style Change	08.04	Sam

Look on the sheet.

Notice that the Revision A is now in the first row.

7.

Revision Number	Numbering	Date	Description	Issued	
A	Alphanumeric	08.13	Door Hardware Change	☐	Joe
A1	A1 A2 A3	08.02	Wall Finish Change	☐	Joe
B1	B1 B2 B3	08.04	Window Style Change	☑	Joe

Place a check on the Issued button for Revision **B1**.

Click **Apply**.

Revision Number	Numbering	Date	Description	Issued
A	Alphanumeric	08.13	Door Hardware Change	☐
A1	A1 A2 A3	08.02	Wall Finish Change	☐
B1	B1 B2 B3	08.04	Window Style Change	☑

Notice that the revision B1 is grayed out in the Revisions dialog.

Sequence	Revision Number	Numbering	Date	Description	Issued
1	A	Alphanumeric	08.13	Door Hardware Change	☐
2	A1	A1 A2 A3	08.02	Wall Finish Change	☐
3	B1	B1 B2 B3	08.04	Window Style Change	☑

Notice once a revision has been issued, you cannot modify the row it has been assigned. You also can no longer Delete it.

Rev.	Description	Date	Issued by
A	Door Hardware Change	08.13	Sam
A1	Wall Finish Change	08.02	Sam
B1	Window Style Change	08.04	Sam

Click **OK**.

Revision B1 is still listed in the Titleblock.

8.

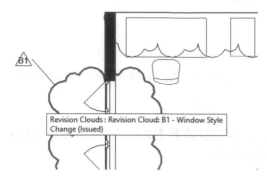

Revision Clouds : Revision Cloud: B1 - Window Style Change (Issued)

Select one of the B1 clouds.

Notice that the revision comment now shows (Issued).

In the Properties pane, the Revision now displays *(Issued).*

Release the selection.

Identity Data	
Revision	Seq. 3 (Issued) - Window Style Change
Revision Number	B1
Revision Date	08.04
Issued to	Joe
Issued by	Sam
Mark	
Comments	Use Assy Code 301

9.

Revisions

Switch to the View tab on the ribbon.

Select **Revisions**.

10.

Add

Click **Add**.

11.

Sequence	Revision Number	Numbering	Date	Description	Issued	Issued to	Issued by	Show
1	A	Alphanumeric	08.13	Door Hardware Change	☐	Joe	Sam	Cloud and Tag
2	A1	A1 A2 A3	08.02	Wall Finish Change	☐	Joe	Sam	Cloud and Tag
3	B1	B1 B2 B3	08.04	Window Style Change	☑	Joe	Sam	Cloud and Tag
4	B2	B1 B2 B3	08.05	Remove a Window	☐	Joe	Sam	Cloud and Tag
5	B3	B1 B2 B3	08.07	Add a Window	☐	Joe	Sam	Cloud and Tag ⌄

Add two more revisions using the B1 B2 B3 numbering schema.

B2 Removes a Window.

B3 Adds a Window.

Click **OK**.

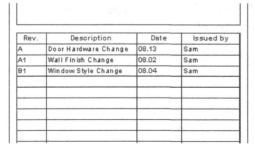

Rev.	Description	Date	Issued by
A	Door Hardware Change	08.13	Sam
A1	Wall Finish Change	08.02	Sam
B1	Window Style Change	08.04	Sam

Notice that the new revisions have not been added to the titleblock.

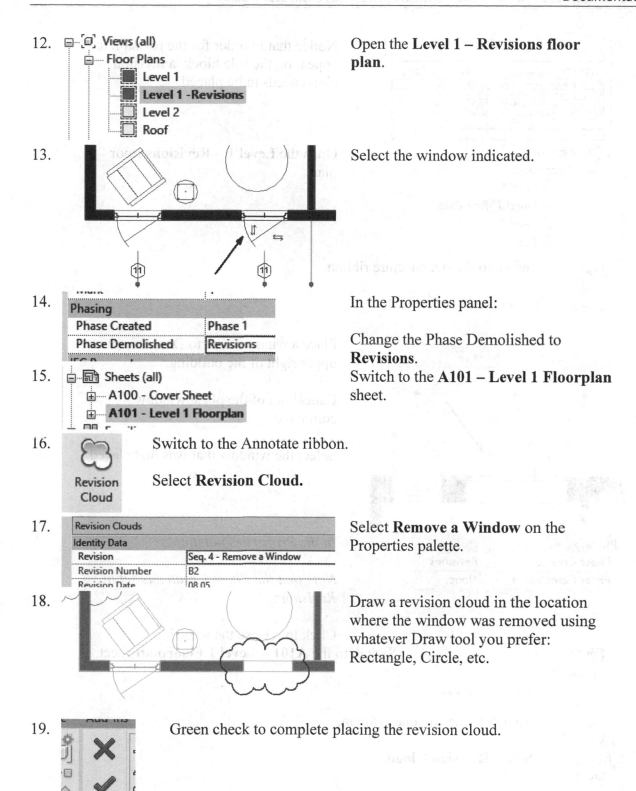

12.   Open the **Level 1 – Revisions floor plan**.

13.   Select the window indicated.

14.   In the Properties panel:

Change the Phase Demolished to **Revisions**.

15.   Switch to the **A101 – Level 1 Floorplan** sheet.

16.   Switch to the Annotate ribbon.

Select **Revision Cloud.**

17.   Select **Remove a Window** on the Properties palette.

18.   Draw a revision cloud in the location where the window was removed using whatever Draw tool you prefer: Rectangle, Circle, etc.

19.   Green check to complete placing the revision cloud.

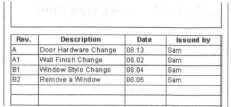

Rev.	Description	Date	Issued by
A	Door Hardware Change	08.13	Sam
A1	Wall Finish Change	08.02	Sam
B1	Window Style Change	08.04	Sam
B2	Remove a Window	08.05	Sam

Notice that in order for the revision to appear on the title block, a revision cloud needs to be placed.

20.

Open the **Level 1 – Revisions** floor plan.

21. Switch to the Architecture ribbon.

Select the **Window** tool.

22.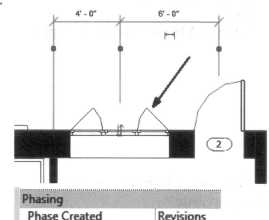

Place a window next to Door 2 in the upper right of the building.

Cancel out of the place window command.

Select the window that was just placed.

In the Properties palette:

Phasing	
Phase Created	Revisions
Phase Demolished	None

Note that the Phase Created is set to Revisions.

Click to release the selection.

23. Switch to the **A101 – Level 1 Floorplan** sheet.

24. 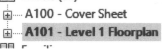 Switch to the Annotate ribbon.

Select **Revision Cloud**.

25.

Revision Clouds	
Identity Data	
Revision	Seq. 5 - Add a Window
Revision Number	B3
Revision Date	08.07
Issued to	Joe
Issued by	Sam

Select **Add a Window** on the Properties palette.

26.

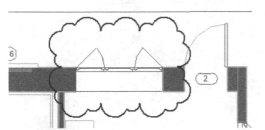

Draw a revision cloud in the location where the window was added using whatever Draw tool you prefer: Rectangle, Circle, etc.

27.

Green check to complete placing the revision cloud.

28.

Rev.	Description	Date	Issued by
A	Door Hardware Change	08.13	Sam
A1	Wall Finish Change	08.02	Sam
B1	Window Style Change	08.04	Sam
B2	Remove a Window	08.05	Sam
B3	Add a Window	08.07	Sam

Check the titleblock to see if the new revision has been added.

29.

Revisions

Switch to the View tab on the ribbon.

Select **Revisions**.

30.

Sequence	Revision Number	Numbering	Date	Description
1	A	Alphanumeric	08.13	Door Hardware Change
2	A1	A1 A2 A3	08.02	Wall Finish Change
3	B1	B1 B2 B3	08.04	Window Style Change
4	B2	B1 B2 B3	08.05	Remove a Window
5	B3	B1 B2 B3	08.07	Add a Window

Highlight Revision B3 and move it above B2.

31.

Revision Number	Numbering	Date	Description
A	Alphanumeric	08.13	Door Hardware Change
A1	A1 A2 A3	08.02	Wall Finish Change
B1	B1 B2 B3	08.04	Window Style Change
B2	B1 B2 B3	08.07	Add a Window
B3	B1 B2 B3	08.05	Remove a Window

The revision number should have updated.

Click **OK**.

32.

Rev.	Description	Date	Issued by
A	Door Hardware Change	08.13	Sam
A1	Wall Finish Change	08.02	Sam
B1	Window Style Change	08.04	Sam
B2	Add a Window	08.07	Sam
B3	Remove a Window	08.05	Sam

Notice that the title block updated as well.

33. Save as *ex3-23.rvt*.

Print SetUp

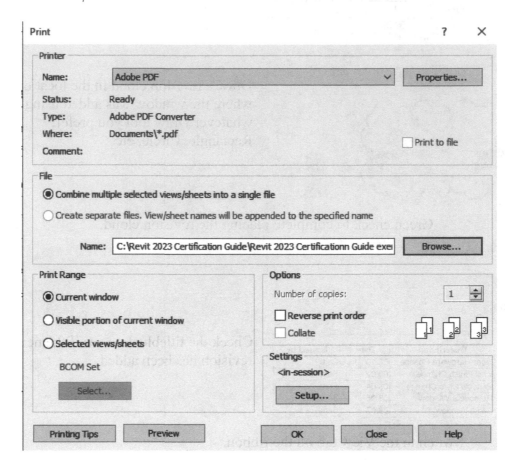

Revit uses standard Windows *print* drivers, so printing from Revit is much like printing from other Windows applications, such as Microsoft Word. Of course, the paper sizes used for architectural drawings are generally much different from those used when you print a Word document, however, so there are some additional settings you need to be aware of.

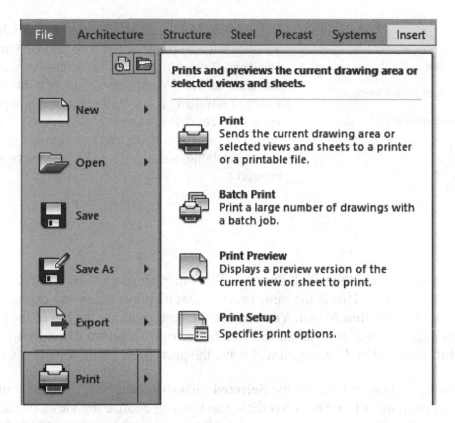

You access all printing functions from the **Application Menu**. When you expand
the **Application Menu** and select **Print**, you can see that there are three options: **Print**, **Print
Preview**, and **Print Setup**. Note that when you click **Print**, you can also access
the **Preview** and **Setup** options from within the **Print** dialog. Click **Print**.

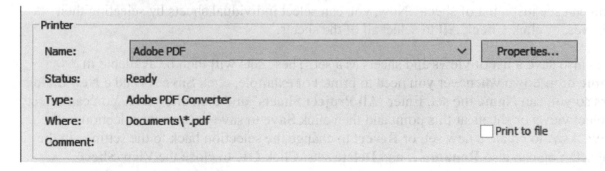

In the **Print** dialog, the first thing you will typically do is choose the printer to which you want to
print. In the **Printer** area, expand the **Name** drop-down. You can print to any available Windows
printer, including drivers that actually create.pdf files, .xps documents, and so on. Your list will
vary, depending on what is available on your system.

When you select **Print to file**, the **Name** field in the **File** area becomes available. You can then
click **Browse** to open a **Browse for Folder** dialog and navigate to the folder in which you want
to save the print file. You can also choose to save the file as a.prn or.plt file. Click **Cancel** to
close this dialog.

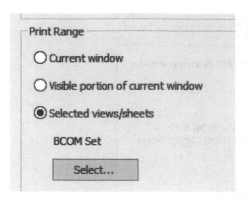

In the **Print Range** area, you can choose what you actually want to print. When you choose **Current window**, Revit will print all of whatever view or sheet is currently active. When you choose **Visible portion of current window**, Revit will print only the portion of the view or sheet that is currently displayed. When either of these options is selected, the two options in the **File** area are unavailable, since only a single view or sheet will be printed.

Select **Current window** and then click **Preview**. Revit displays a preview showing what the printed sheet will look like. This is the same preview that displays when you choose **Print Preview** from the **Application Menu**. You can use the buttons along the top of the preview to switch between pages, if available, and also zoom in and out. To return to the **Print** dialog, you must click **Print**. If you click **Close**, you will leave the print function all together. Click **Print**.

Under **Print Range**, when you choose the **Selected views/sheets** option, you must then click **Select…** to open the **View/Sheet Set** dialog so you can choose the views or sheets you want to print. Note that if you are printing to a file, Revit will create a separate file for each view or sheet when you select this option.

In the **View/Sheet Set** dialog, you can see a list of all the views and sheets in the current project. Realize that checkboxes below the list let you display just **Views** or just **Sheets**. Clear **Views** so that you can see just a list of sheets. Now, you can select individual sheets by selecting their checkboxes, or click **Check All** to select all of the sheets.

You can also save a list of views and sheets as a set. These sets will then be available in the **Name** drop-down whenever you need to print. For example, click **Save As** and a **New** dialog appears so you can **Name** the set. Enter "**All Project Sheets**" and then click **OK**. You can select or deselect views or sheets at this point and then click **Save** to save the current selection to the set, **Save As…** to create a new set, or **Revert** to change the selection back to the settings in the current set. You can also **Rename…** and **Delete** sets. Click **OK** to close the **View/Sheet Set** dialog.

Back in the **Print** dialog, in the **Print Range** area, the **All Project Sheets** set is shown, indicating that it is selected. Also, when printing selected views or sheets, the **Preview** button is unavailable.

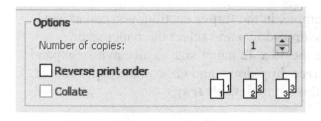

In the **Options** area, you can specify the **Number of copies** as long as **Print to file** is not selected. You can also select **Reverse print order**, and when printing multiple copies, **Collate** is available if you want to organize those copies into finished sets.

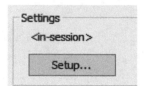

In the **Settings** area, you can click **Setup** to display the **Print Setup** dialog.

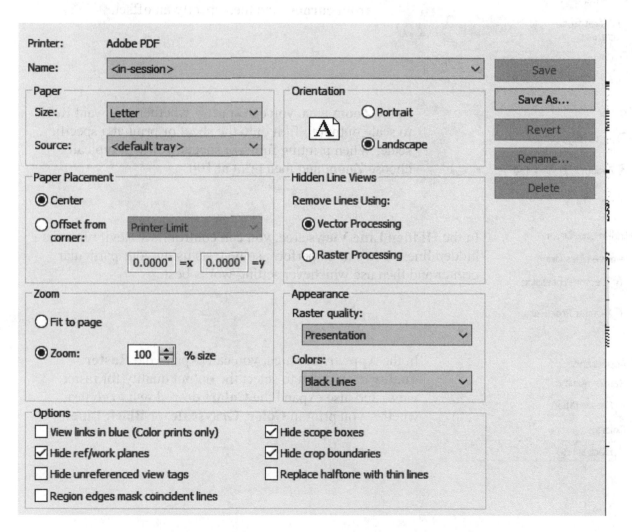

This is the same dialog that displays when you choose **Print Setup** in the **Application Menu**. Here, you can control various settings for paper size, print orientation, and so on. Some of the options available will depend on your particular printer.

For example, in the **Paper** section, you can expand the **Size** drop-down and select the paper size. Choose the **30 x 42** paper size. You can also expand the **Source** drop-down and choose the paper source. Leave this set to **default tray**.

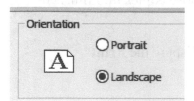

In the **Orientation** area, you can select a **Portrait** or **Landscape** orientation.

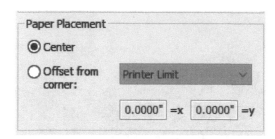

In the **Paper Placement** area, you can choose to either **Center** the print on the sheet or choose **Offset from corner:** and then specify an offset.

In the **Zoom** area, you can specify whether you want Revit to scale output to fit it onto the sheet or print at a specific scale. When printing full-size sheets, you will typically choose **Zoom** and then print at 100%.

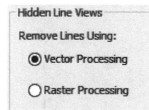

In the **Hidden Line Views** area, you can control how Revit removes hidden lines. You should perform some tests using your particular printer and then use whichever setting works best.

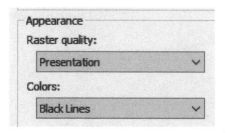

In the **Appearance** area, you can expand the **Raster quality** drop-down to select the output quality for raster views and also expand the **Colors** drop-down to control whether you print in **Color**, **Grayscale**, or **Black Lines**.

Options
- [ ] View links in blue (Color prints only)
- [x] Hide ref/work planes
- [ ] Hide unreferenced view tags
- [ ] Region edges mask coincident lines
- [x] Hide scope boxes
- [x] Hide crop boundaries
- [ ] Replace halftone with thin lines

Lastly, in the **Options** area, you can select various print options. For example, by default, **Hide ref/work planes**, **Hide scope boxes**, and **Hide crop boundaries** are selected. Those objects will not print unless you clear these checkboxes. You can also hide view tags that are unreferenced and replace halftone shading with thin lines.

Once you have adjusted these settings, you can save them as a named setup so that you can easily restore these settings in the future. Click **Save As...**. In the **New** dialog, **Name** this "**Landscape Full Size**" and then click **OK**. Once you do, the new print setup appears in the **Name** field and you will be able to select it again in the future. If you make any changes at this point, you can **Save** them to the current setup, or you can **Revert** the changes. You can easily save multiple print setups, rename them, and delete any that you no longer need.

Exercise 3-24
Configure Custom Print Setup

Drawing Name: **custom print.rvt**
Estimated Time to Completion: 20 Minutes

Scope
Set up a project for printing.

Solution

1. 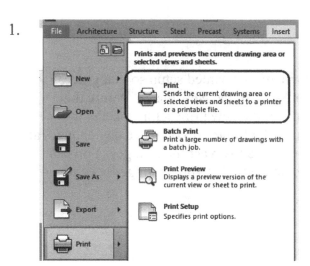 Go to **File→Print→Print**.

2. 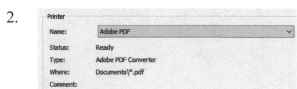 Select **Adobe PDF** as the printer to print a PDF file.

3. 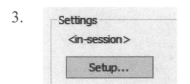 Click the **Setup** button.

4. 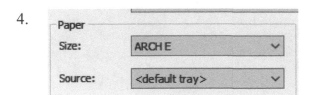 Set the Paper Size: to **ARCH E**.

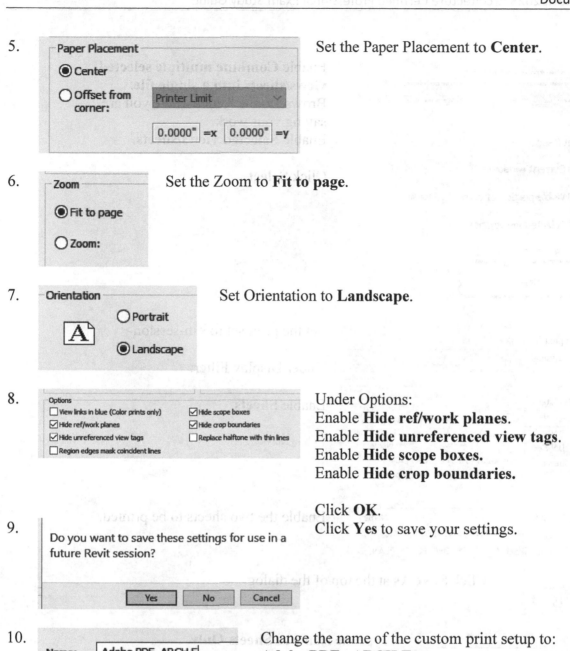

5.   Set the Paper Placement to **Center**.

6.   Set the Zoom to **Fit to page**.

7.   Set Orientation to **Landscape**.

8.   Under Options:
Enable **Hide ref/work planes**.
Enable **Hide unreferenced view tags**.
Enable **Hide scope boxes**.
Enable **Hide crop boundaries.**

Click **OK**.

9.   Click **Yes** to save your settings.

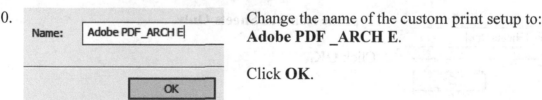

10.   Change the name of the custom print setup to:
Adobe PDF _ARCH E.

Click **OK**.

11.   If you look in the Settings area, you should see the saved setting name.

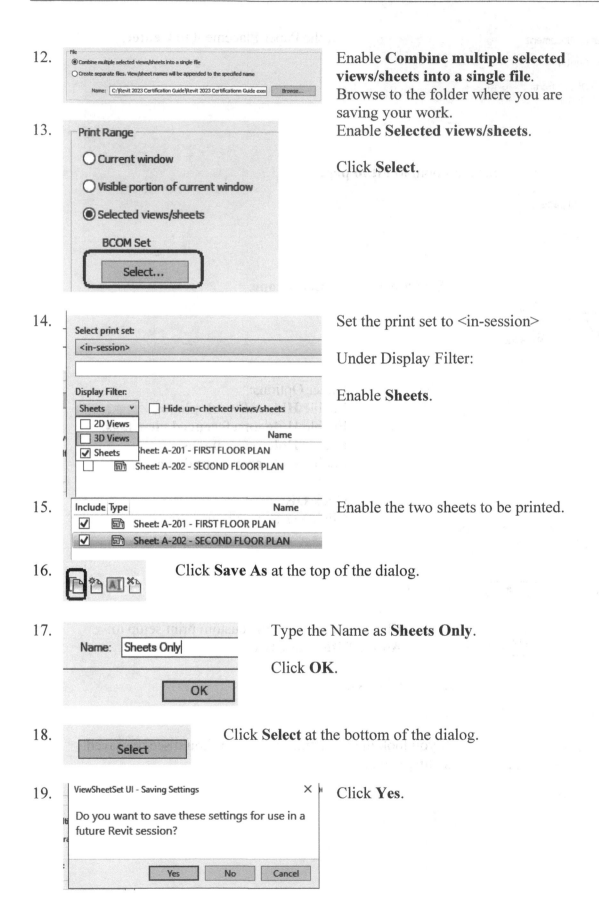

12. Enable **Combine multiple selected views/sheets into a single file**. Browse to the folder where you are saving your work.
Enable **Selected views/sheets**.

13. Click **Select**.

14. Set the print set to <in-session>

Under Display Filter:

Enable **Sheets**.

15. Enable the two sheets to be printed.

16. Click **Save As** at the top of the dialog.

17. Type the Name as **Sheets Only**.

Click **OK**.

18. Click **Select** at the bottom of the dialog.

19. Click **Yes**.

20.

Print Range

○ Current window

○ Visible portion of current window

◉ Selected views/sheets

[Sheets Only]

[Select...]

Notice that Sheets Only is the active set under Print Range.

Click **OK** to create the PDF file.

21.

File name: sheets only

Save as type: PDF Files (*.pdf)

Browse to where you want to save the file.

Name the file *sheets only.pdf*.

Color Schemes

Colors schemes are useful for graphically illustrating categories of spaces, rooms or areas.

For example, you can create a color scheme by room name, area, occupancy, or department. If you want to color rooms in a floor plan by department, set the Department parameter value for each room to the necessary value, and then create a color scheme based on the values of the Department parameter. You can then add a color fill legend to identify the department that each color represents.

Color schemes can be placed in floor plans, ceiling plans, and sections.

Color schemes can be applied to:

- Rooms
- Areas
- Spaces or Zones
- Pipes or ducts

Exercise 3-25

Color Schemes

Drawing Name: **i_Color_Scheme.rvt**
Estimated Time to Completion: 15 Minutes

Scope
Create an area plan using a color scheme.
Place a color scheme legend.
Modify the appearance of a color scheme legend.

Solution

1.
Activate the **01- Entry Level** floor plan.

2.
In the Properties pane, scroll down to the **Color Scheme** field.
Click **<none>**.

3.
Select the **Rooms** category.
Highlight **Name**.

4.
In the Title field, enter **Area Legend**.
Under Color, select the **Area** scheme.

5.
Colors are not preserved when changing which parameter is colored. To color by a different parameter, consider making a new color scheme.

Click **OK**.

6.

Note that different colors are applied depending on the square footage of the rooms.

Click **OK**.

7.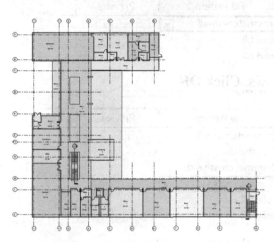

The rooms fill in according to the square footage.

8.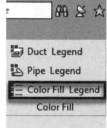

Activate the **Annotate** ribbon.
Select the **Color Fill Legend** tool on the Color Fill panel.

9.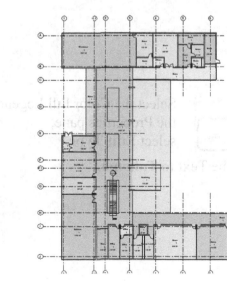

Place the legend in the view.

10.  Select the legend that was placed in the display window.
Select **Edit Scheme** from the Scheme panel on the tab on ribbon.

11. Change the Fill Pattern for 47 SF **to Crosshatch-Small**.

	Value	Visible	Color	Fill Pattern	Preview	
1	47 SF	☑	RGB 156-185	Crosshatch-small		
2	58 SF	☑	PANTONE 3	Solid fill		
3	64 SF	☑	PANTONE 6	Solid fill		

12. Change the hatch patterns for the next few rows. Click **OK**.

	Value	Visible	Color	Fill Pattern	Preview	
1	47 SF	☑	RGB 156-185	Crosshatch-small		
2	58 SF	☑	PANTONE 3	Crosshatch		
3	64 SF	☑	PANTONE 6	Diagonal crosshatch		
4	76 SF	☑	RGB 139-166	Diagonal down		
5	90 SF	☑	PANTONE 6	Solid fill		
6	94 SF	☑	RGB 096-175	Solid fill		

13.   *Note that the legend updates.*

Area Legend

☐ 47 SF

☐ 58 SF

☐ 64 SF

☐ 76 SF

14. Color Fill Legend
    1

    Legends (1)   [Edit Type]

    Select the Color Fill Legend. On the Properties pane, select **Edit Type.**

15.
Text	
Font	Tahoma
Size	3/16"
Bold	☑
Italic	☐
Underline	☐

Change the font for the Text to **Tahoma.**
Enable **Bold.**

16.

Change the font for the Title Text to **Tahoma.**
Enable **Bold**.
Enable **Italic.**
Click **OK**.

17. Note how the legend updates.

Area Legend
☐ 47 SF
☐ 58 SF
☐ 64 SF

18. Close without saving.

Exercise 3-26

Color Scheme by Department

Drawing Name: **i_Color_Scheme_2.rvt**
Estimated Time to Completion: 10 Minutes

Scope
Create a floor plan using a color scheme.
Place a color scheme legend using the Department parameter.

Solution

1.

Activate the **GROUND FLOOR** floor plan.

2.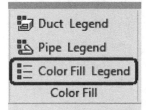

Switch to the Annotate ribbon.

Click **Color Fill Legend** on the Color Fill panel.

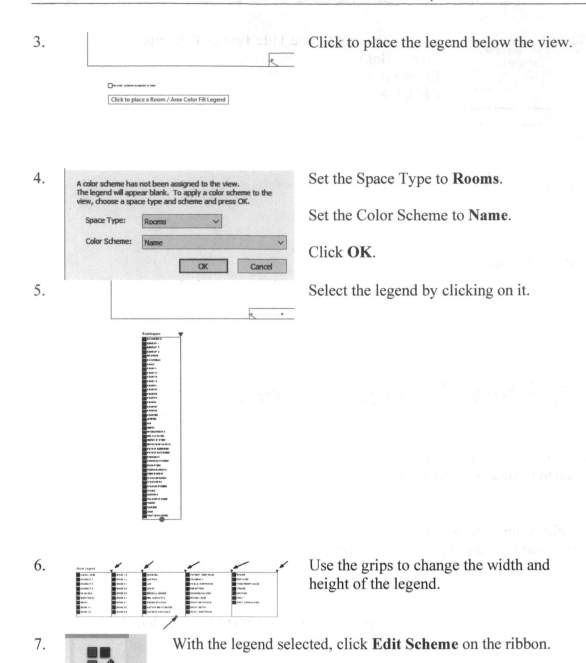

3. Click to place the legend below the view.

4. A color scheme has not been assigned to the view. The legend will appear blank. To apply a color scheme to the view, choose a space type and scheme and press OK.

    Space Type: Rooms

    Color Scheme: Name

    OK   Cancel

    Set the Space Type to **Rooms**.

    Set the Color Scheme to **Name**.

    Click **OK**.

5. Select the legend by clicking on it.

6. Use the grips to change the width and height of the legend.

7. With the legend selected, click **Edit Scheme** on the ribbon.

    Edit Scheme

    Scheme

8. Highlight **Department**.

Click **OK**.

Schemes
Category:
Rooms

(none)
Name
Department

9. The legend updates.

Department Legend

The color fills changes on the floor plan.

☐ Circulation

☐ Core

☐ Exam

☐ Nursing

10. Open the **ROOM SCHEDULE** view.

Schedules/Quantities (all)
 Area Schedule (Gross Building)
 Area Schedule (Rentable)
 DOOR SCHEDULE
 ROOM SCHEDULE

11. The last column in the schedule is Department.

J	K
n Style	Department
	Circulation
	Circulation
	Core
	Circulation
	Core
	Core
	Core
	Circulation
	Nursing
	Core
	Circulation
	Core
	Nursing

This is what controls how the color fill schedule is organized and how colors are applied.

Department is a default parameter for Rooms. You do not need to create it, but you do need to fill in the values you want to use in your schedule.

12. Save as *ex3-26.rvt*.

Export to DWG

Many companies continue to use AutoCAD. AutoCAD uses a dwg file extension. Your company may find itself working with an outside vendor that would prefer the drawing sheets or building model in AutoCAD format to make it easier for them to modify and use. If you find yourself in this situation, the hardest thing to control is any revisions to the project. Be sure to have strong communication in place so that you can be informed of any changes in the project regardless of what software is being used. AutoCAD can use 3D models, but all the metadata (parameters and properties will be lost).

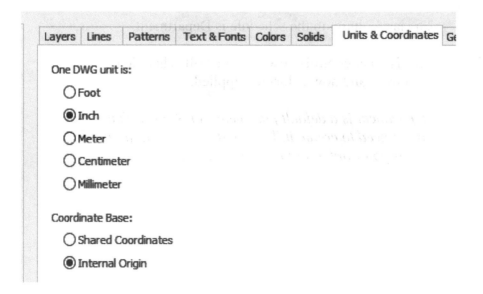

You can assign each Revit category to a corresponding AutoCAD layer using different standards or create your own standard and load from a *.txt file.

Before you export, make sure you set the correct units for the DWG file and where you want the internal origin to be specified.

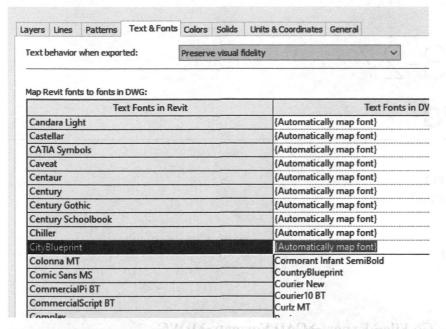

| Layers | Lines | Patterns | Text & Fonts | Colors | Solids | Units & Coordinates | General |

Text behavior when exported: | Preserve visual fidelity ⌄ |

Map Revit fonts to fonts in DWG:

Text Fonts in Revit	Text Fonts in DW
Candara Light	{Automatically map font}
Castellar	{Automatically map font}
CATIA Symbols	{Automatically map font}
Caveat	{Automatically map font}
Centaur	{Automatically map font}
Century	{Automatically map font}
Century Gothic	{Automatically map font}
Century Schoolbook	{Automatically map font}
Chiller	{Automatically map font}
CityBlueprint	{Automatically map font}
Colonna MT	Cormorant Infant SemiBold
Comic Sans MS	CountryBlueprint
CommercialPi BT	Courier New
CommercialScript BT	Courier10 BT
Complex	Curlz MT

Revit uses Microsoft ttf fonts. AutoCAD uses shx fonts and ttf fonts.

To ensure text and dimensions look proper, you can map the font you are using to a known AutoCAD font.

| Layers | Lines | Patterns | Text & Fonts | Colors | Solids | Units & Coordinates | General |

Room, space and area boundaries:
☐ Export rooms, spaces and areas as polylines

Non-plottable layers:
☑ Make layers containing the following text not plottable:

NPLT

Options:
☑ Hide scope boxes
☑ Hide reference planes
☑ Hide unreferenced view tags
☐ Preserve coincident lines

Default export options:
☑ Export views on sheets and links as external references

Export to file format: | AutoCAD 2018 Format ⌄ |

By default, rooms, spaces, and areas are not included in any export to DWG.

When you export the 3D model, you export any views that have been placed on sheets, including drafting views, as external references.

Exercise 3-27

Export to DWG

Drawing Name: **export dwg.rvt**
Estimated Time to Completion: 5 Minutes

Scope
Export sheets to DWG format.

Solution

1. 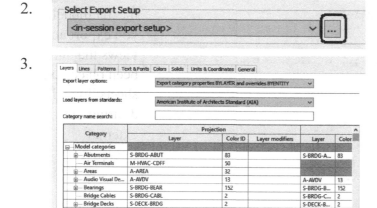 Go **File→Export→CAD Formats→DWG**.

2. Select **<in-session export setup>**.
   Click the **...** button.

3. Each category is assigned an AutoCAD layer.

   Load layers from standards: **American Institute of Architects Standard (AIA)**.

   Click **OK**.

4. 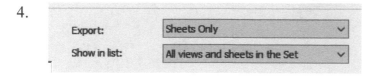 Set Export to **Sheets Only**.

5.

Include	Type	Name ▲
☑	🗐	Sheet: A-201 - FIRST FLOOR P...
☐	🗐	Sheet: A-202 - SECOND FLOO...

Enable to export only the first sheet A-201.

Click **Next**.

6.

File name/prefix:	A-201
Files of type:	AutoCAD 2018 DWG Files (*.dwg)
Naming:	Automatic - Short ⌄

☐ Export views on sheets and links as external references

Set the Naming to **Automatic-Short**.

Disable **Export views on sheets and links as external references**.

This will embed any views/links into the drawing file.

Browse to where you want to save the file.

Click **OK**.

7.

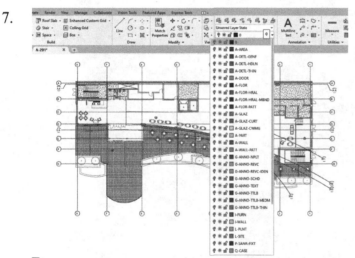

If you have access to AutoCAD, you can open the exported drawing to see how the layers look and how the views are placed.

The floor plan is on the model tab.

8.

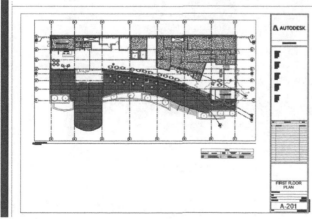

The titleblock and schedule are available on the layout tab.

The floorplan is on the layout tab as a view.

9.

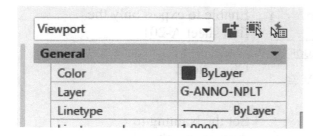

The viewport was automatically placed on the G-ANNO-NPLT layer.

This means the viewport outline will not be plotted.

Practice Exam

1. When you add text notes to a view using the Text tool, you can control all of the following except:

   A. The display of leader lines
   B. Text Alignment
   C. Text Font
   D. Whether the text is bold or underlined

2. To use a different text font, you need to:

   A. Use the Text tool.
   B. Create a new Text family and change the Type Properties.
   C. Create a new Text family and change the Instance Properties.
   D. Modify the text.

3. _____ are the measurements that display in the drawing when you select an element. These dimensions disappear when you complete the action or deselect the element.

   A. Temporary
   B. Permanent
   C. Listening
   D. Constraints

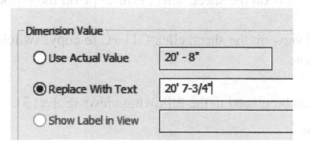

4. You click on a permanent dimension to change the value. You enable Replace With Text and type in the desired value. Then you click OK. What happens next?

   A. An error message appears advising that you cannot modify a dimension value using text
   B. The dimension updates with the new value you entered
   C. The dimension updates and the elements shift location
   D. The dimension updates, but the element remains in the same position

5. To add a new sheet to a project: (select all that apply)

   A. Right click on the Sheets category in the Project Browser and select New Sheet

B. Select the New Sheet tool on the View tab on the ribbon
C. Select New Sheet tool on the Annotation tab on the ribbon
D. Select New Sheet tool on the Manage tab on the ribbon

6. To change the size of a sheet used in a Revit project:

A. Right click on the sheet in the Project Browser and select Properties
B. Highlight the sheet in the Project Browser and select Edit Type in the Properties panel
C. Select the title block on the sheet, select a new title block using the Type Selector
D. Highlight the sheet in the Project Browser and use the Type Selector to change the size

7. To add a plan view to a sheet: (select all that apply)

A. Drag and drop a view from the Project Browser on to a sheet
B. Select the Place View tool from the View tab on the ribbon
C. Select the Insert View tool from the Insert tab on the ribbon
D. Select the Place View tool from the Insert tab on the ribbon

8. You want to place the Level 1 floor plan view on more than one sheet. The best method to use is:

A. Duplicate the view so it can be placed on more than one sheet
B. Drag the view on each desired sheet
C. Highlight the view on the sheet, select Edit Type on the Properties Panel, select Duplicate
D. Highlight the view on the sheet, click CTL+C to copy, switch to the other sheet, click CTL+V to paste

9. Color Schemes can be placed in the following views (Select 3):

A. Floor plan
B. Ceiling Plan
C. Section
D. Elevation
E. 3D

10. To define the colors and fill patterns used in a color scheme legend, select the legend, and Click Edit Scheme on the:

    A. Properties palette
    B. View Control Bar
    C. Ribbon
    D. Options Bar

11. A legend is a view that can be placed on:

    A. A plan view
    B. A drafting view
    C. Multiple Plan Regions
    D. Multiple Sheets

12. To change the font used in a Color Scheme Legend:

    A. Select the legend, then select Edit Scheme from the ribbon.
    B. Select the text in the legend, double click to edit.
    C. Select the legend, then select Edit Type from the Properties pane.
    D. Select the legend, right click and select Edit Family.

13. Filled regions can be placed in the following views: (Select all that apply)

    A. Floor Plan
    B. Elevation
    C. Section
    D. Detail View

14. Filled regions: (select all that apply)

    A. Are two-dimensional elements
    B. View-specific (only visible in the view where they are created and placed)
    C. Use fill patterns
    D. Can be placed in 3D sections

15. Detail Components are:

    A. System Families
    B. Mass Families
    C. Annotation Families
    D. Loadable Families

16. The Repeating Detail tool places an array of _____

    A. Annotations
    B. Filled Regions
    C. Detail Views
    D. Detail Components

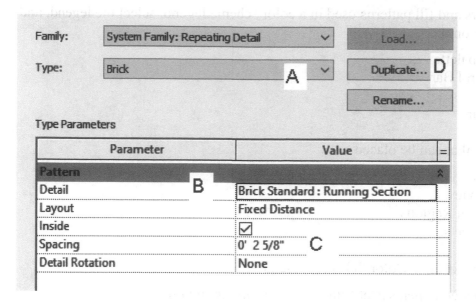

17. Specify the location to select to use a different detail component for the repeating detail.

18. Tagging elements by category:
    A. Allows you to select which elements will be tagged by selecting the element
    B. Allows you to select all the elements in a category (doors/windows/rooms/etc) by simply selecting which tags to be applied in a view
    C. Allows you to select which tag to be placed and then select the element to be tagged one at a time
    D. Applies the category tag to all the selected elements in the entire project

19. T/F Phases allow you to display different versions of a model.

20. T/F Phase filters changes the graphic display of a view depending on the phase.

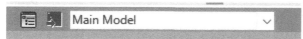

21. This section below the View Control Bar is used to manage:

    A. Phases
    B. Design Options
    C. Worksets
    D. Links

22. T/F Phases can be applied to schedules.

23. To set the Phase to be applied to a view:

    A. Activate the view and select the desired filter in the Properties panel
    B. Go to the Manage tab on the ribbon and set the Phase assigned to each view
    C. Go to the View Template and assign the phase for that view
    D. Go to the View Control Bar and assign the desired phase

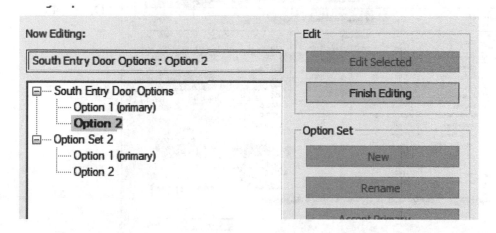

24. You click on Finish Editing. This:

    A. Deletes that design option
    B. Moves elements from one design option to another
    C. Closes the design option you are working in
    D. Incorporates the design option into the main model

25. What happens when you change the sequence of revisions in the Sheet Issues/Revisions dialog?

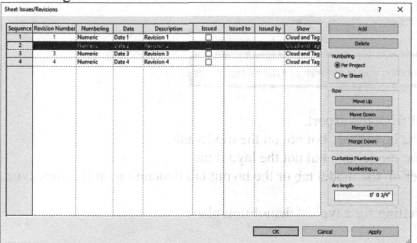

B. The Revision Number only changes for revisions marked as Issued.

C. The Revision Number for revisions NOT on sheets changes accordingly.

D. The Revision Number assigned to each revision changes accordingly.

A. The Revision Number does NOT change because the Numbering is Per Project.

26. A designer is exporting a model to DWG. The designer needs to export the model based on the internal origin. Where should the designer click to adjust this setting?

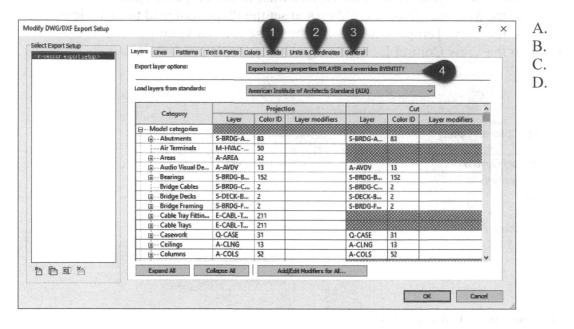

A.   1
B.   2
C.   3
D.   4

27. A designer is exporting a Revit sheet to a DWG file. She disables the Export views on sheets and links as external references as shown. There are two views on the sheet. What will happen to those views when you open the DWG file in AutoCAD?

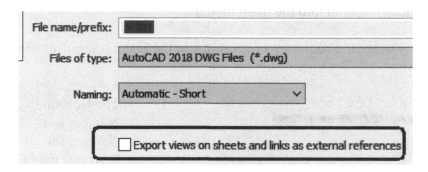

A. They will be excluded from the export.
B. They will appear on the layout tab, but not on the model tab.
C. They will appear on the model tab, but not the layout tab.
D. They will appear either on the model tab or the layout tab depending on the view type.

28. What is the phase filter setting for a typical demolition plan?

        A. Show Demo + New
        B. Show Demo
        C. Show Previous + Demo
        D. Show Previous + New

29. What happens when you click the icon below a temporary dimension?

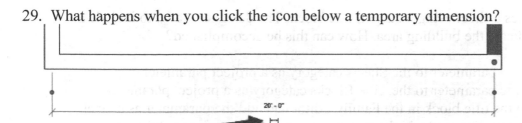

A. Makes the temporary dimension permanent.
B. Deletes the dimension.
C. Locks the dimension.
D. Activates the dimension value to modify it.

30. A designer is creating a repeating detail family. The designer wants the spacing between each CMU block to be the same in the repeating detail. Which Layout method should be selected?

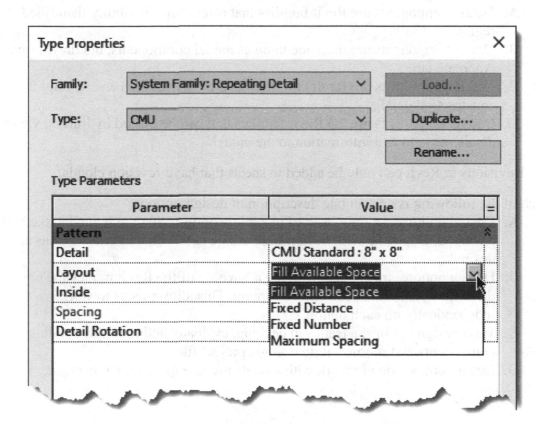

A. Fixed Distance
B. Maximum Spacing
C. Fixed Number
D. Fill Available Space

31. A designer creates a sheet using a custom title block. The designer wants to include a parameter to indicate the building area. How can this be accomplished?

    A. Add the parameter to the Sheets category as a project parameter.
    B. Add the parameter to the Title Blocks category as a project parameter.
    C. Open the title block in the Family Editor and add the parameter as a label.
    D. Open the sheet and add the parameter as text inside of angle brackets.

32. Which of the following methods can be used to organize a color fill legend?

    A. Name, Department, Space, Room.
    B. Wall face, floor face, or ceiling face.
    C. Spaces only. Color fill plans are used for HVAC zoning.
    D. Rooms, either by Name or Department

33. What is a detail component?
    A. Detail components are Revit families that offer more flexibility than filled regions.
    B. Detail components are the same thing as model components, but drawn from the Annotate tab.
    C. Detail components are Revit families that are only used when creating construction details.
    D. Detail components are 2D Revit families that can be placed in drafting views or detail views to add information to the model.

34. T/F   Revisions in Revit can only be added to sheets that have revision clouds.

35. Which of the following is an accurate description of design options?
    A. Design options can only be used to explore options that are related to the building envelope. Duplicate with detailing should be used when interior options are desired.
    B. Design options are created by doing a Save As of the Revit model and creating different files to track each design option. This allows users to work independently on each option.
    C. With design options, a team can develop, evaluate, and redesign building components and rooms within a single project file
    D. Design options do not work with models that use more than one phase.

ANSWERS: 1) D; 2) B; 3) A; 4) A; 5) A & B; 6) C; 7) A & B; 8) A 9) A,B, C, & D; 10) C; 11) D; 12) C; 13) A.B.C.&D; 14) A,B, & C; 15) D; 18) D; 19) F; 20) T; 21) B; 22) T; 23) A; 24) C; 25) D; 26) B; 27) D; 28) C; 29) A; 30) A; 31) C; 32) D; 33) D; 34) T; 35) C

Lesson

04

Views

This lesson addresses the following exam questions:

- View Properties
- Object Visibility Settings
- Section Views
- Elevation Views
- 3D Views
- View Templates
- Scope Boxes
- Schedules
- Legends
- Rendering
- Revision Clouds

The Project Browser lists all the views available in the project. Any view can be dragged and dropped onto a sheet. Once a view is used or consumed on a sheet, it cannot be placed a second time on a sheet – even on a different sheet. Instead, you must create a duplicate view. You can create as many duplicate views as you like. Each duplicate view may have different annotations, line weight settings, detail levels, etc. Annotations are specific to a view. If a view is deleted, any annotations are also deleted.

Revit has bidirectional associativity. This means that changes in one view are automatically reflected in all views. For example, if you modify the dimensions or locations of a window in one view, the change is reflected in all the views, including the 3D view. How elements are displayed in each view (line color, pattern, and line weight) is controlled using Graphic Overrides.

You can control the appearance of Revit elements using Object Visibility Settings. These settings control line color, line type, and line weight. You can create templates which have different Object Visibility Settings for different project types.

Browser Organization

Many users find it more productive to organize or sort the project browser using view parameters.

Use the Browser Organization tool to group and sort views, sheets, and schedules/quantities in the way that best supports your work. You can specify six levels of grouping. Within groups, items are sorted in ascending or descending order of a selected property.

By default, the Project Browser displays all views (by view type), all sheets (by sheet number and sheet name), and all schedules and quantities (by name).

In addition to grouping and sorting views, you can limit the views that display in the Project Browser by applying a filter. This approach is useful when the project includes many views, sheets, and schedules/quantities, and you want the Project Browser to list a subset. You can specify up to 3 levels of filtering.

The combination of criteria used to filter, group, and sort items in the Project Browser is an organization scheme. You can create separate schemes for views, sheets, and schedules/quantities in the project. Create as many schemes as needed to support different phases of your work.

When creating an organization scheme, you can use any properties of the view, sheet, or schedule/quantity as the criteria for grouping, sorting, and filtering. You can also use project parameters and shared parameters for grouping and filtering.

When working on a project, you can quickly change the organization schemes applied to the Project Browser at any time. Switch between organization schemes whenever needed based on your current work.

Exercise 4-1
Browser Organization

Drawing Name: **browser.rvt**
Estimated Time to Completion: 20 Minutes

Scope

Creating a new browser organization schema.
Using parameters in the project browser.

Solution

1. 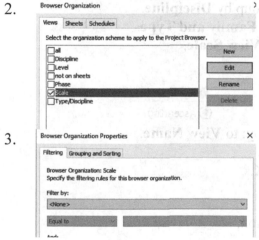   Highlight the Views in the Project Browser.

   Right click and select **Browser Organization**.

2.   Notice that you can organize Views, Sheets, and Schedules. Each of these can be organized a different way, if you like.

   Currently the views are organized by Scale.

   Click **Edit**.

3.   No filtering has been applied to this organization schema.

   Click the **Grouping and Sorting** tab.

4.

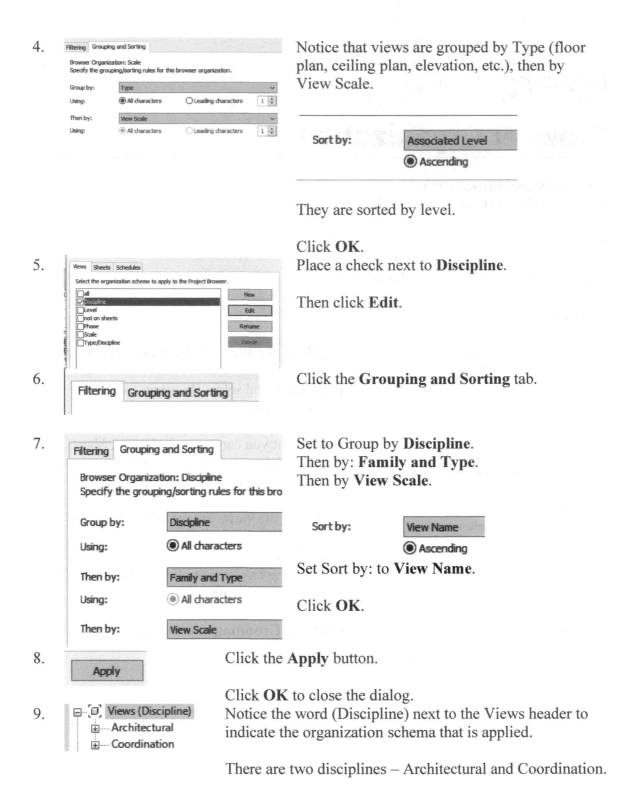

Notice that views are grouped by Type (floor plan, ceiling plan, elevation, etc.), then by View Scale.

They are sorted by level.

Click **OK**.
Place a check next to **Discipline**.

Then click **Edit**.

5.

6.

Click the **Grouping and Sorting** tab.

7.

Set to Group by **Discipline**.
Then by: **Family and Type**.
Then by **View Scale**.

Set Sort by: to **View Name**.

Click **OK**.

8.

Click the **Apply** button.

Click **OK** to close the dialog.

9.

Notice the word (Discipline) next to the Views header to indicate the organization schema that is applied.

There are two disciplines – Architectural and Coordination.

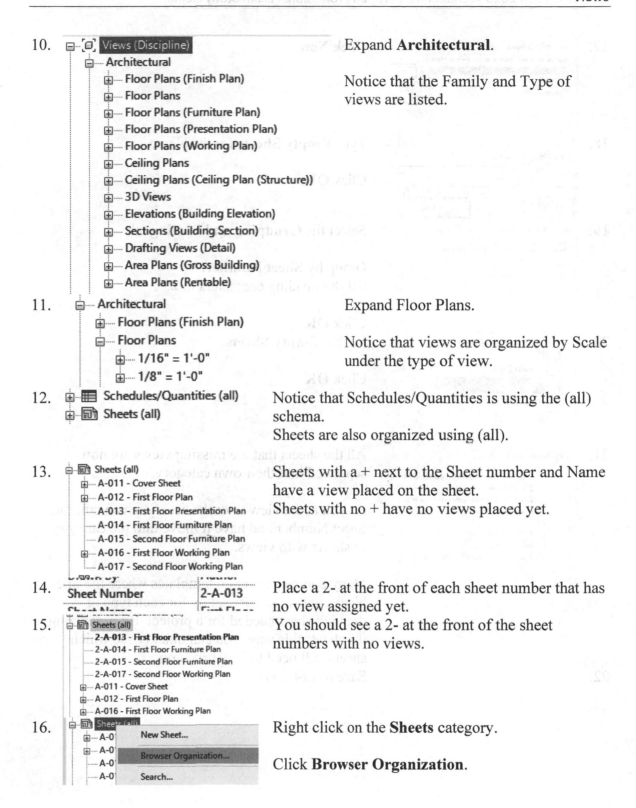

10. **Views (Discipline)**
     Architectural
       Floor Plans (Finish Plan)
       Floor Plans
       Floor Plans (Furniture Plan)
       Floor Plans (Presentation Plan)
       Floor Plans (Working Plan)
       Ceiling Plans
       Ceiling Plans (Ceiling Plan (Structure))
       3D Views
       Elevations (Building Elevation)
       Sections (Building Section)
       Drafting Views (Detail)
       Area Plans (Gross Building)
       Area Plans (Rentable)

Expand **Architectural**.

Notice that the Family and Type of views are listed.

11. Architectural
       Floor Plans (Finish Plan)
       Floor Plans
         1/16" = 1'-0"
         1/8" = 1'-0"

Expand Floor Plans.

Notice that views are organized by Scale under the type of view.

12. Schedules/Quantities (all)
     Sheets (all)

Notice that Schedules/Quantities is using the (all) schema.
Sheets are also organized using (all).

13. Sheets (all)
       A-011 - Cover Sheet
       A-012 - First Floor Plan
       A-013 - First Floor Presentation Plan
       A-014 - First Floor Furniture Plan
       A-015 - Second Floor Furniture Plan
       A-016 - First Floor Working Plan
       A-017 - Second Floor Working Plan

Sheets with a + next to the Sheet number and Name have a view placed on the sheet.
Sheets with no + have no views placed yet.

14. Sheet Number   2-A-013

Place a 2- at the front of each sheet number that has no view assigned yet.

15. Sheets (all)
       2-A-013 - First Floor Presentation Plan
       2-A-014 - First Floor Furniture Plan
       2-A-015 - Second Floor Furniture Plan
       2-A-017 - Second Floor Working Plan
       A-011 - Cover Sheet
       A-012 - First Floor Plan
       A-016 - First Floor Working Plan

You should see a 2- at the front of the sheet numbers with no views.

16. Sheets (all)
       A-0   New Sheet...
       A-0
       A-0   Browser Organization...
       A-0   Search...

Right click on the **Sheets** category.

Click **Browser Organization**.

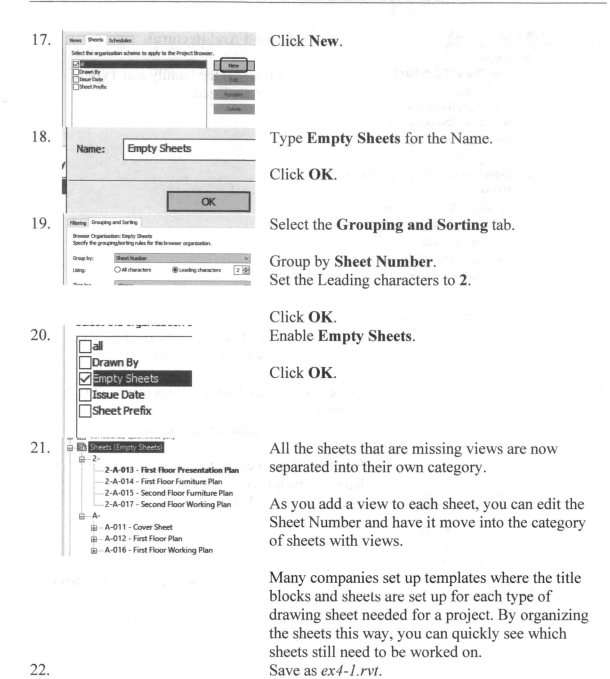

17. Click **New**.

18. Type **Empty Sheets** for the Name.

    Click **OK**.

19. Select the **Grouping and Sorting** tab.

    Group by **Sheet Number**.
    Set the Leading characters to **2**.

20. Click **OK**.
    Enable **Empty Sheets**.

    Click **OK**.

21. All the sheets that are missing views are now separated into their own category.

    As you add a view to each sheet, you can edit the Sheet Number and have it move into the category of sheets with views.

    Many companies set up templates where the title blocks and sheets are set up for each type of drawing sheet needed for a project. By organizing the sheets this way, you can quickly see which sheets still need to be worked on.

22. Save as *ex4-1.rvt*.

View Scale

The view scale controls the scale of the view as it appears on the drawing sheet.

You can assign a different scale to each view in a project. You can also create custom view scales.

Many view types in Revit contain a "View Scale" property- such a Floor Plans, Ceiling Plans, Sections, Elevations, Callouts, and Drafting Views. The "View Scale" parameter allows you to set the scale at which each particular view will be printed out. You can also modify the view scale to ensure that a view fits on a sheet.

Key Points

- Each view has its own "View Scale" property.
- You can change the scale of a view at any time- using the View Control Bar or View Properties.
- Revit maintains annotations (tags, dimensions, etc.) at their actual printed size, regardless of the scale of the view.

If you assign a template to a view, the view scale may be defined by the template.

Exercise 4-2
Change the View Scale

Drawing Name: **view_scale.rvt**
Estimated Time to Completion: 10 Minutes

Scope
Changing a view scale.
Adding a custom view scale.

Solution

1. Activate the **01 FLOOR PLAN**.

2. Zoom into the area with Rooms 2504 and 2503.

3. Look down at the bottom left hand corner of the active view window. Locate the View Control Bar.

    The first button on the View Control Bar is the View Scale.

    You can see that this view is indeed set to **1:96**.

    Click on the View Scale button.

    A pop-up displays a list of all the available scales.

    At the top of the list is Custom. This is used to define a

scale that is not available in the default list.

4.   1 : 100   Change the view scale to **1:100**.

Did you notice how the room tags adjusted size in relation to the view scale?

5.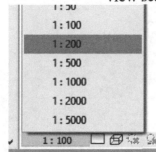

Change the view scale to **1:200**.

Did the room tags get bigger or smaller?

In fact, the room tags remained the same size. The model elements increased in scale.

6. In the browser, locate the Sheets category.

Highlight and right click to select **New Sheet**.

7.   Load...   Click **Load** to load a new titleblock.

8. Browse to the *Titleblocks* folder under English\US.

9. Select *A2 metric.rfa*.
Click **Open**.
Click **OK** to create a new sheet.

10.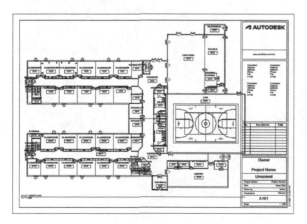

Drag and drop the 01 FLOOR PLAN onto the sheet.

11.

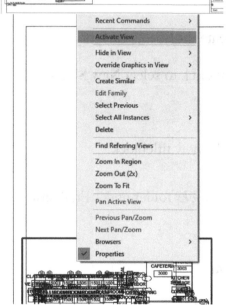

Select the view on the sheet.

On the Properties pane:

Set the View Scale to 1:500.

Click **Apply.**

12.

The view updates.

13.

Select the view.
Right click and select **Activate View**.

The View Control bar is now visible.

14.

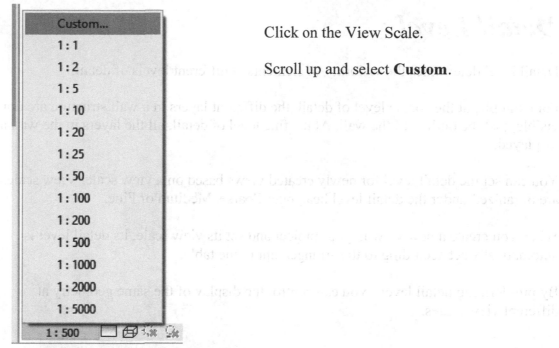

Click on the View Scale.

Scroll up and select **Custom**.

15.

Type **250** for the ratio.

Click **OK**.

16.

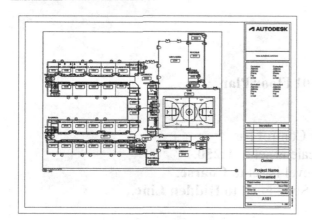

Right click on the view.

Select **Deactivate View**.

17.

The view can now be adjusted to place on the sheet.

Close without saving.

Detail Level

Detail level determines the visibility of elements at different levels of detail.

For example, at the coarse level of detail, the different layers in a wall structure are not visible, just the outline of the wall. At the fine level of detail, all the layers in the wall are displayed.

You can set the detail level for newly created views based on a view scale. View scales are organized under the detail level headings: Coarse, Medium or Fine.

When you create a new view in your project and set its view scale, its detail level is automatically set according to the arrangement in the table.

By pre-defining detail levels, you can control the display of the same geometry at different view scales.

Exercise 4-3

Change the Detail Level of a View

Drawing Name: **detail_level.rvt**
Estimated Time to Completion: 15 Minutes

Scope
Defining Detail Levels for View Scales

Solution

1. Floor Plans
    Architectural
      01 FLOOR PLAN

    Activate the **01 Floor Plan**.

2. 1 : 250

    On the View Control bar:
    The View Scale is set to **1:250**.
    The Detail Level is set to **Coarse**.
    The Display Style is set to **Hidden Line**.

    You should be able to identify the different icons associated with Detail Level and Display Style for the certification exam.

3. 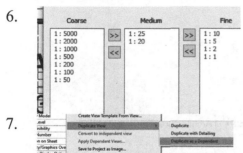 Zoom into the lower left area of the floor plan – where Rooms 2010 and 2008 are located.

4. Switch to the Manage ribbon.

   Select **Additional Settings** on the ribbon.

5. Select **Detail Level** from the drop-down list.

6.

Coarse	Medium	Fine
1 : 5000	1 : 25	1 : 10
1 : 2000	1 : 20	1 : 5
1 : 1000		1 : 2
1 : 500		1 : 1
1 : 200		
1 : 100		
1 : 50		

   Note that each View Scale has been assigned a Detail Level setting.

   Click **OK**.

7. Highlight the **01 FLOOR PLAN**.

   Right click and select **Duplicate View→Duplicate as a Dependent.**

8. 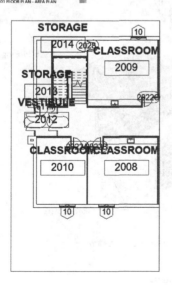 Crop the view using the crop region so only Rooms 2008, 2009, and 2010 are displayed.

   Change the view scale to **1:10**.

   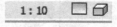

   The Detail Level is set to **Coarse**.
   The View Display is set to **Hidden Line**.

   You may recall on the Detail Level settings, the Detail Level for a View Scale of 1:10 should be Fine.

9. 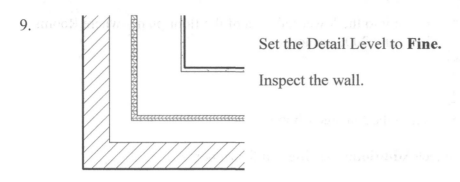 Set the Detail Level to **Fine.**

Inspect the wall.

10. Change the View Scale to **1:500.**
Change the View Scale to **Coarse.**

1 : 500

Notice how the wall display changes.

11. Close without saving.

Drafting Views

The following is a sample drafting view created using the 2D detailing tools in Revit Architecture. This is not a 3D view.

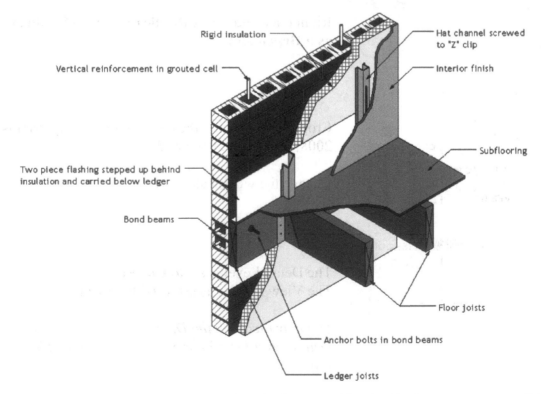

Use a drafting view to create unassociated, view-specific details that are not part of the modeled design.

Rather than create a callout and add details to it, you may want to create detail conditions where the model is not needed (for example, a carpet-transition detail which shows where carpet switches to tile, or roof-drain details not based on a callout on the roof). For this purpose, create a drafting view.

In a drafting view, you create details at differing view scales (coarse, medium, or fine) and use 2D detailing tools: detail lines, detail regions, detail components, insulation, reference planes, dimensions, symbols, and text. These are the exact same tools used in creating a detail view. However, drafting views do not display any model elements. When you create a drafting view in a project, it is saved with the project.

> When using drafting views, consider the following:
>
> - Similar to other views, drafting views are listed in the Project Browser under Drafting Views.
>
> - All of the detailing tools used in detail views are available to you in drafting views.
>
> - Any callouts placed in a drafting view must be reference callouts.
>
> - Although not associated with the model, you can still drag the drafting views from the browser onto a drawing sheet.
>
> - Name drafting views so they are organized neatly in the Project Browser

Many companies will create a library of drafting views that are used across projects. For example, threshold details for doors or sill details for windows. These details provide valuable construction information for the subcontractor.

The following are some examples of drafting view type naming:

- 00-Accessibility-ADA
- 01-General Requirements
- 02-Site
- 03-Concrete
- 04-Masonry
- 05-Metals
- 06-Walls
- 07-Roof
- 08-Door & Windows
- 09-Finishes
- 10-Specialties
- 11-Misc
- 12-Casework
- 13-Stairs Ramps Elevators
- 14-Mechanical
- 15-Plumbing
- 16-Electrical

Exercise 4-4

Create a Drafting View

Drawing Name: **drafting_view.rvt**
Estimated Time to Completion: 5 Minutes

Scope
Create a drafting view.
Load detail items.
Add text.

Solution

1.   Activate the **View ribbon.**

   Click **Drafting View.**

2. Type **06- Interior Wall GWB** in the Name field.
   Set the Scale to **1:25**.

   Click **OK**.

3. 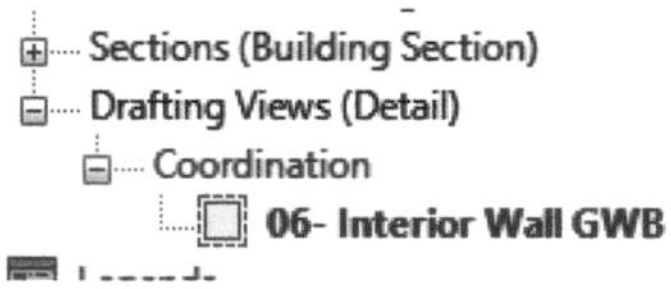 Note that the Drafting View category is added to the Project Browser and the view is listed under Coordination.

   The view is blank. *Remember a drafting view is a 2D view that is independent of the model.*

4. 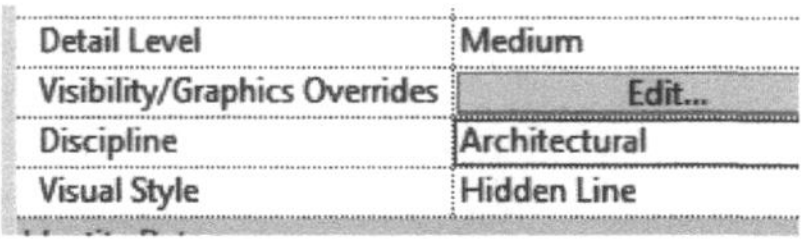 Change the Discipline to **Architectural** in the Properties palette.

   The Project Browser will update.

5.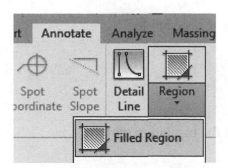

Switch to the **Annotate** ribbon.

Select the **Filled Region** tool.

6.

Select the **Filled Region: Insulation** type from the Properties palette.

7.

Select the **Rectangle** tool from the Draw panel.

8.

Set the Line Style to **Medium Lines**.

9.

Draw a 0.6 x 2100 rectangle.

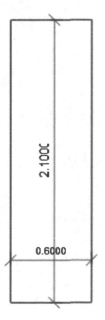

10.  Click the Green Check to complete the rectangle.

11. A rectangle is placed with the insulation pattern.

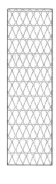

12. 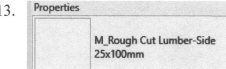 Select **Detail Component**.

13. Select **M_Rough Cut Lumber – Side 25 x 100 mm** using the Type Selector.

Properties

M_Rough Cut Lumber-Side 25x100mm

14. Draw the rough lumber by selecting the left bottom and top end points of the filled region.

15. Select the rough lumber detail item.

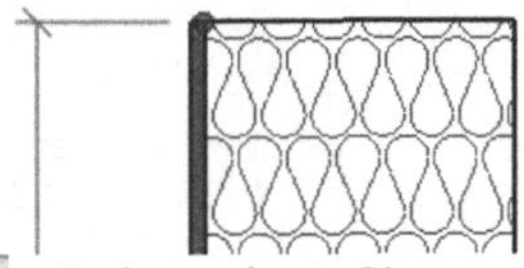

16.  Select the **Mirror – Draw Axis** tool.

17. Select the Midpoint of the top of the filled region to start the axis.

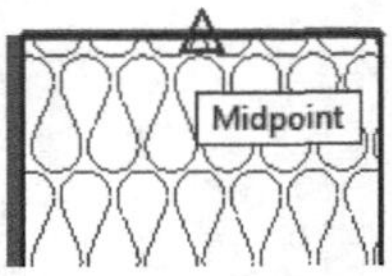

Midpoint

18. Draw the axis straight down the middle of the filled region.

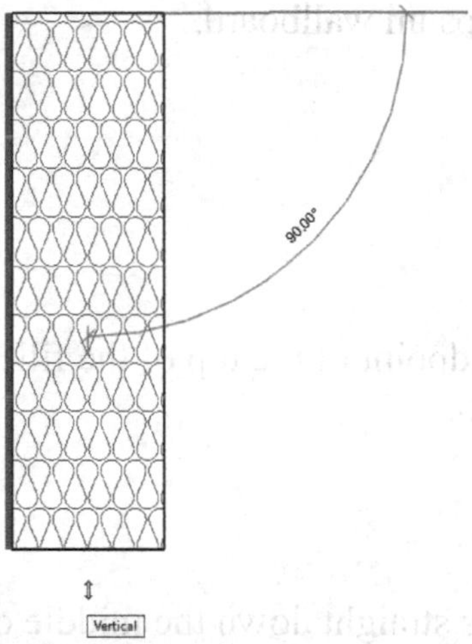

There are now pieces of rough-cut lumber on both sides of the insulation.

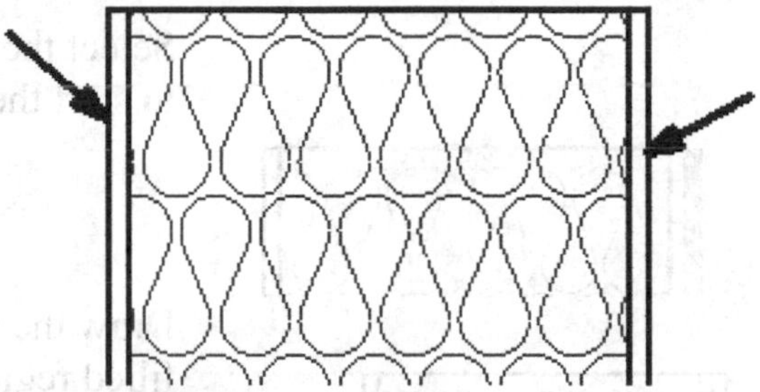

90.00°

Vertical

19. 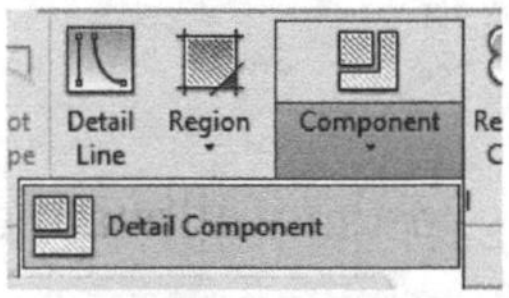 Select **Detail Component**.

ot Detail Region Component Re
pe Line

Detail Component

20.  Select **M_Gypsum Wallboard-Section 25 mm** using the Type Selector.

M_Gypsum Wallboard-Section 25mm

21. Place the gypsum wallboard by selecting the left bottom and top end points of the rough-cut lumber.

22. Select the gypsum wallboard.

23. Select the **Mirror – Draw Axis** tool.

24. Select the Midpoint of the top of the filled region to start the axis.

25. Draw the axis straight down the middle of the filled region.

There are now sections of gypsum wallboard on both sides of the insulation.

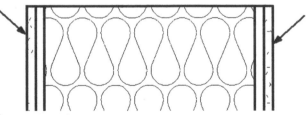

26. Select **Detail Component**.

27.

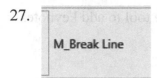

Select **M_Break Line** using the Type Selector.

28.

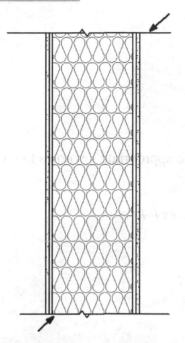

Place a break line at the top and the bottom of the wall section.

29.

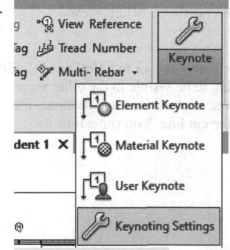

Go to **Keynote→Keynoting Settings**.

Verify that the **RevitKeynotes_Metric** are loaded. If they are not, then browse to the Libraries under RVT2023 and locate the correct file to load.

30.

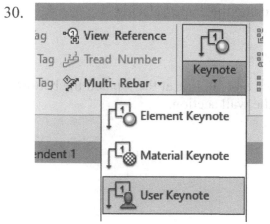

Use the **User Keynote** tool to add keynotes to the view.

31.

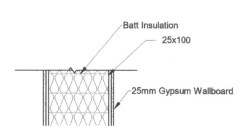

Add the appropriate keynotes to the drafting view.

Save as *ex4-4.rvt*.

Scope Boxes

Scope boxes are created in plan views but are visible in other view categories.

To create a Scope Box, you must be in either a Floor Plan View or in a Reflected Ceiling Plan. However, once a scope box is created, it is going to be visible in the other view categories: sections, callouts, elevations and 3D views. In elevations and sections, the scope box is only going to be visible if it intersects the cut line. You can adjust the extents of the scope box in all view categories.

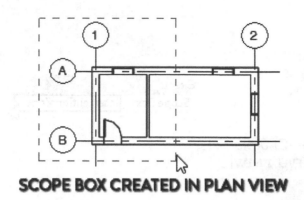

SCOPE BOX CREATED IN PLAN VIEW

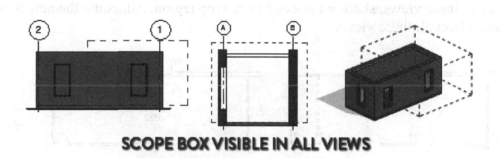

SCOPE BOX VISIBLE IN ALL VIEWS

Consider this office building renovation project. The area affected is in the middle of the building. You want the views to be cropped to fit the red rectangle.

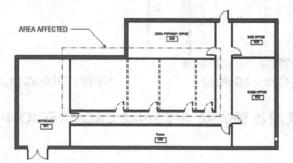

The thing is: you have a lot of views to create. Existing floor plan. Demolished floor plan. New floor plan. Ceilings. Finishes. Layout. All in all, you'll have about 10 views that need the exact same crop region.

An archaic workflow would be to manually adjust the crop region of each view. That would probably work. But what if the project changes and the area affected gets bigger? You have to adjust all the cropped regions again?

That's where the power of scope boxes come into play. Go to the View tab and create a Scope Box. Match it to your intervention area. Give it a name.

Now, apply the scope box to all the views that will be using this cropped region. To save time, select all the views in the project browser by holding the CTRL key.

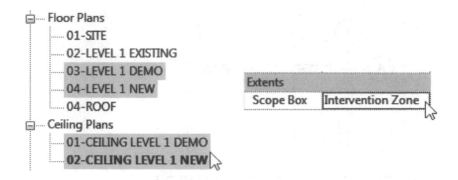

Look at all these views, sharing the exact same crop region. Adjusting the new Scope Box will affect all these views.

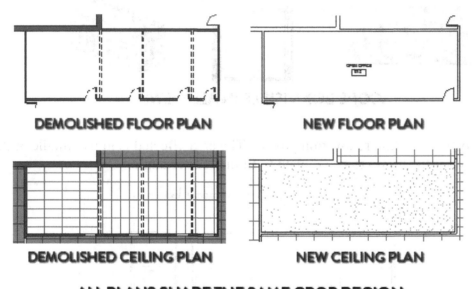

DEMOLISHED FLOOR PLAN   NEW FLOOR PLAN

DEMOLISHED CEILING PLAN   NEW CEILING PLAN

ALL PLANS SHARE THE SAME CROP REGION

Even though Scope Boxes can only be applied in plan view, they are 3D objects. That means they have an assigned height.

| Name: | Scope Box 1 | Height: | 12000 |

The Height is assigned in the Option bar.

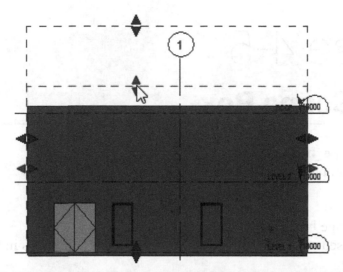

DRAG BLUE ARROWS TO ADJUST SCOPE BOX HEIGHT

Scope boxes can also be used to control the extents and visibility of elements like grids, levels and reference planes. Each of these elements can be assigned to a specific scope box, limiting the 3D extents to the dashed green line limit.

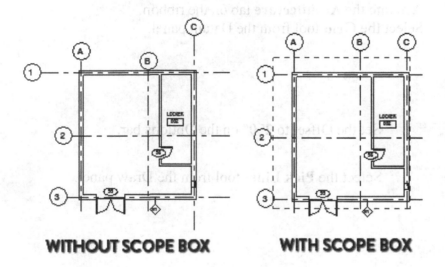

WITHOUT SCOPE BOX   **WITH SCOPE BOX**

Exercise 4-5

Create a Scope Box

Drawing Name: **i_scope_box.rvt**
Estimated Time to Completion: 15 Minutes

Scope
Create and apply a scope box.
Scope boxes can be used to control the visibility of grid lines and levels in views.

Solution

1. 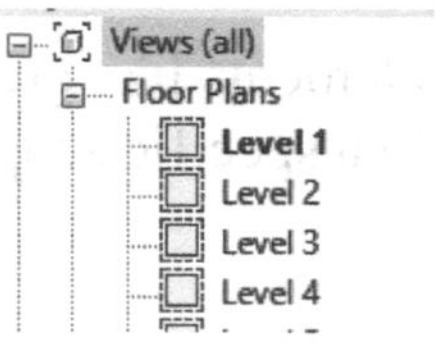 Activate the **Level 1** floor plan.

2. 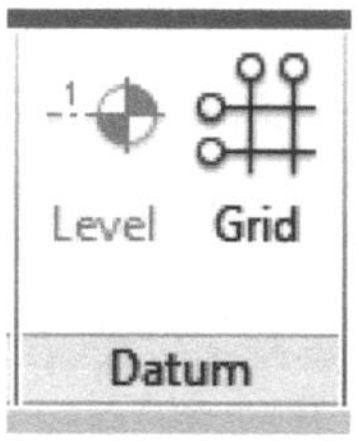 Activate the **Architecture** tab on the ribbon.
    Select the **Grid** tool from the Datum panel.

3. 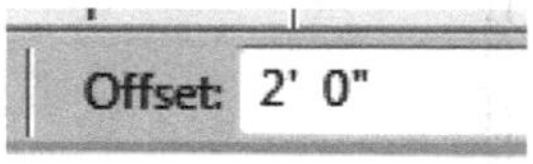 Set the Offset to **2′ 0″** on the Options bar.

4. 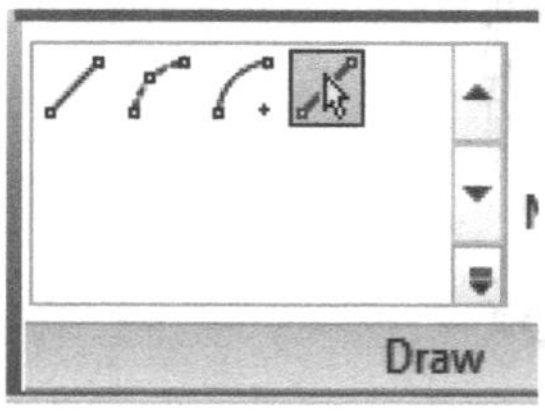 Select the **Pick Lines** tool from the Draw panel.

5.

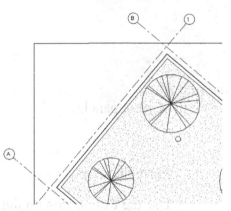

Place three grid lines using the exterior side of the walls to offset.

6.

Re-label the grid bubbles so that the two long grid lines are A and B and the short grid line is 1.

7.

Select the **Grid** tool from the Datum panel.

8. Offset: 180' 0"   Set the Offset to **180' 0"** on the Options bar.

9.

Select the **Pick Lines** tool from the Draw panel.

10.

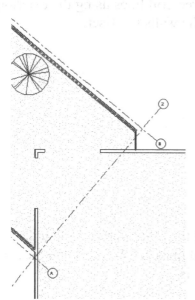

Place the lower grid line by selecting the upper wall and offsetting 180′.
Re-label the grid line **2**.

11.

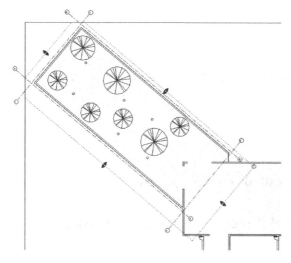

Activate the **View** tab on the ribbon.
Select the **Scope Box** tool from the Create panel.

Scope
Box

12.

Place the scope box.

Use the Rotate icon on the corner to rotate the scope box into position.

Use the blue grips to control the size of the scope box.

13.

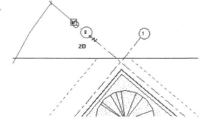

Select the grid line labeled **B**.

14. 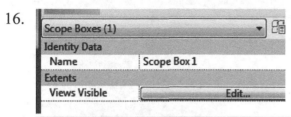 In the Properties pane:
Set the Scope Box to Scope Box 1; the scope box which was just placed.

15. Repeat for the other three grid lines.
You can use the Control key to select more than one grid line and set the Properties of all three at the same time.

16. 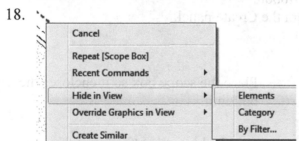 Select the Scope Box.
Select **Edit** next to Views Visible in the Properties pane.

17. 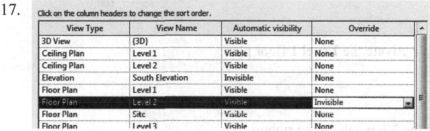 Set the Level 2 Floor Plan Override Invisible.
Click **OK**.

View Type	View Name	Automatic visibility	Override
3D View	{3D}	Visible	None
Ceiling Plan	Level 1	Visible	None
Ceiling Plan	Level 2	Visible	None
Elevation	South Elevation	Invisible	None
Floor Plan	Level 1	Visible	None
Floor Plan	Level 2	Visible	Invisible
Floor Plan	Site	Visible	None
Floor Plan	Level 3	Visible	None

Click on the column headers to change the sort order.

18. Select the scope box.
Right click and select **Hide in View→ Elements**.

The scope box is no longer visible in the view.

Cancel

Repeat [Scope Box]
Recent Commands

Hide in View   Elements
Override Graphics in View   Category
                              By Filter...
Create Similar

19. Floor Plans
····· Level 1
····· **Level 2**
····· Level 3

Activate Level 2.

The grid lines and scope box are not visible.

20. Close the file without saving.

Tip: To make the hide/isolate mode permanent to the view: on the View Control bar, click the glasses icon and then click Apply Hide/Isolate to the view.

Exercise 4-6
Use a Scope Box to Crop Multiple Views

Drawing Name: **i_scope_multiple.rvt**
Estimated Time to Completion: 5 Minutes

Scope
Create and apply a scope box to crop multiple views.

Solution

1.   Activate the **Level 1** floor plan.

2.   Activate the **View** tab on the ribbon.
   Select the **Scope Box** tool from the Create panel.

3.   Place the scope box so it encloses the building.

   Use the blue grips to control the size of the scope box.

4.   In the Properties panel:

   Change the Name of the Scope Box to **Building Model**.

5.   Activate the Level 1 floor plan.
   In the Properties panel:
   Set the Scope Box to: **Building Model**.

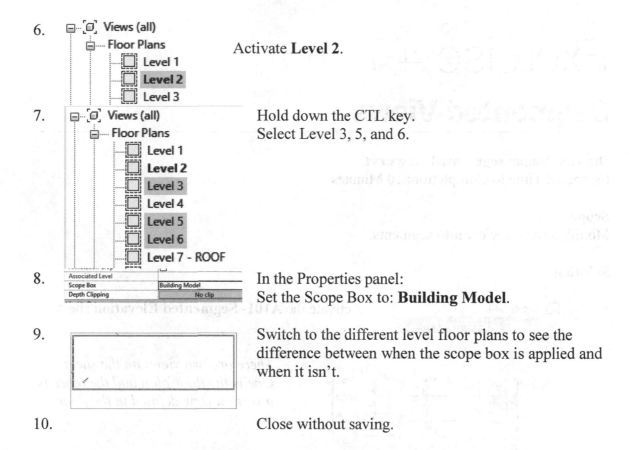

6.   Activate **Level 2**.

7.   Hold down the CTL key.
Select Level 3, 5, and 6.

8.   In the Properties panel:
Set the Scope Box to: **Building Model**.

9.   Switch to the different level floor plans to see the difference between when the scope box is applied and when it isn't.

10.   Close without saving.

Segmented Views

Split a section or elevation view to permit viewing otherwise obscured parts of the view.

This function allows you to vary a section view or an elevation view to show disparate parts of the model without having to create a different view. For example, you may find that landscaping obscures the parts of the model that you would like to see in an elevation view. Splitting the elevation allows you to work around these obstacles.

Exercise 4-7

Segmented Views

Drawing Name: **segmented views.rvt**
Estimated Time to Completion: 10 Minutes

Scope
Modify a section view into segments.

Solution

1. 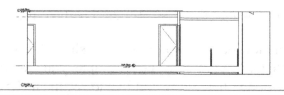 Activate the **A101- Segmented Elevation** sheet.

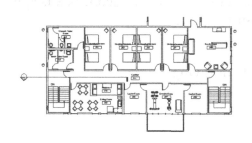

There are two views on the sheet. One is the floor plan and the other is a section view defined in the floor plan.

2. 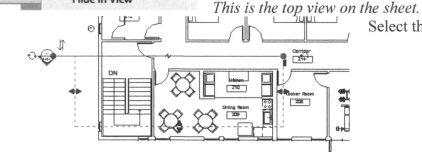 Right click on the **Main Floor** floor plan view. Select **Activate View**.

This is the top view on the sheet.

3. Select the section line.

4. 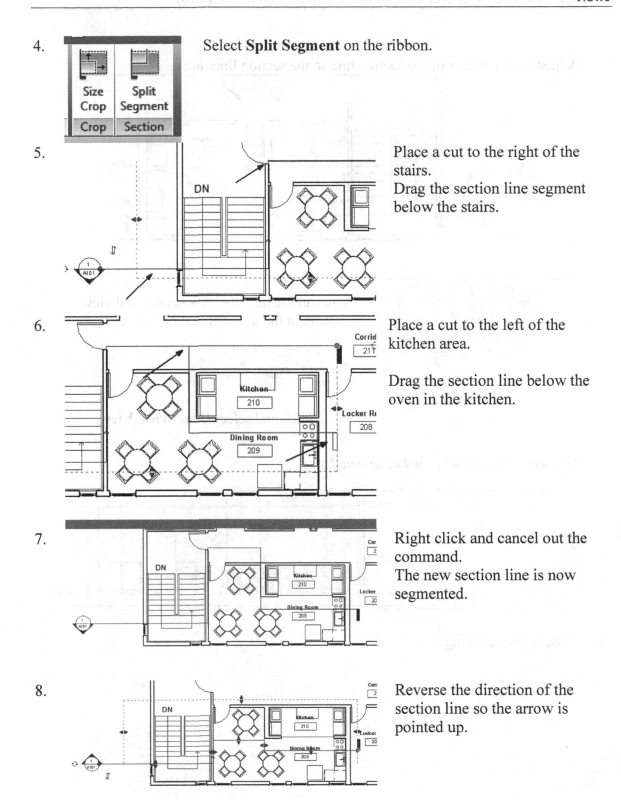 Select **Split Segment** on the ribbon.

5. Place a cut to the right of the stairs.
Drag the section line segment below the stairs.

6. Place a cut to the left of the kitchen area.

Drag the section line below the oven in the kitchen.

7. Right click and cancel out the command.
The new section line is now segmented.

8. Reverse the direction of the section line so the arrow is pointed up.

9. Adjust the segments of the section line so the section lines are as shown.

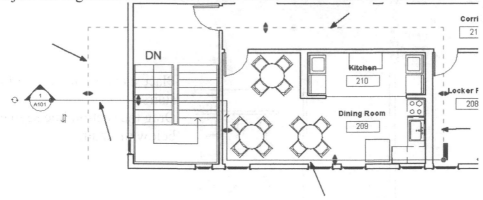

10.

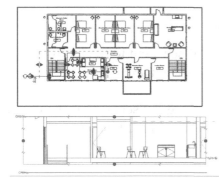

You can see how the new segmented view appears in the lower elevation view.

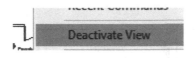

Right click and select **Deactivate View**.

Note how the section view has updated.

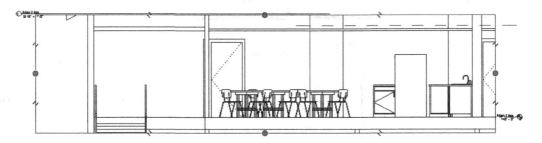

11. Close without saving.

Exercise 4-8

Rotate a View

Drawing Name: **rotate_view.rvt**
Estimated Time to Completion: 10 Minutes

Scope
To rotate a plan view

Solution

1. Open the **Main Floor – Kitchen** floor plan view.

2. Verify the Crop Region is enabled as **Visible**.

3. Select the crop region outline.

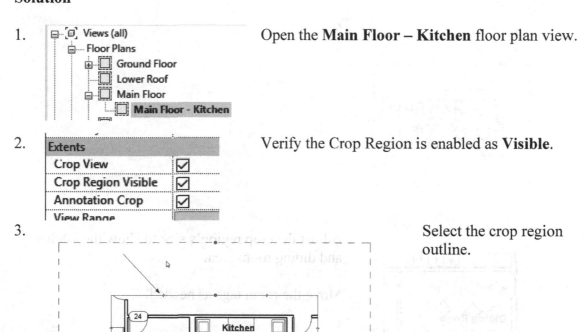

4. Select the **Rotate** tool on the Modify panel of the ribbon.

5.

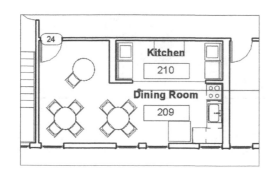

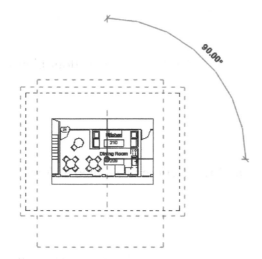

Set the starting point at 0 degrees on the Options bar.

Click the Place button.

Click the center of the view.

6.

Rotate the view 90 degrees.

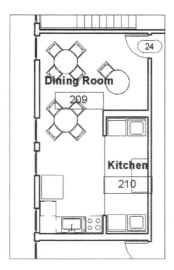

7.

Adjust the crop region's size to show the kitchen and dining room area.

Move the room tags as needed.

Did you notice that the tags rotated with the view?

8.

Close without saving.

Graphic Overrides

Graphic Overrides allow you to control the appearance of elements or categories in specific views. You can also use graphic overrides to identify linked files or worksets. By changing the appearance of elements, categories, linked files, and/or worksets, you are provided with a visual cue about the model.

You can override the cut, projection, and surface display for model categories and filters. For annotation categories and imported categories, you can edit the projection and surface display. In addition, for model categories and filters, you can apply transparency to faces. You can also specify visibility, half-tone display, and detail level of an element category, filter, or individual element.

Exercise 4-9

Graphic Overrides of Linked Files

Drawing Name: **graphic overrides.rvt**
Estimated Time to Completion: 15 Minutes

Scope
Manage Links
Apply Graphic Overrides to a linked file

Solution

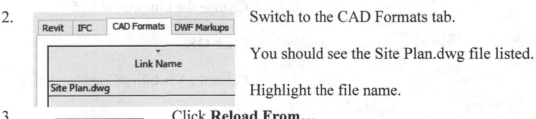

1.   Open the **Insert** ribbon.

     Click **Manage Links**.

2.   Switch to the CAD Formats tab.

     You should see the Site Plan.dwg file listed.

     Highlight the file name.

3.   Click **Reload From...**

4.

| File name: | Site Plan.dwg |
| Files of type: | DWG Files (*.dwg) |

Locate the file in the folder where you have downloaded the exercise files.
Select it and click **Open**.

This will reset the path so Revit can load the file into the project.

5.

Save Positions

☐ Preserve graphic overrides

At the bottom of the dialog, disable **Preserve graphic overrides**.

Click **OK** to close the dialog.

6.

⊟ [O] Views (all)
  ⊟ Floor Plans
    ☐ Enlarged Dining Room Plan
    ☐ Enlarged Loft Plan
    ☐ Foundation
    ☐ Level 1
    ☐ Level 2
    ☐ Presentation View
    ☐ Roof
    ☐ **Site Plan**

Open the **Site Plan** view under Floor Plans.

7.

Floor Plan: Site Plan	⊞ Edit Type
Graphics	☆ ^
View Scale	1" = 40'-0"
Scale Value 1:	480
Display Model	Normal
Detail Level	Coarse
Parts Visibility	Show Original
Visibility/Graphics Overrides	Edit...
Graphic Display Options	Edit...

Click **Edit** next to Visiblity/Graphics Overrides in the Properties palette.

You can also use the shortcut VV or VG.

8.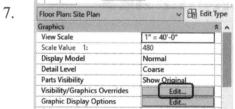

| Model Categories | Annotation Categories | Analytical Model Categories | Imported Categories | Filters |

☑ Show imported categories in this view   If a category is

Visibility	Projection/Surface		Halftone
	Lines	Patterns	
⊟ ☑ Imports in Families			☐
☑ 0			
⊟ ☑ Site Plan.dwg	Override...		☐
☑ 0			
☑ C-Curb			
☑ C-Prop-Line			
☑ C-Road-Cntl			
☑ C-Site-Prkg			
☑ Defpoints			

Switch to the Imported Categories tab.

Highlight the *Site Plan.dwg*.

Click **Override** in the Lines column.

9.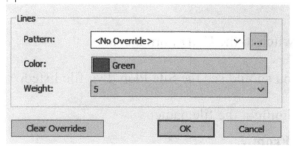

Lines

Pattern: <No Override>   ...

Color: Green

Weight: 5

Clear Overrides   OK   Cancel

Change the Color to **Green**.
Change the Lineweight to **5**.

Click **OK**.

Close the Visibility/Graphics dialog.

10. 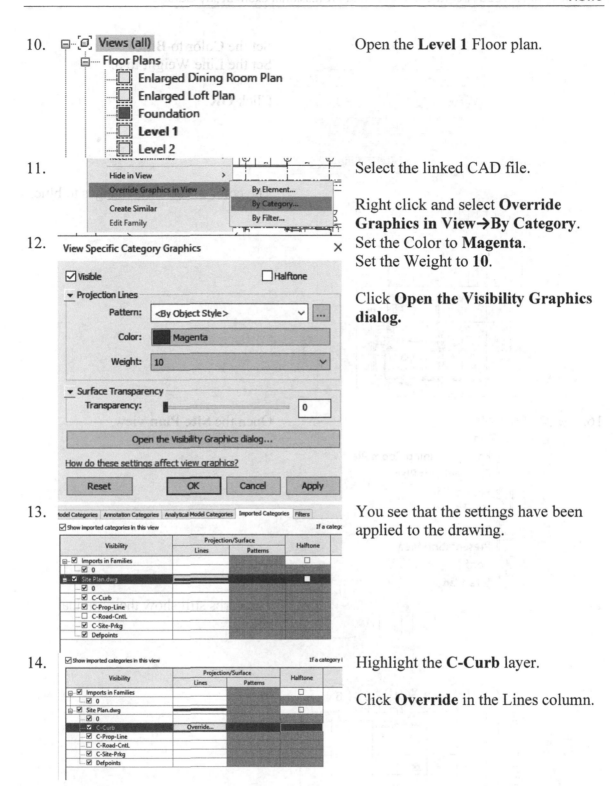 Open the **Level 1** Floor plan.

11. Select the linked CAD file.

Right click and select **Override Graphics in View→By Category**.
Set the Color to **Magenta**.
Set the Weight to **10**.

Click **Open the Visibility Graphics dialog.**

12.

13. You see that the settings have been applied to the drawing.

14. Highlight the **C-Curb** layer.

Click **Override** in the Lines column.

15.

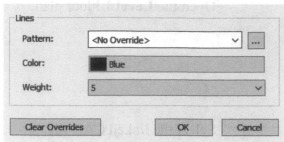

Set the Color to **Blue**.
Set the Line Weight to **5**.

Click **OK**.

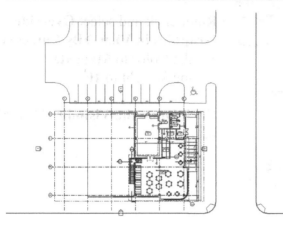

The color of the curb changes to blue.

16.

Open the **Site Plan** view.

17.

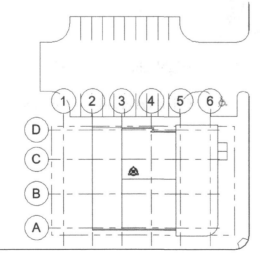

The curbs still show their original color.

18.   Save as *ex4-9.rvt*.

Duplicating Views

In Revit you can duplicate a view such as a floor plan or section. Once a view is placed on a sheet, it is "consumed" and cannot be re-used. So, the best way to place a different version of the same view is to duplicate views. For example, you might have one version of a floor plan that displays the fire ratings of the walls. You might have one version of the floor plan that displays a color scheme for area or department use. You might have a version of the floor plan that displays how furniture will be placed. You might want to create a view of the floor plan that focuses on a specific section of the floorplan, like the lavatories or stairs.

There are three methods for copying a view. The three types are as follows:

- Duplicate
- Duplicate with detailing
- Duplicate as Dependent

Duplicate

This is the most common way to duplicate a view. This option:

- Copies visibility settings
- No detailing
- Independent from original view

This creates multiple copies of the same view, such as a floor plan, but use them for different reasons such as fire escape plans, room area plans & dimensions. Each view will be of the same floor plan, but each view will have different annotations, tags, and displays.

Duplicate with Detailing

This is similar to duplicate; however, as the name suggests, it will also copy any detailing such as dimensions and text. The new duplicated view is still independent from the original view.

- Copies visibility settings
- Copies detailing
- Independent from original view

This can be useful if you want to preserve any tags or notes that have been added to the view. For example you want to use the same view but at different scales or colour schemes.

Duplicate as Dependent

Duplicating as a dependent will create identical copies of the original view; the new views are also tied to the original view as child objects. What this means is if you add a dimension in the original view, all dependent views will also have the new dimension.

- Shares visibility settings
- Copies detailing
- Is synchronised to the original view
-

This can be extremely useful if you have a large floor plan and you want to make several views which are cropped to specific regions.

Exercise 4-10

Duplicating Views

Drawing Name: **duplicating_views.rvt**
Estimated Time to Completion: 15 Minutes

Scope
Duplicate view with Detailing
Duplicate view as Dependent
Duplicate view
Understand the difference between the different duplicating views options

Solution

1. 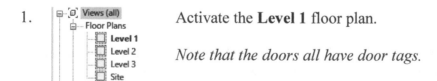 Activate the **Level 1** floor plan.

   Note that the doors all have door tags.

2. 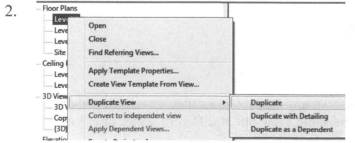 Highlight Level 1.
   Select **Duplicate View→Duplicate**.

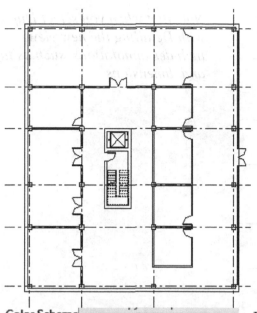

When you select Duplicate View→Duplicate, any annotations, such as tags and dimensions, are not duplicated.

3. Highlight the copied level 1.

Right click and select **Rename**.

Name: Level 1 - No Annotations

Type **Level 1- No Annotations**.

Click **OK**.

4. Highlight Level 1.
Right click and select **Duplicate View→Duplicate with Detailing**.

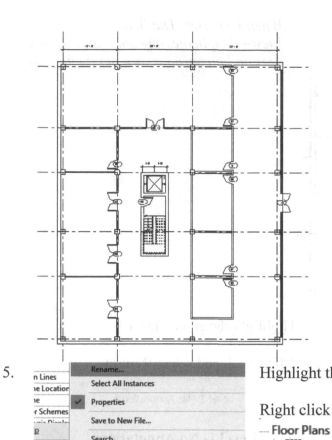

Note that when you select Duplicate with Detailing the new view includes annotations, such as tags and dimensions.

5. Highlight the copied level 1.

Right click and select **Rename**.

---- Floor Plans
    Level 1
    Level 1- Original Annotations
    Level 1- No Annotations
    Level 2

Type **Level 1- Original Annotations**.

Click **OK**.

6.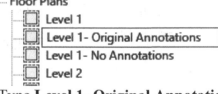

Highlight Level 1.
Right click and select **Duplicate View→Duplicate as a Dependent**.

7.

Notice that the duplicated view includes the annotations.

Floor Plans
    Level 1
        Level 1 - Dependent 1
    Level 1- No Annotations
    Level 1- Original Annotations

In the Project Browser, the dependent view is listed underneath the parent view.

8.

Floor Plans
    Level 1
        Level 1 -Stairs Area

Rename the dependent view **Level 1- Stairs area**.

9.

Floor Plans
    Level 1
        Level 1 -Stairs Area

Activate **Level 1- Stairs area**.

10.

Extents	
Crop View	☑
Crop Region Vis...	☑
Annotation Crop	☑

In the Properties pane,
Enable **Crop View**.
Enable **Crop Region Visible**.

11.

Adjust the crop region to focus the view on the stairs area.

12.

Floor Plans
    Level 1
        Level 1 -Stairs Area
    Level 1- No Annotations
    Level 1- Original Annotations

Activate **Level 1**.

13. 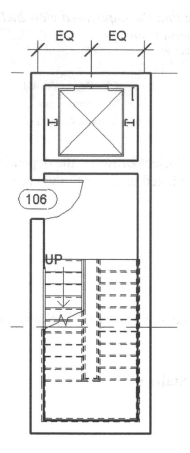 Zoom into the stairs area.

14. [Annotate] Activate the **Annotate** ribbon.

15. Select the **Tread Number** tool on the Tag panel.

Tread
Number

Tread Numbers can only be applied to Component-based stairs.

16. Select the middle line that highlights when your mouse hovers over the left side of the stairs.

17.

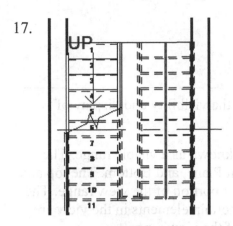

Select the middle line on the right side of the stairs. Right click and select CANCEL to exit the command.

18.

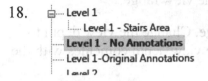

Activate **Level 1- Original Annotations**.

Notice that the new annotations - the tread numbers - are not visible in this view.

19.

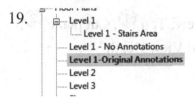

Activate **Level 1 - No Annotations**.

Notice that the new annotations - the tread numbers - are not visible in this view.

20.

Activate the **Level 1- Stairs area** dependent view.

Notice that any annotations added to the parent view are added to the dependent view.

21. Close without saving.

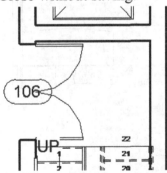

Extra: Change the door labeled 106 to a double flush door. Which Level 1 views display the new door type?

View Range

The view range is a set of horizontal planes that control the visibility and display of objects in a plan view.

Every plan view has a property called **view range**, also known as a visible range. The horizontal planes that define the view range are Top, Cut Plane, and Bottom. The top and bottom clip planes represent the topmost and bottommost portion of the view range. The cut plane is a plane that determines the height at which certain elements in the view are shown as cut. These 3 planes define the primary range of the view range.

View depth is an additional plane beyond the primary range. Change the view depth to show elements below the bottom clip plane. By default, the view depth coincides with the bottom clip plane.

The following elevation shows the view range ⑦ of a plan view: Top ①, Cut plane ②, Bottom ③, Offset (from bottom) ④, Primary Range ⑤, and View Depth ⑥.

The plan view on the right shows the result for this view range.

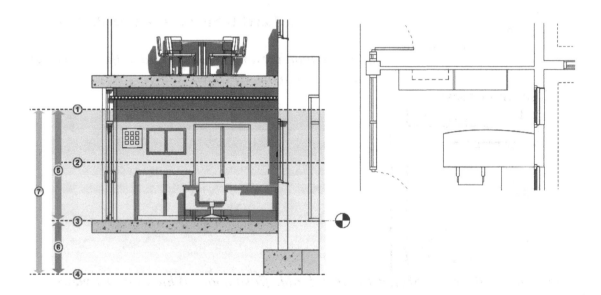

You will have at least one question on the Professional exam where you need to answer a question regarding one of the settings in the dialog.

Exercise 4-11

View Range

Drawing Name: **i_view_range.rvt**
Estimated Time to Completion: 5 Minutes

Scope
Determine the view depth of a view.

Solution

1. 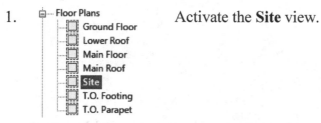 Activate the **Site** view.

2. 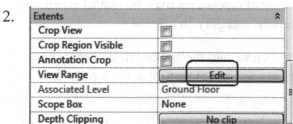 In the Properties pane:
Scroll down to **View Range** located under the Extents category.
Select the **Edit** button.

3. 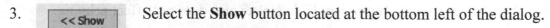 Select the **Show** button located at the bottom left of the dialog.

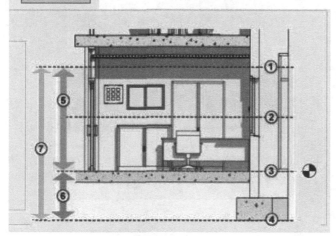

A representation of how view range settings work is displayed to assist you in defining the view range of a view.

Each number represents a field in the dialog box.

The elevation shows the view range ⑦ of a plan view.

4.

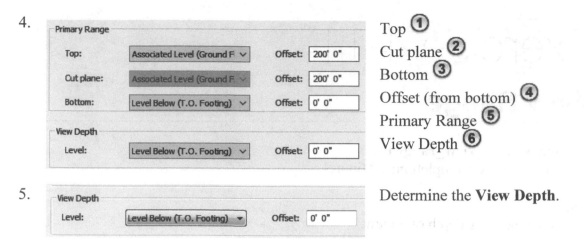

Top ①
Cut plane ②
Bottom ③
Offset (from bottom) ④
Primary Range ⑤
View Depth ⑥

5.

Determine the **View Depth**.

6.   Click **OK**.

7.   Close without saving.

Hidden Elements

Temporarily hiding or isolating elements or element categories may be useful when you want to see or edit only a few elements of a certain category in a view.

The Hide tool hides the selected elements in the view, and the Isolate tool shows the selected elements and hides all other elements in the view. The tool affects only the active view in the drawing area.

Element visibility reverts back to its original state when you close the project, unless you make the changes permanent. Temporary Hide/Isolate also does not affect printing.

Exercise 4-12

Reveal Hidden Elements

Drawing Name: **i_visibility.rvt**
Estimated Time to Completion: 5 Minutes

Scope
Turn on the display of hidden elements

Solution

1. Activate **the Ground Floor Admin Wing** floor plan.

   Views (all)
   └─ Floor Plans
      ├─ Ground Floor
      ├─ **Ground Floor Admin Wing**
      ├─ Lower Roof
      └─ Main Floor

2. 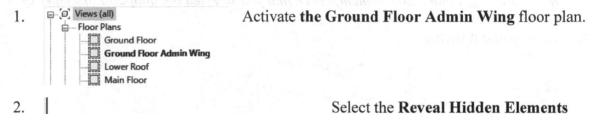 Select the **Reveal Hidden Elements** tool.

3. Items highlighted in magenta are hidden. Window around the two tables while holding down the CONTROL key to select them.

4. 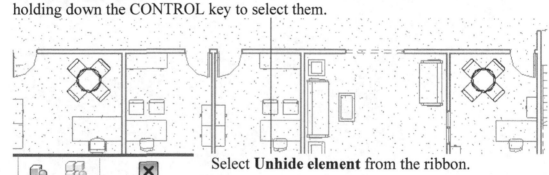 Select **Unhide element** from the ribbon.

The tables will no longer be displayed as magenta (hidden elements).

5. 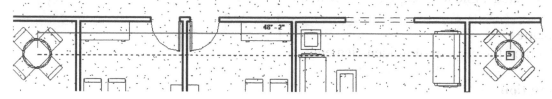 Select the **Close Hidden Elements** tool.

6. The view will be restored.

7. Select the **Measure** tool from the Quick Access toolbar.

8. Determine the distance between the center of the two tables.
    Did you get 48' 2"?

    If you didn't get that measurement, check that you selected the midpoint or center of the two tables.

9. Close without saving.

View Templates

A view template is a collection of view properties, such as view scale, discipline, detail level, and visibility settings.

Use view templates to apply standard settings to views. View templates can help to ensure adherence to office standards and achieve consistency across construction document sets.

Before creating view templates, first think about how you use views. For each type of view (floor plan, elevation, section, 3D view, and so on), what styles do you use? For example, an architect may use many styles of floor plan views, such as power and signal, partition, demolition, furniture, and enlarged.

You can create a view template for each style to control settings for the visibility/graphics overrides of categories, view scales, detail levels, graphic display options, and more.

Filters

Filters provide a way to override the graphic display and control the visibility of elements that share common properties in a view.

For example, if you need to change the line style and color for different conduit types, you can create a filter that selects all conduits in the view that have the color 'red' in the description parameter. You can then select the filter, define the visibility and graphic display settings (such as line style and color), and apply the filter to the view. When you do this, all conduits that meet the criteria defined in the filter update with the appropriate visibility and graphics settings. You need to set up the view filters and then apply those filters to each view in order to display the conduits with the correct colors and linetypes.

Exercise 4-13
Create a View Template

Drawing Name: **view_templates.rvt**
Estimated Time to Completion: 30 Minutes

Scope
Apply a wall tag.
Create a view template.
Create a view filter.
Apply view settings to a view.

Solution

1. 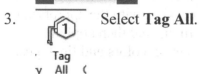 Activate Level 1.

2. 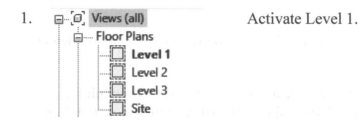 Activate the **Annotate** ribbon.

3. Select **Tag All**.

4. Highlight the **Wall tag - fire rating** as the tag to be used and click **OK**.

Structural Framing Tags	Structural Framing Tag : Standard
Wall Tags	Wall Tag : 1/2"
Wall Tags	Wall Tag : 1/4"
Wall Tags	wall tag- fire rating
Window Tags	Window Tag

The wall tag - fire rating is a custom family. It was pre-loaded into this exercise but is included with the exercise files on the publisher's website for your use.

5. 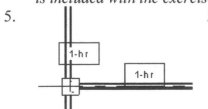 Zoom in to inspect the tags.

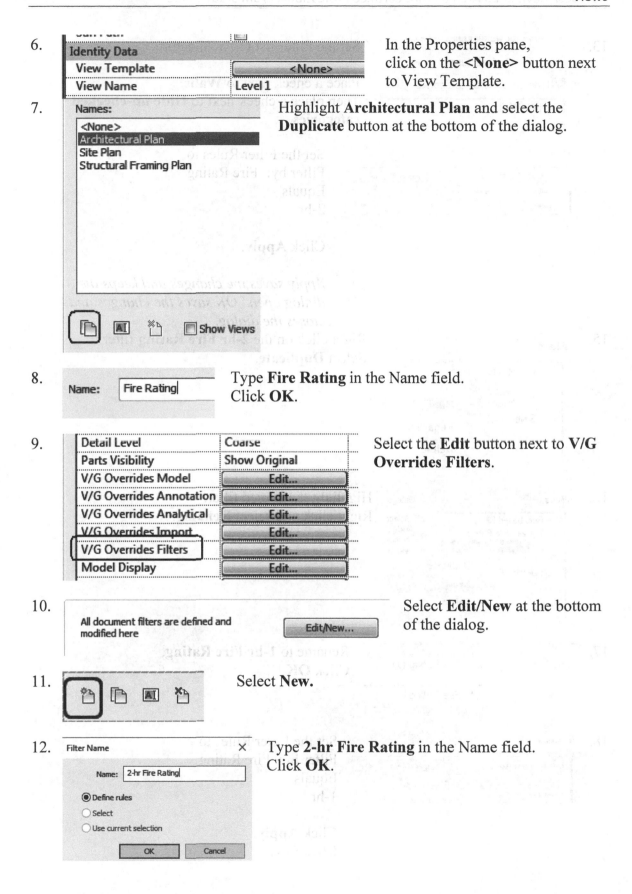

6. In the Properties pane, click on the **<None>** button next to View Template.

7. Highlight **Architectural Plan** and select the **Duplicate** button at the bottom of the dialog.

8. Type **Fire Rating** in the Name field.
Click **OK**.

9. Select the **Edit** button next to **V/G Overrides Filters**.

10. Select **Edit/New** at the bottom of the dialog.

11. Select **New.**

12. Type **2-hr Fire Rating** in the Name field.
Click **OK**.

13. 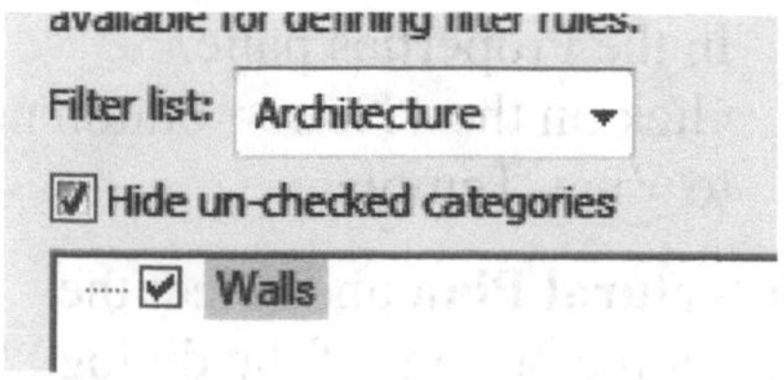 Set the Filter list to Architecture.

Place a check next to **Walls**.
Then place a check next to **Hide un-checked categories**.

14. 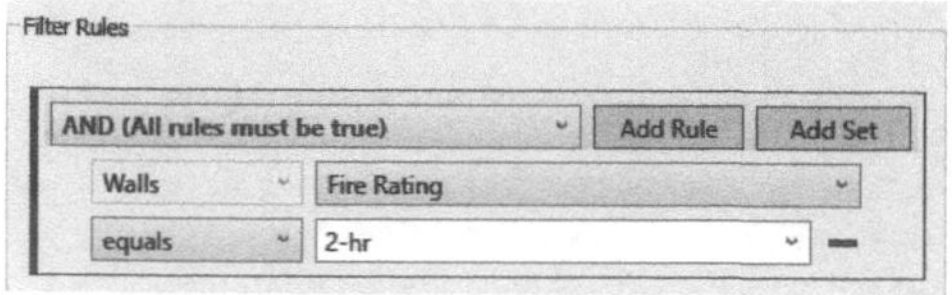 Set the Filter Rules to
Filter by: Fire Rating
Equals
2-hr.

Click **Apply**.

Apply saves the changes and keeps the dialog open. OK saves the changes and closes the dialog.

15. 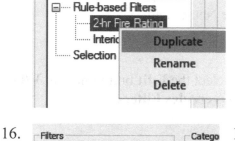 Right click on the **2-hr Fire Rating** filter.
Select **Duplicate**.

16. 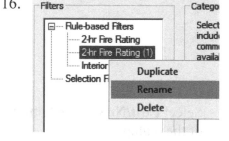 Highlight the copied filter.
Right click and select **Rename**.

17. 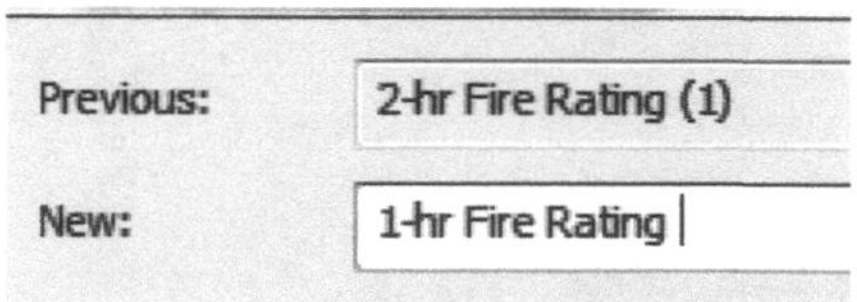 Rename to **1-hr Fire Rating**.
Click **OK**.

18. 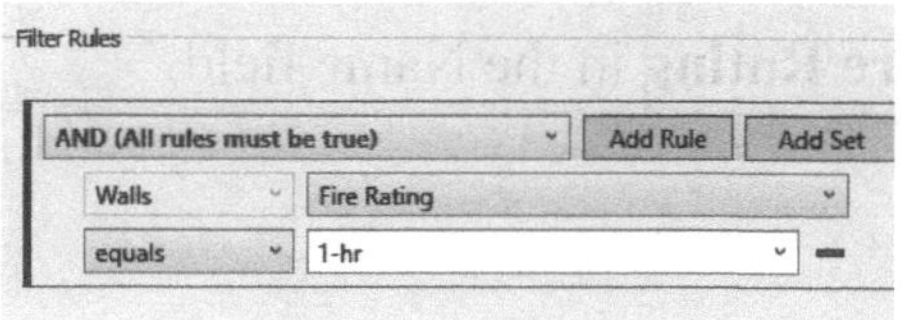 Set the Filter Rules to
Filter by: Fire Rating
Equals
1-hr.

Click **Apply**.
Click **OK**.

19.

Select the **Add** button at the bottom of the Filters tab.

Name	Vi...
No filters have been applied to this vi...	

Add   Remove

20.

Hold down the Control key.
Highlight the 1-hr and 2-hr fire rating filters and click **OK**.

Select one or more filters to insert.

- Rule-based Filters
  - 1-hr Fire Rating
  - 2-hr Fire Rating
  - Interior
- Selection Filters

21.

You should see the two fire rating filters listed.

| Model Categories | Annotation Categories | Analytical Model Categories | Imported Categories | Filters |

Name	Visibility	Projection/Surface			Cut		Halftone
		Lines	Patterns	Transparen...	Lines	Patterns	
2-hr Fire Rating	☑	Override...	Override...	Override...	Override...	Override...	☐
1-hr Fire Rating	☑						☐

If the interior filter was accidentally added, simply highlight it and select Remove to delete it.

22.

Highlight the **2-hr Fire Rating** filter.

Name	Visibility	Projection/Surface		
		Lines	Patterns	Tr...
2-hr Fire Rating	☑	Override...	Override...	C...
1-hr Fire Rating	☑			

23. Select the **Pattern Override** under Projection/Surface.

24.

Select the **2 Hour** fill pattern for the Foreground.

Pattern Overrides
Foreground ☑ Visible
Pattern: | 2 Hour | ||||||||||||||||||| ∨ | ... |

25.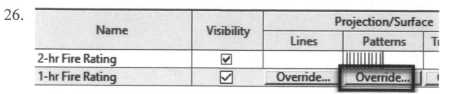

Set the Color to **Blue.**
Click **OK.**

26.

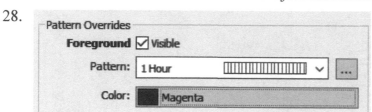

Highlight the **1-hr Fire Rating** filter.

27. Select the **Pattern Override** under Projection/Surface.

28.

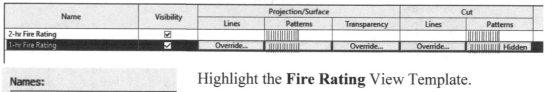

Set the fill pattern for the Foreground to **1 Hour** and the Color to **Magenta.**
Click **OK.**

29. Apply the same settings to the Cut Overrides. Click **OK**.

Name	Visibility	Projection/Surface			Cut	
		Lines	Patterns	Transparency	Lines	Patterns
2-hr Fire Rating	☑		IIIIIIIIIIII			IIIIIIIIIIII
1-hr Fire Rating	☑	Override...	IIIIIIIIIIII	Override...	Override...	IIIIIIIIIIII Hidden

30.

Names:

<None>
Architectural Plan
Fire Rating
Site Plan
Structural Framing Plan

Highlight the **Fire Rating** View Template.

Click **OK**.

31. 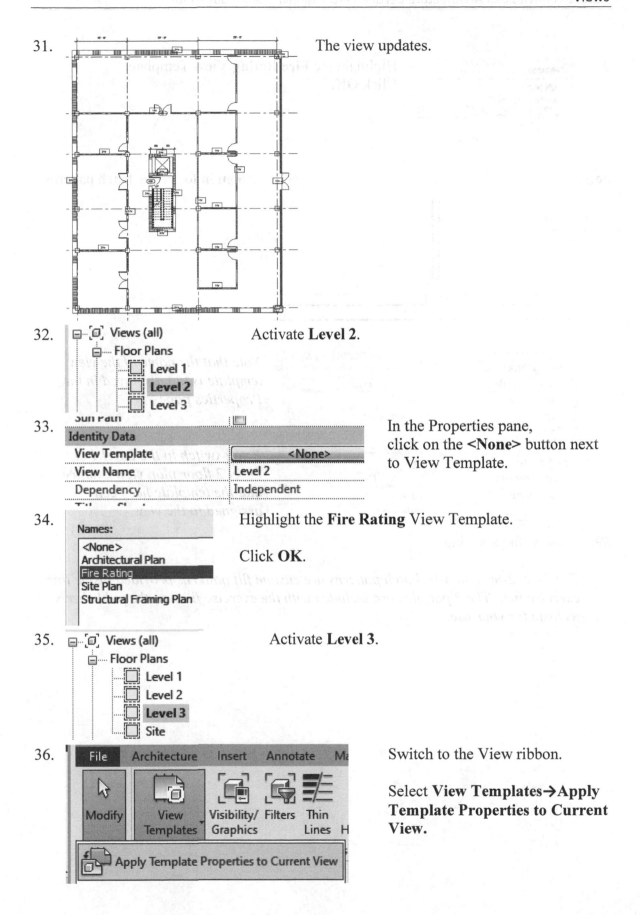 The view updates.

32. Activate **Level 2**.

33. In the Properties pane, click on the **<None>** button next to View Template.

34. Highlight the **Fire Rating** View Template.

    Click **OK**.

35. Activate **Level 3**.

36. Switch to the View ribbon.

    Select **View Templates→Apply Template Properties to Current View.**

37.

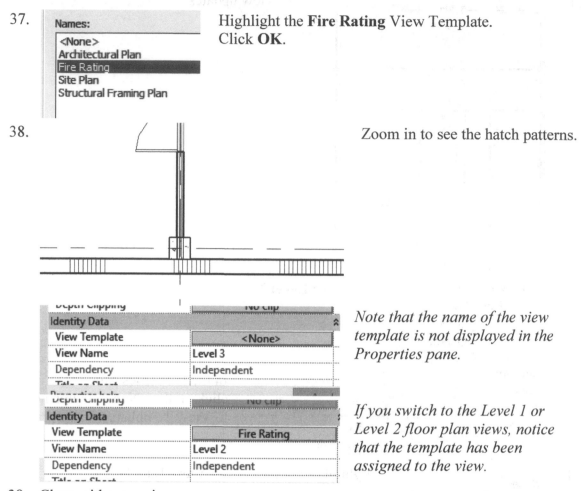

Highlight the **Fire Rating** View Template.
Click **OK**.

38.

Zoom in to see the hatch patterns.

Note that the name of the view template is not displayed in the Properties pane.

If you switch to the Level 1 or Level 2 floor plan views, notice that the template has been assigned to the view.

39. Close without saving.

The 1-hr, 2-hr, and 3-hr hatch patterns are custom fill patterns provided inside this exercise file. The *.pat files are included with the exercise files on the publisher's website for your use.

Exercise 4-14

Apply a View Template to a Sheet

Drawing Name: **view templates2.rvt**
Estimated Time to Completion: 10 Minutes

Scope

Use a view template to set all the views on a sheet to the same view scale.
View templates are used to standardize project views. In large offices, different people are working on the same project. By using a view template, everybody's sheets and views will be consistent.

Solution

1. 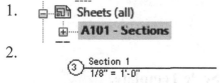   Open the **A101 – Sections** sheet.

2.   Zoom into each title bar and see the scale for the view.

3.   Open the **Section 1** view.

4. 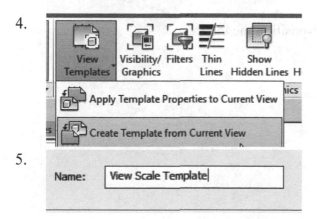   Activate the View ribbon.
   Under View Templates, select **Create Template from Current View.**

5.   Name the new template **View Scale Template**.
   Click **OK**.

6.   Uncheck all the boxes EXCEPT View Scale. Click **OK**.

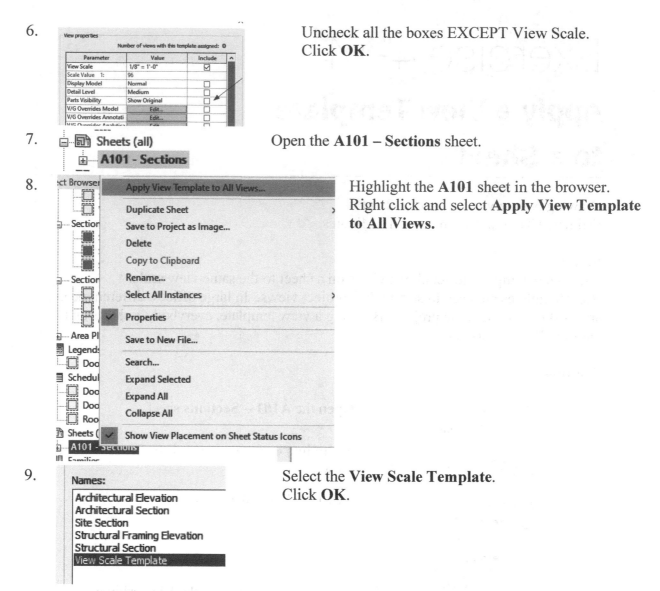

7.   Sheets (all)   Open the **A101 – Sections** sheet.

                  A101 - Sections

8.   Highlight the **A101** sheet in the browser. Right click and select **Apply View Template to All Views.**

9.   Select the **View Scale Template**. Click **OK**.

10.   All the views on the sheet adjusted. Re-position the views.

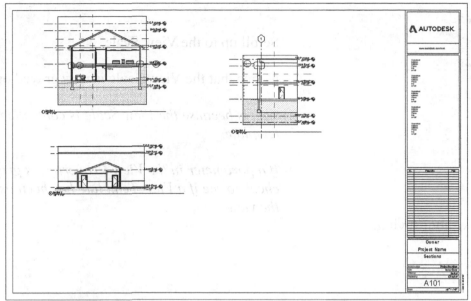

11.   ⊟ Sections (Building Section)   Open **Section 2** view.
             Section 1
             Section 2
             Section 3

12.

Graphics	
View Scale	**1/8" = 1'-0"**
Scale Value 1:	96
Display Model	**Normal**

Note that the view scale has been modified for the view.

13.

Identity Data	
View Template	<None>
View Name	Section 2
Dependency	Independent
Title on Sheet	

Scroll down the Property panel.

The View Template was not assigned.

Click **<None>.**

14.

Names:

<None>
Architectural Elevation
Architectural Section
Site Section
Structural Framing Elevation
Structural Section
View Scale Template

Select the **View Scale Template**.

Click **OK**.

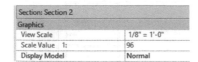

Section: Section 2	
Graphics	
View Scale	1/8" = 1'-0"
Scale Value 1:	96
Display Model	Normal

Scroll up to the View Scale.

Notice that the View Scale is now grayed out.

This is because the View Scale is controlled by the template.

If a parameter in the View Properties is grayed out, check to see if a View Template has been assigned to the view.

15. Close without saving.

Call-outs

The standard 'Callout' command places a rectangle defining the callout view extents. This will create a new view, and place the view reference rectangle on the active view. For reasons listed below, this method is not preferred.

- You cannot change the reference.
- You cannot move the callout to a different parent view.
- You cannot change the callout view family between Plan/Detail (Plans only).
- The callout rectangle can be stretched or rotated in the parent view – this will make the same change of extents or rotation to the callout view itself, which is not always desirable.
- You cannot have the callout extents slightly different on the parent view (to make callouts readable) - they have to match the view cropping exactly.
- If you copy and paste a Callout it creates an entirely new view with a different reference.

Reference Other View Callouts can be created by:

Create a view to be referenced by the callout – eg. duplicate another view and crop it
Make sure the view is cropped (unless it is a drafting view).
On the view to place the callout, select the 'Callout' command, then select 'Reference Other View', and select a relevant view name from the drop-down menu

This method has many advantages:

- Callouts can be moved to another view
- Callout references can be changed to refer to different view
- References will update if the view/sheet number is changed
- Callouts can be copied to another parent view

It also has a few disadvantages:
It is possible for the user to select the wrong view in the list - there is no automatic check for this.

If you want to have "Sim" showing on a callout for similar details on multiple callouts, that is a Type property of the view being referred to, not a property of the callout symbol.

That means it is all or nothing – i.e. all instances of the callout must have "Sim" or not. It also changes the view type so it may move in your project browser, depending on your browser organization scheme.

Exercise 4-15

Create a Call-out View

Drawing Name: **callouts.rvt**
Estimated Time to Completion: 10 Minutes

Scope
Place a callout to a drafting view

Solution

1.

    Sections (Building Section)
       Architectural
          Section 1
          Section 2
          Section 3
          Section 4

Open the **Section 3** view under Sections (Building Section).

2.

Collaborate | View

Section | Callout

Rectangle

Switch to the View ribbon.

Select the **Callout→Rectangle** tool.

3.

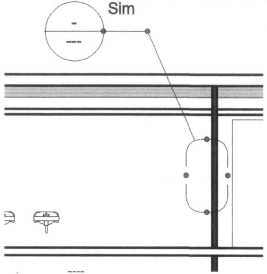

On the ribbon:

Enable **Reference Other View**.

In the drop-down: select **Drafting View: 06-Interior Wall GWB**.

4.

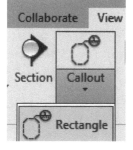

Sketch the rectangle over the interior wall.

Use the grips to adjust the position of the callout.

5.

Open the **Section 4** view under Sections (Building Section).

6.

Switch to the View ribbon.

Select the **Callout→Rectangle** tool.

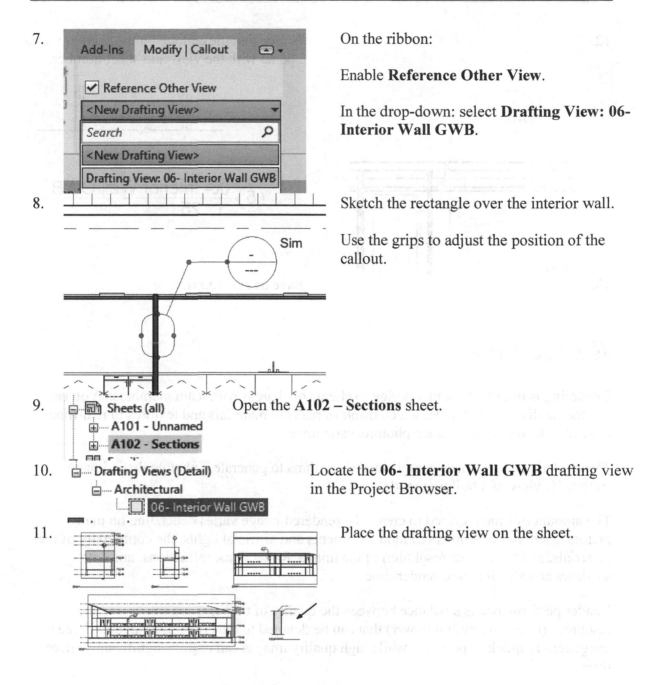

7. On the ribbon:

Enable **Reference Other View**.

In the drop-down: select **Drafting View: 06-Interior Wall GWB**.

8. Sketch the rectangle over the interior wall.

Use the grips to adjust the position of the callout.

9. Open the **A102 – Sections** sheet.

10. Locate the **06- Interior Wall GWB** drafting view in the Project Browser.

11. Place the drafting view on the sheet.

12.

Notice that the callouts in the section views update with the view and sheet number.

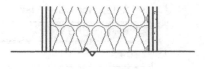

5 | 06- Interior Wall GWB
1 : 25

13.

Save as *ex4-15.rvt*.

Rendering

Rendering is used to present a design to clients or share it with team members. You can use the Realistic visual style, which displays realistic materials and textures in real-time or render the model to create a photorealistic image.

The rendering engine uses complicated algorithms to generate a photorealistic image from a 3D view of a building model.

The amount of time required to create the rendered image varies depending on many factors, such as the numbers of model elements and artificial lights, the complexity of the materials, and the size or resolution of the image. Reflections, refractions, and soft shadows can also increase render time.

Render performance is a balance between the quality of the resulting image and the resources (time, computing power) that can be devoted to the effort. Low quality images are generally quick to produce, while high quality images can require significantly more time.

Before rendering an image, consider whether you need a high-quality image or a draft quality image. In general, start by rendering a draft quality image to see the results of the initial settings. Then refine materials, lights, and other settings to improve the image. As you get closer to the desired result, you can use the medium quality setting to produce a more realistic image. Use the high-quality setting to produce a final image only when you are sure that the material render appearances and the render settings will give the desired result.

Exercise 4-16
Render a View

Drawing Name: **render.rvt**
Estimated Time to Completion: 15 Minutes

Scope
Create a rendering.
Control rendering options.
Save a rendering to the project.

Solution

1. 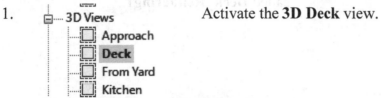 Activate the **3D Deck** view.

2. 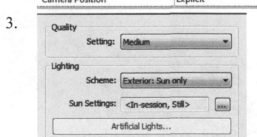 On the Properties pane,
under Camera,
select **Edit** for Rendering Settings.

3. Set the Quality Setting to **Medium**.
Click **OK**.

4.  Select the **Rendering** tool located on the Display
Control bar.

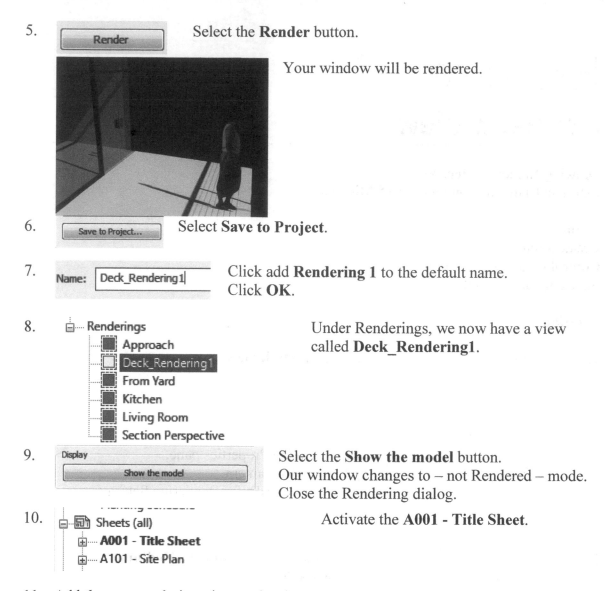

5. Select the **Render** button.

Your window will be rendered.

6. Select **Save to Project**.

7. Click add **Rendering 1** to the default name.
Click **OK**.

Name: Deck_Rendering1

8. Under Renderings, we now have a view called **Deck_Rendering1**.

9. Select the **Show the model** button.
Our window changes to – not Rendered – mode.
Close the Rendering dialog.

10. Activate the **A001 - Title Sheet**.

11. Add the new rendering view to the sheet.

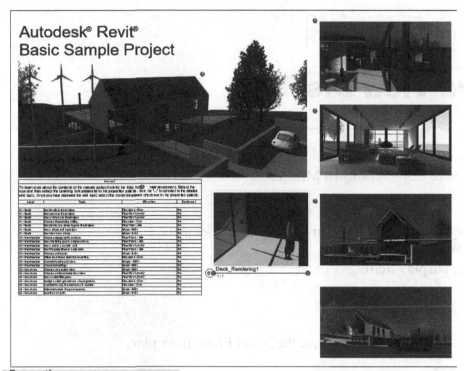

12.   Select the view that was just placed.

On the Properties panel:

Select **Viewport: No Title** using the Type Selector.

13.   Save the file as *ex4-16.rvt*.

Exercise 4-17
Change View Display

Drawing Name: **i_views.rvt**
Estimated Time to Completion: 10 Minutes

Scope
Use Temporary Hide/Isolate to control visibility of elements.
Change Line Width Display.
Change Object Display Settings.

Solution

1. Activate the **Main Floor** floor plan.

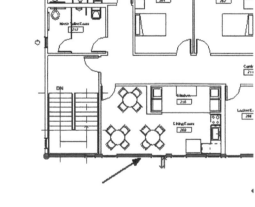

2. Select one of the exterior walls so it is highlighted.

3. 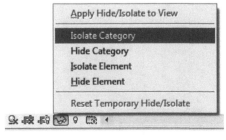 Select the **Temporary Hide/Isolate** tool.
Right click and select **Isolate Category**.

4.

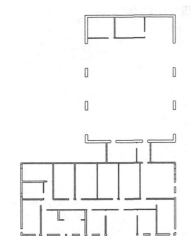

Only the exterior walls are visible.

5.

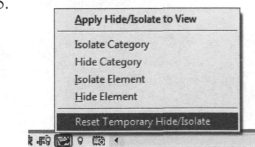

Select the **Temporary Hide/Isolate** tool. Right click and select **Reset Temporary Hide/Isolate**.

This restores the view.

6.

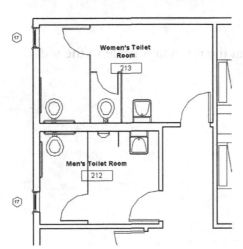

Zoom into the region where the lavatories are located.

7.
Custom...
12" = 1'-0"
6" = 1'-0"
3" = 1'-0"
1 1/2" = 1'-0"
1" = 1'-0"
3/4" = 1'-0"
1/2" = 1'-0"
3/8" = 1'-0"
1/4" = 1'-0"
3/16" = 1'-0"
1/8" = 1'-0"
1" = 10'-0"
3/32" = 1'-0"
1/16" = 1'-0"
1" = 20'-0"
3/64" = 1'-0"
1" = 30'-0"
1/32" = 1'-0"
1" = 40'-0"
1" = 50'-0"
1" = 60'-0"
1/64" = 1'-0"
1" = 80'-0"
1" = 100'-0"
1" = 160'-0"
1" = 200'-0"
1" = 300'-0"
1" = 400'-0"
1/4" = 1'-0"

Change the view scale to **1/8″ = 1′-0″**.

8.
Women's Toilet Room
213
Men's Toilet Room
212
17
17

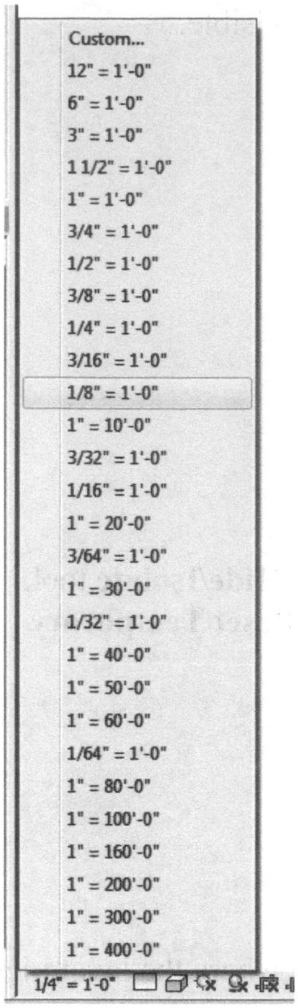

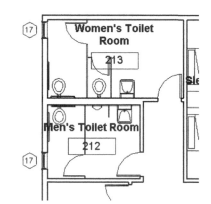

Note that the room tags scale to the view.

9.
Modify

Activate the **Modify** ribbon.

10.
View

Select the **Linework** tool on the View panel.

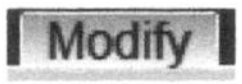

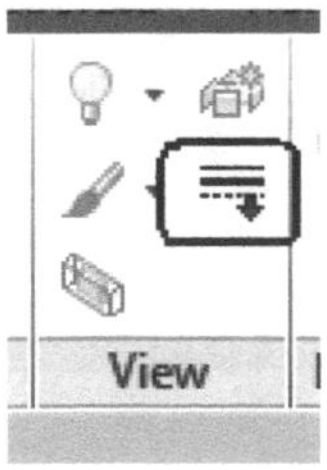

11.   Line Style:

<Overhead>

Line Style

Set the Line Style to **Overhead**.

12.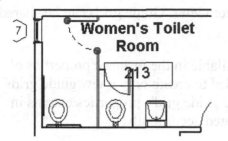

Select the door swing on the toilet cubicle.

Note that the door swing's appearance changes.

13.   Sections (Building Section)
 — **Section 1**
 — Section 2
 — Section 3

Activate **Section 1**.

14.   Materials Object Snaps
            Styles

Activate the Manage ribbon.

Select **Settings→Object Styles**.

15.

Detail Items	1		■ Black
Doors	2	2	■ Black
Elevation Swing	1	1	■ Black
Frame/Mullion	3	3	■ Magenta
Glass	1	4	■ Black
Hidden Lines	2	2	■ Blue
Opening	1	3	■ Black
Panel	3	5	■ Blue
Plan Swing	1	1	■ Black

Expand the **Doors** category on the Model Objects tab.

Change the Line Weight, Line Color, and Line Pattern for the Panel and Frame.

Click **Apply** to see the changes.

16.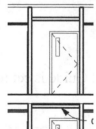

You can move the dialog over so you can see how the display is changed.

Click OK to close the dialog.

Note that linework changes are specific to the view, but object settings changes affect all views.

17.   Close without saving.

Guide Grids

Guide grids help arrange views so that they appear in the same location from sheet to sheet.

You can display the same guide grid in different sheet views. Guide grids can be shared between sheets.

When new guide grids are created, they become available in the instance properties of sheets and can be applied to sheets. It is recommended to create only a few guide grids and then apply them to sheets. When you change the guide grid's properties/extents in one sheet, all the sheets which use that grid are updated accordingly.

Exercise 4-18
Using Guide Grids

Drawing Name: **aligning_views.rvt**
Estimated Time to Completion: 30 Minutes

Scope
Using Grid Guides

Solution

1. Activate the sheet with **Level 1** floor plan.

2. In the Properties pane:
   Scroll down and note that Grid Guide is set to <None>.

3. Activate the **View** ribbon.
   Select **Guide Grid** on the Sheet Composition panel.

4.

Assign Guide Grid

Choose existing:

Create new:
Name: Guide Grid 1

OK   Cancel

Click **OK**.

5.

Dimensions	
Guide Spacing	2"
Identity Data	
Name	Guide Grid 1

Select the Guide Grid.
Select by left clicking on an edge.
In the Properties pane,
set the Guide Spacing to **2"**.
Click **Apply**.

6.

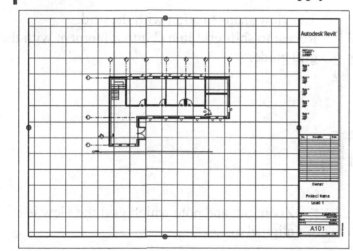

The grid updates.
Use the blue grips to adjust
the grid so it lies entirely
inside the title block.

7.

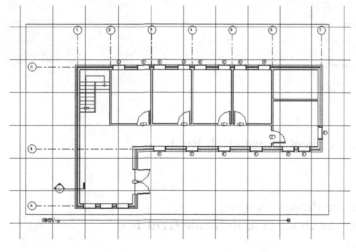

Select the viewport.

Select the **Move** tool from
the Modify Panel.

8.

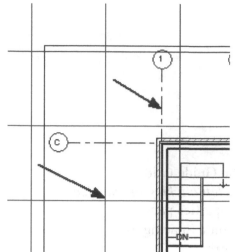

Select on the grid line 1 in the view.
Then select the guide grid.

9.

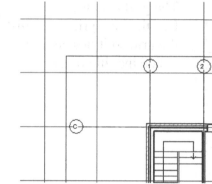

The view snaps to align with the grid.
Repeat to shift grid line C into alignment with the
grid guide.

10.

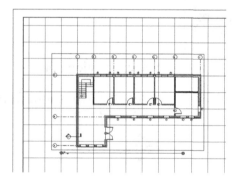

Zoom out and note which grid the view is
aligned to.

11.

Schedules/Quantities
Sheets (all)
⊞— A101 - Level 1
⊞— **A102 - High Roof**
⊞— A103 - Low Roof

Activate the sheet named **High Roof**.

12.

Other
File Path   C:\Revit 2023 Certification...
Guide Grid   Guide Grid 1

In the Properties pane,
set the Guide Grid to **Guide Grid 1**.

13.

Select the viewport.

Select the **Move** tool from the Modify Panel.

14. 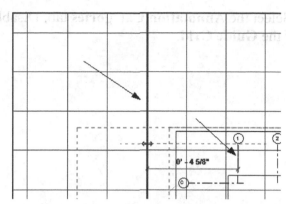 Select on the grid line 1 in the view. Then select the guide grid.

15. 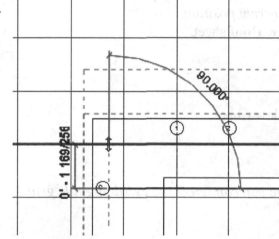 Repeat to shift grid line C into alignment with the grid guide.

16. 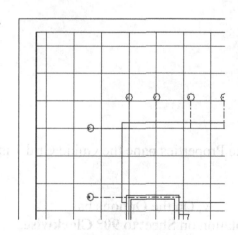 Verify that the view is aligned to the same guide grid cells as sheet A101.

17. 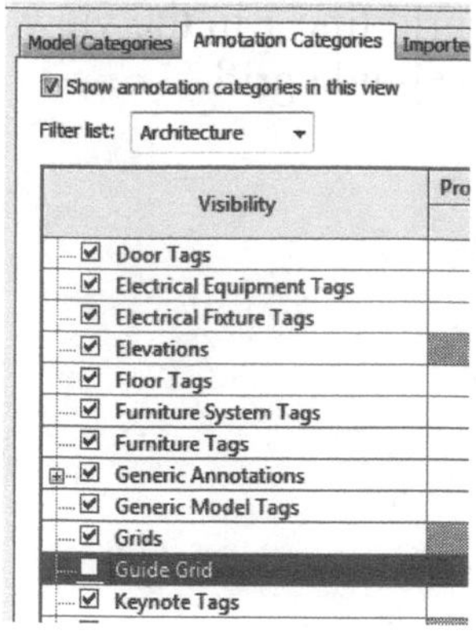 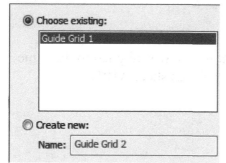 Type **VV**. Select the **Annotation Categories** tab. Disable visibility of the **Guide Grid**.
Click **OK**.

18. 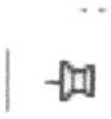 Select the viewport. Select **Pin** from the Modify panel.
This will lock the view to its current position.

19. 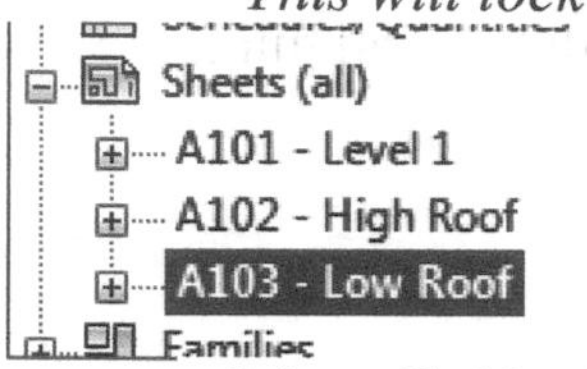 Activate the **Low Roof** sheet.

Sheets (all)
- A101 - Level 1
- A102 - High Roof
- A103 - Low Roof

Families

20.  Select **Guide Grid** from the View ribbon.

Guide Grid

21. *Note you can select the existing guide grid.*

Assign Guide Grid

○ Choose existing:

Guide Grid 1

Click **OK**.

○ Create new:

Name: Guide Grid 2

22. 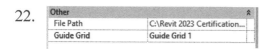 Note in the Properties pane the Guide Grid 1 is listed.

23.  Select the view. On the Options bar, set the Rotation on Sheet to **90° Clockwise**.

24. 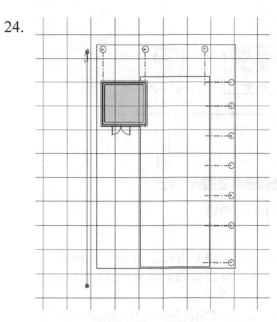   The view rotates.

25. Close the file without saving.

Schedules

Schedules are spreadsheets connected to your model. If you modify an element in the model, it will be updated in the schedule and the other way around. Autodesk refers to schedules as "tabulated views", so look for those words in exam questions. If you see those words, the exam is asking about schedules.

Revit offers four different schedule types:

Basic Schedule – used to list and quantify model elements, such as walls, doors, and windows

Sheet & View Lists – used to list views and which sheets they are placed on

Material Takeoff – allows you to calculate the materials used for construction. For example, the amount of concrete needed for a floor slab, or the amount of plywood used in a wall.

Note Block - used to organize plan and general notes

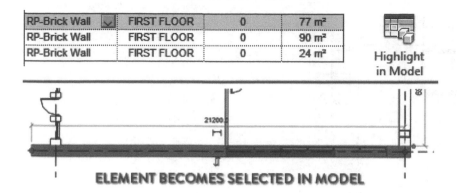

ELEMENT BECOMES SELECTED IN MODEL

Sometimes, you see an element in a schedule and you need to locate it in the model. Click **Highlight in Model** and a view will open with the element appearing in blue.

When creating a Material Takeoff schedule type, use the Calculated Parameter to calculate values together. For example, multiply the Material Cost with the Material Area to get the total price for each material.

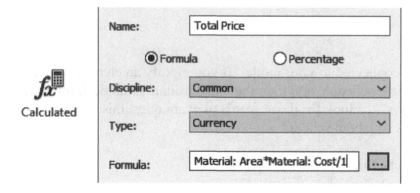

Click **Calculated** in the ribbon, select **Currency** and set the **Name**. Then select the two fields in the formula and put * between them. Add a **/1** at the end of the formula to fix units, else you will get a warning.

In this example we calculated price, but this tool can be used to calculate anything or create percentages.

Use the Filters tab to exclude specific elements from the schedule. For example, if you want a schedule with walls that are at least 1000mm long, add a «*is greater than or equal*» Length filter. As you see in the resulting schedule, walls below 1000mm are hidden.

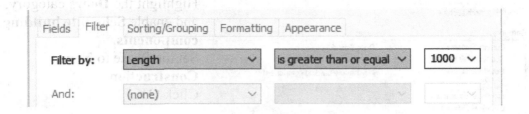

<RP-Wall Schedule>				
A	**B**	**C**	**D**	**E**
Type	Base Constraint	Base Offset	Area	Length
Curtain Wall Ext	FIRST FLOOR	0	5 m²	1034
Curtain Wall Ext	FIRST FLOOR	0	6 m²	1061
WOOD WALL	FIRST FLOOR	0	10 m²	7550

Exercise 4-19

Define Element Properties in a Schedule

Drawing Name: **i_schedules.rvt**
Estimated Time to Completion: 5 Minutes

Scope
Create a door schedule.
Use the sort and group feature to determine how many doors of a specific type are on a level.

Solution

1.

Activate the **View** tab on the ribbon.
Select **Schedule/Quantities** from the Create panel.

2.

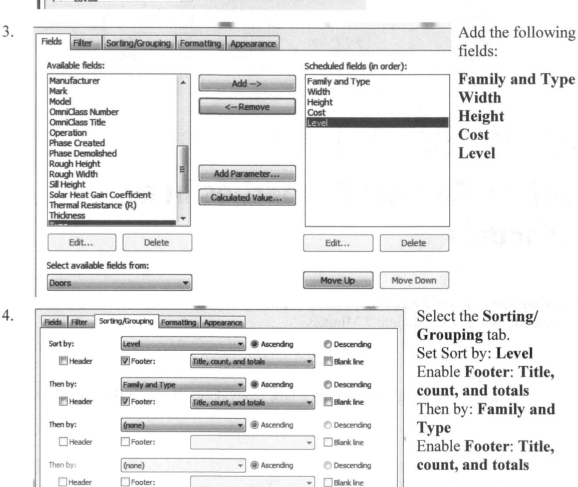

Highlight the **Doors** category, and enable **Schedule building components**.
Set the Phase to **New Construction**.
Click **OK**.

3.

Add the following fields:

Family and Type
Width
Height
Cost
Level

4.

Select the **Sorting/ Grouping** tab.
Set Sort by: **Level**
Enable **Footer: Title, count, and totals**
Then by: **Family and Type**
Enable **Footer: Title, count, and totals**

Enable **Itemize every instance.**

5.

☐ Hidden field   Conditional Format...

☐ Show conditional format on sheets

| Standard | ⌄ |

| Standard |
| Calculate totals |
| Calculate minimum |
| Calculate maximum |
| Calculate minimum and maximum |

Select the **Formatting** tab. Highlight **Cost**. Disable **Show conditional format on sheets**. Select **Calculate totals** from the drop-down list.

6.

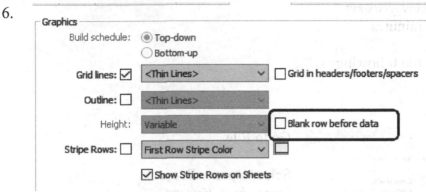

Select the **Appearance** tab. Clear the **Blank row before data** checkbox.

7. Click **OK**.

8.

Overhead-Sectional Glass: 14 x 14: 0			74320.00	
Single-Flush Vi	3' - 0"	7' - 0"	240.15	Main Floor
Single-Flush Vi	3' - 0"	7' - 0"	240.15	Main Floor
Single-Flush Vi	3' - 0"	7' - 0"	240.15	Main Floor
Single-Flush Vi	3' - 0"	7' - 0"	240.15	Main Floor
Single-Flush Vi	3' - 0"	7' - 0"	240.15	Main Floor
Single-Flush Vi	3' - 0"	7' - 0"	240.15	Main Floor
Single-Flush Vision: 36" x 84": 6			1440.90	

Locate how many **Single-Flush Vision: 36″ x 84″** doors are placed on the Main Floor level.

9. Close without saving.

Exercise 4-20

Sheet Lists

Drawing Name: *sheet_lists.rvt*
Estimated Time: 5 minutes

This exercise reinforces the following skills:
  ❑ Schedules
  ❑ Sheets

1. 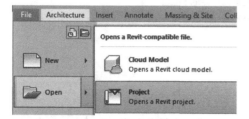   Go to **File**.

    Select **Open→Project**.
    Open *sheet_lists.rvt.*

2.   Activate the **View** ribbon.

3.   Go to **Schedules→Sheet List.**

    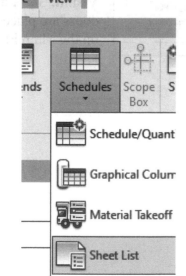

4.   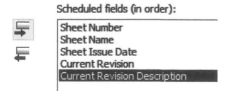   Select the following fields in order:
      • Sheet Number
      • Sheet Name
      • Sheet Issue Date
      • Current Revision
      • Current Revision Description
    Click **OK**.

5.  The schedule view opens.
Save as *ex4-20.rvt*.

Legends

A Legend is used to display a list of various model components and annotations used in a project.

Legends fall into three main types: Component, Keynote and Symbol.

Component Legends display symbolic representations of model components. Examples of elements that would appear in a component legend are electrical fixtures, wall types, mechanical equipment and site elements.

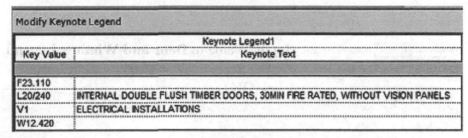

Keynote Legends are schedules which list the keynotes applied in a view on the sheet and what the different key values designate.

Symbol Legends are similar to keynote legends. They list the different annotation symbols used in a view and what they mean.

Legends can be placed on more than one sheet. If you use the same legends across projects, you can use Transfer Project Standards to copy legends from one project to another.

Exercise 4-21
Create a Legend

Drawing Name: **i_Legends.rvt**
Estimated Time to Completion: 30 Minutes

Scope
Create a Legend.
Add to a sheet.

Solution

1.  Activate the **View** ribbon.
   Select **Legend** from the Create panel.

2. Set the Name to **Door and Window Legend**.
   Set the Scale to **¼″ = 1′-0″**.
   Click **OK**.

3. The Legend is listed in the browser. An empty view window is opened.

4. 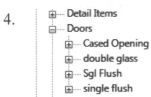 In the browser, locate all the door families.

5. 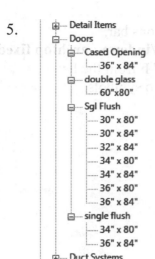 Expand each door family to see which types are available.

6. 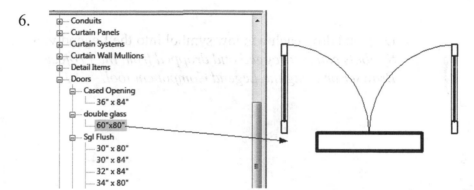 Highlight the **60" x 80" Double Glass** door.
Drag and drop it into the Legend view.

Right click and select **Cancel** to exit the command.

7. 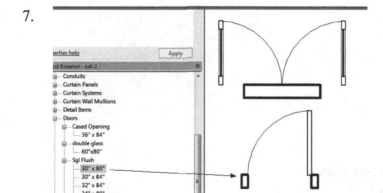 Highlight the **30" x 80" Sgl Flush** door.

Drag and drop it into the Legend view.

Right click and select **Cancel** to exit the command.

8. 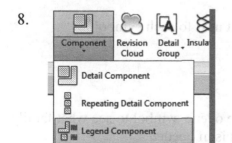 Activate the Annotate ribbon.

Select the **Legend Component** tool from the Detail panel.

9.

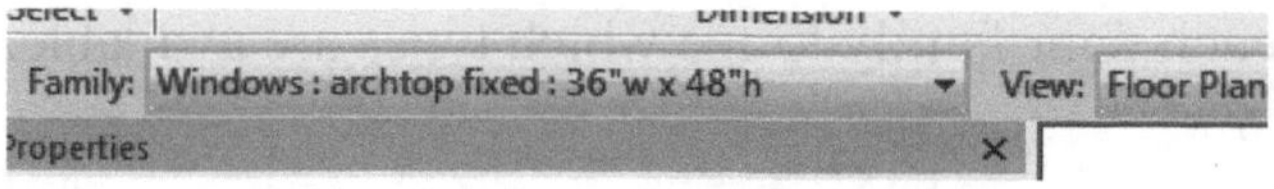

Family: Windows : archtop fixed : 36"w x 48"h ▼   View: Floor Plan

Properties   ×

On the Options bar, select the **Windows : archtop fixed : 36″ w x 48″ h** from the list.

10.

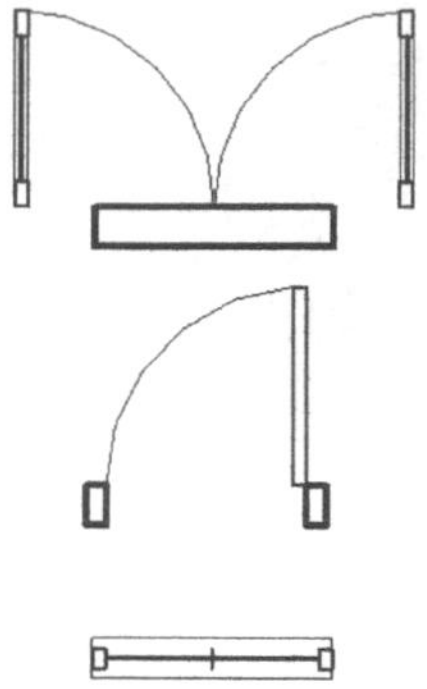

Place the symbol below the other two symbols.

11.

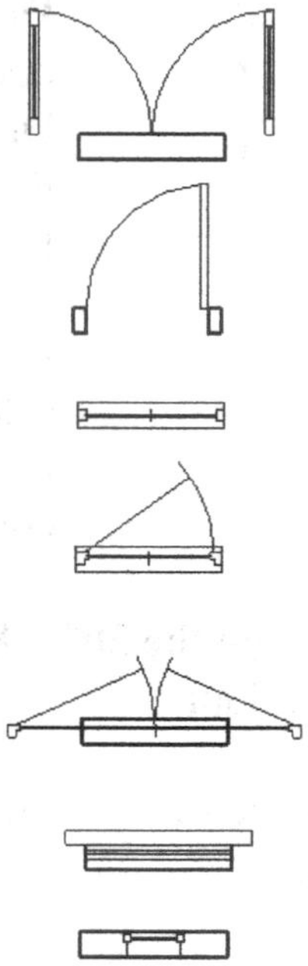

Drag and drop each window symbol into the Legend view. *Symbols can be dragged and dropped from the Project Browser or using the Legend Component tool.*

12.

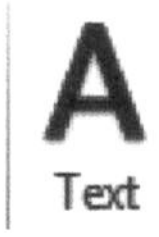

A

Text

Activate the **Annotate** ribbon.
Select the **Text** tool.

13.

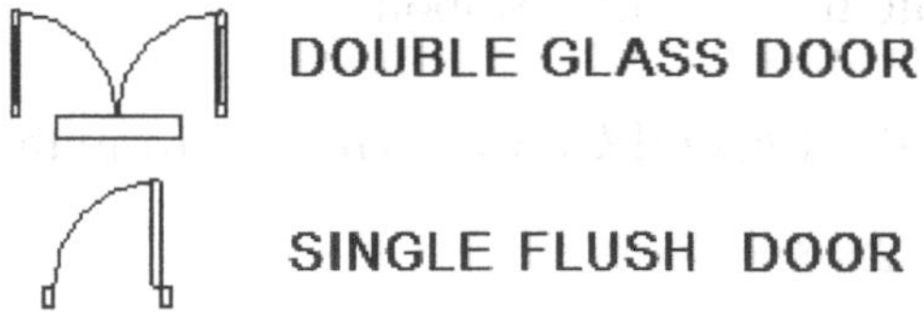

DOUBLE GLASS DOOR

SINGLE FLUSH DOOR

Add text next to each symbol.

14.

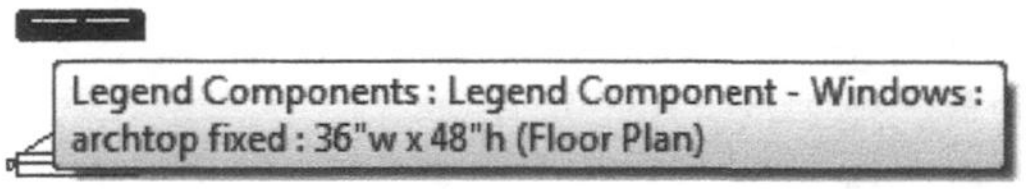

Legend Components : Legend Component - Windows :
archtop fixed : 36"w x 48"h (Floor Plan)

Mouse over a symbol to see what family type it is, if needed.

15.
DOUBLE GLASS DOOR
SINGLE FLUSH DOOR
ARCHTOP FIXED
CASEMENT
DOUBLE CASEMENT WITH TRIM
DOUBLE HUNG WITH TRIM
FIXED

Add text as shown.

16.
Detail
Line

Select the **Detail Line** tool from the Detail panel on the Annotate ribbon.

17.
SYMBOL
DESCRIPTION
DOUBLE GLASS DOOR
SINGLE FLUSH DOOR
ARCHTOP FIXED
CASEMENT
DOUBLE CASEMENT WITH TRIM
DOUBLE HUNG WITH TRIM
FIXED

Add a rectangle.
Add a vertical and horizontal line as shown.
Add the header text.

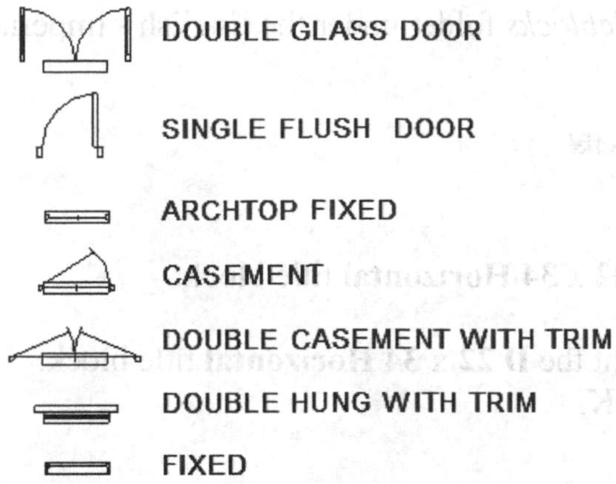

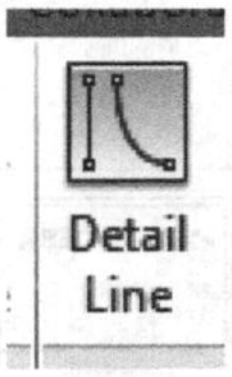

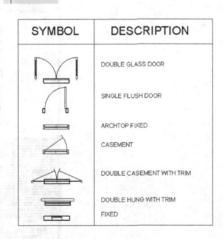

18.
Schedules/Quantities
Sheets (all)
A100
Familie
Anno
Cable
Ceiling
New Sheet...
Browser Organization...
Search...

Locate the **Sheets** category in the browser.
Right click and select **New Sheet**.

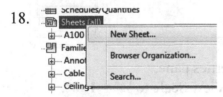

19.
Load...

Select **Load**.

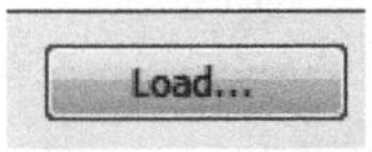

20. 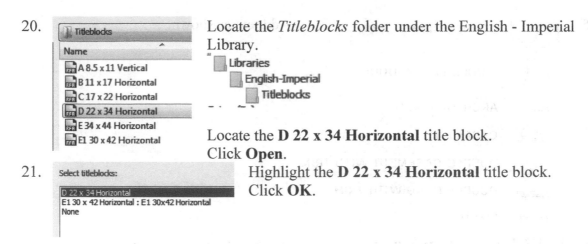 Locate the *Titleblocks* folder under the English - Imperial Library.

Locate the **D 22 x 34 Horizontal** title block.
Click **Open**.

21. Highlight the **D 22 x 34 Horizontal** title block.
Click **OK**.

22. Drag and drop the Level 1 floor plan onto the sheet.

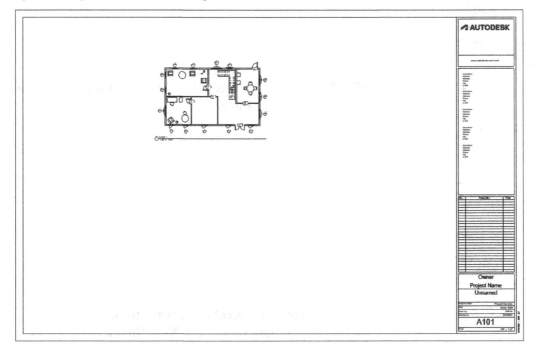

23.

Graphics	
View Scale	1/4" = 1'-0"
Scale Value 1:	48
Display Model	Normal

In the Properties pane,
change the View Scale to ¼″ = 1′-0″.

24. Drag and drop the Door and Window Legend onto the sheet.
Place next to the floor plan view.

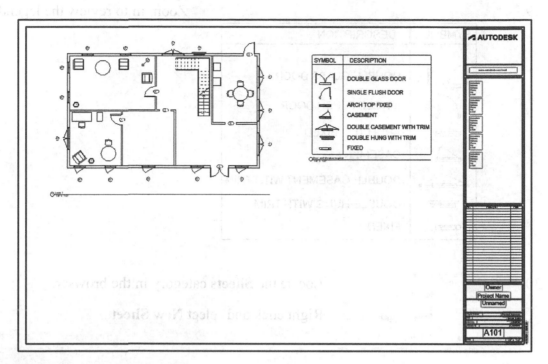

25. Select the legend so it highlights.

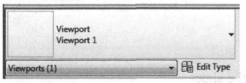

Select **Edit Type** on the Properties pane.

26. Duplicate...   Select **Duplicate**.

27. Name: Viewport with no Title   Type **Viewport with no Title**. Click **OK**.

28.

Type Parameters	
Parameter	
Graphics	
Title	view title
Show Title	No
Show Extension Line	☐
Line Weight	1
Color	■ Black
Line Pattern	Solid

Uncheck **Show Extension Line**.
Set Show Title to **No**.
Click **OK**.

29.

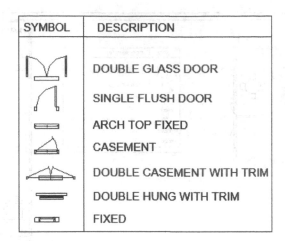

Zoom in to review the legend.

SYMBOL	DESCRIPTION
	DOUBLE GLASS DOOR
	SINGLE FLUSH DOOR
	ARCH TOP FIXED
	CASEMENT
	DOUBLE CASEMENT WITH TRIM
	DOUBLE HUNG WITH TRIM
	FIXED

30.

Locate the **Sheets** category in the browser.

Right click and select **New Sheet**.

31.

Highlight the **D 22 x 34 Horizontal** title block.

Click **OK**.

Select titleblocks:

D 22 x 34 Horizontal
E1 30 x 42 Horizontal : E1 30x42 Horizontal
None

32. Drag and drop the **Level 2** floor plan onto the sheet.

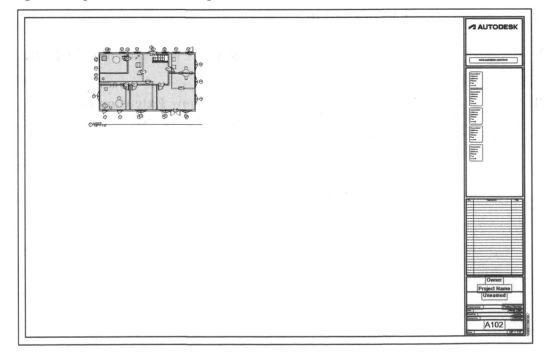

33.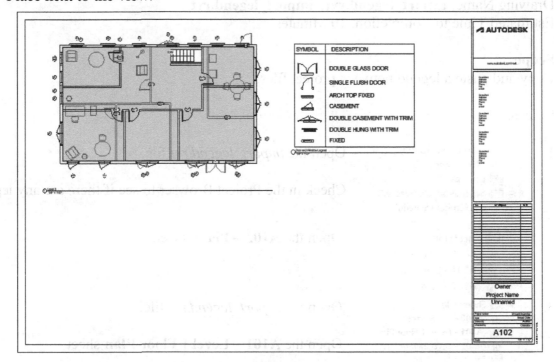

In the Properties pane,
change the View Scale to **¼″ = 1′-0″**.

34. Drag and drop the Door and Window Legend onto the sheet.
Place next to the view.

35. Select the legend view and then use the type selector in the Properties pane to set
whether the view should have a title or not.

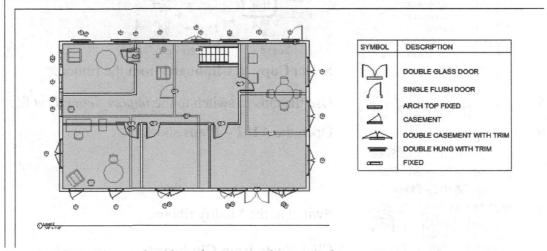

Note that legends can be placed on more than one sheet.

36. Save as *ex4-21.rvt*.

Exercise 4-22
Import a Legend

Drawing Name: **export_legend.rvt, import_legend.rvt**
Estimated Time to Completion: 10 Minutes

Scope
Copy and paste a legend between project files.

Solution

1.
   - Detail Views (Detail)
   - Renderings
   - Legends
   - Schedules/Quantities (all)
     - Furniture Schedule

   Open the *import_legend.rvt* file.

   Check in the Project Browser to see if there are any legends.

2.
   - Sheets (all)
     - A001 - Title Sheet
     - A101 - Site Plan
     - **A102 - Plans**

   Open the **A102 – Plans** sheet.

3.
   - Sheets (all)
     - A100 - Cover Sheet
     - **A101 - Level 1 Floor Plan**
     - A102 - Level 2 Floor Plan

   Open the *export_legend.rvt* file.

   Open the **A101 – Level 1 Floor Plan** sheet.

4.

   Select the legend view.

   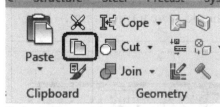

   Select **Copy to Clipboard** from the ribbon.

5. 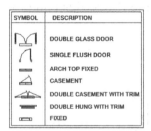   Use the tabs to switch to the *import_legend.rvt* file.

6.
   - Sheets (all)
     - A001 - Title Sheet
     - A101 - Site Plan
     - **A102 - Plans**

   Open the **A102 – Plans** sheet.

7.

   Switch to the Modify ribbon.

   Click **Paste from Clipboard**.

8.   Click **OK**.

9.   Place the legend view on the sheet.

10.   Notice that the legend is now available in the Project Browser as well.

11.   Save as *ex4-22.rvt*.

Grids

Grids are system families. Grid lines are finite planes. Grids can be straight lines, arcs, or multi-segmented.

Revit automatically numbers each grid. To change the grid number, click the number, enter the new value, and Click ENTER. You can use letters for grid line values. If you change the first grid number to a letter, all subsequent grid lines update appropriately. Each grid ID must be unique. If you have already assigned an ID, it cannot be used on another grid.

As you draw grid lines, the heads and tails of the lines can align to one another. If grid lines are aligned and you select a line, a lock appears to indicate the alignment. If you move the grid extents, all aligned grid lines move with it.

Grids are Annotation elements. But, unlike most annotation elements, they DO appear across different views. For example, you can draw a grid on your ground floor plan, and it would then appear on the subsequent floors (levels) of your model. You can control the display of grids on different levels using a scope box. Grids are datum elements.

A grid line consists of two main parts: the grid line itself and the Grid Header (i.e. the bubble at the end of the grid line). The default setting is for the grid line to have a grid header at one end only.

On the Professional exam, expect a grid question relating to how to control the view display of grids using 2D extents, 3D extents and/or scope boxes.

Exercise 4-23

Create and Modify Grids

Drawing Name: **grids.rvt**
Estimated Time to Completion: 20 Minutes

Scope
Create and Modify Grids

Solution

1. Activate the **Level 1** floor plan.

2. Select the **Grid** tool from the Architecture ribbon.

3.

Draw a grid on the left side of the display.

Left pick on the bottom of the screen to start the grid line.

Move the mouse up.

Left pick to place the end point of the grid line.

Cancel out of the command.

Notice that by default the grid line only displays a bubble on one end.

4. 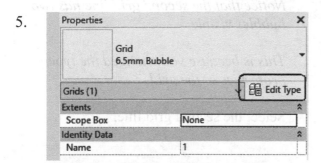 Select the grid that was just placed.

A small box appears at the end of the grid line.

Left click inside the box.

This enables the visibility of the grid bubble.

5. 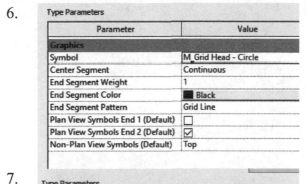 Select the grid line.
On the Properties panel, select **Edit Type**.

6. Study the properties that are controlled by the grid type.

Parameter	Value
Graphics	
Symbol	M_Grid Head - Circle
Center Segment	Continuous
End Segment Weight	1
End Segment Color	■ Black
End Segment Pattern	Grid Line
Plan View Symbols End 1 (Default)	☐
Plan View Symbols End 2 (Default)	☑
Non-Plan View Symbols (Default)	Top

7.

Parameter	Value
Graphics	
Symbol	M_Grid Head - Circle
Center Segment	Custom
Center Segment Weight	1
Center Segment Color	■ Green
Center Segment Pattern	Long Dash
End Segment Weight	1
End Segment Color	■ Black
End Segment Pattern	Grid Line
End Segments Length	0.0250
Plan View Symbols End 1 (Default)	☑
Plan View Symbols End 2 (Default)	☑
Non-Plan View Symbols (Default)	Top

Change the Center Segment to **Custom**.
Set the Center Segment Color to **Green**.
Set the Center Segment Pattern to **Long Dash**.
Enable Plane View Symbols End 1.
Enable Plane View Symbols End 2.

Click **OK**.

Left click in the window to release the selection.

Notice how the appearance of the grid line changes.

8. 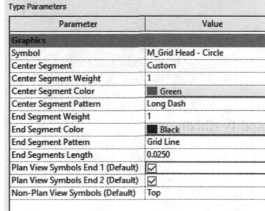 Select the **Grid** tool from the Architecture ribbon.

9.

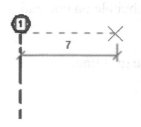

Start a second grid line 7 m to the right of the first grid line.

Use Object Tracking and the Temporary dimension to help you locate the grid line.

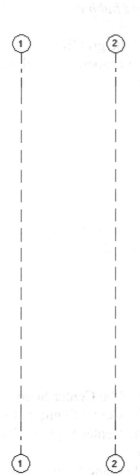

Notice that the second grid line has two bubbles visible.

This is because you changed the type properties of the grid.

Select the second grid line.

10. Select the **Copy** tool from the Modify ribbon.

11.

On the Options bar:

Enable **Constrain**.
Enable **Multiple**.

Constrain is similar to using ORTHO mode in AutoCAD. It constrains movement in the horizontal/vertical direction.

Multiple allows you to place more than one copy.

12. Select the endpoint below the top bubble on the second grid line.

13.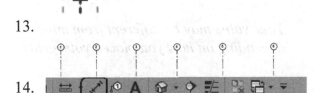

Place four more grids to the right of the existing grids. Don't worry about the dimensions yet.

14.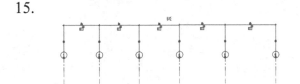

Select the **Aligned Dimension** tool from the Quick Access toolbar located at the top of the window.

15. Select each grid, starting with grid 1.

Then left click above the grids to place the dimension.

This creates a multi-segmented dimension, also known as a dimension string.

16. EQ Left click on the EQ symbol displayed above the dimension.

This toggles the dimension string to set the distance between the grids as equal. This is considered a constraint in Revit.

Click ESC to exit the dimension command or right click and select CANCEL.

17.

Linear Dimension Style
Diagonal - 2.5mm Arial

Dimensions (1)		Edit Type
Graphics		
Leader	☑	
Baseline Offset	0.0000 mm	
Other		
Label	<None>	
Equality Display	Equality Text	

Value
Equality Text
Equality Formula

Select the multi-segmented dimension.

On the Properties palette, you can set the Equality Display to Value, Equality Text, or Equality Formula.

Select **Value** from the drop-down.

The dimension now displays a numerical value.

18.

Select Grid 2.

Select the dimension value located between grid 1 and grid 2.

Your value may be different from mine depending on how you placed your grids.

19.

Change the value to **7.**

Left click anywhere in the display to release the selection.

20.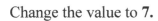

Notice that all the dimensions update as they have been set equal.

21.

1 : 100

Click on the **Reveal Constraints** tool located on the display bar at the bottom of the window.

22. 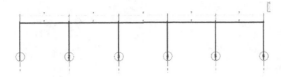 The dimensions are displayed as red and bold.

This is because they were defined using an EQ (equal) constraint and are set to always be equal.

23. Click on the Reveal Constraints tool located on the display bar at the bottom of the window to toggle the display off.

24. Save as *ex4-23.rvt*.

Levels

Levels are finite horizontal planes that act as a reference for level-hosted elements, such as roofs, floors, and ceilings. You can resize their extents so that they do not display in certain views.

You can modify level type properties, such as Elevation Base and Line Weight, in the Type Properties dialog.

If the elevation base is set as the project base point, the level elevation will be displayed to the Origin 0,0. If it is set as the survey point, the level will be displayed according to the defined relative coordinates. (This is a possible question on the Professional exam.)

Modify instance properties to specify the level's elevation, computation height, name, and more.

On the certification exam, you may need to identify which level properties are instance properties and which level properties are type properties.

Levels are finite horizontal planes that act as a reference for level-hosted elements, such as roofs, floors, and ceilings.

Create a level for each known story or other needed reference of the building (for example: first floor, top of wall, or bottom of foundation).

To add levels, you must be in a section or elevation view. When you add levels, you can create an associated plan view.

You can resize the extents of a level so that they do not display in certain views.

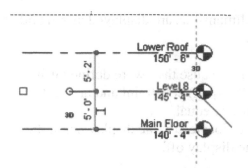

Levels which are blue are story levels. They have floor and ceiling plan views associated to the levels. Levels displayed in black are reference levels and have no associated views.

You should be able to identify the different level components and properties. There will be one question regarding levels win the Professional exam.

You also should be able to identify the components of a level element.

For example, the elevation dimension shown is a permanent dimension when the level is not selected because it is always displayed as long as the level is displayed. If the level is selected, the dimension displayed is a temporary dimension and can be modified.

Exercise 4-24

Create and Modify Levels

Drawing Name: **i_levels.rvt**
Estimated Time to Completion: 5 Minutes

Scope
Placing a level.

Solution

1.   Elevations (Building Elevation)
   - East
   - North
   - **South**
   - West

   Activate the **South Elevation**.

   The level names have been turned off.

2.   Main Floor
   140' - 4"

   Select each level and place a check in the square that appears. This will turn on visibility of the level name.

3.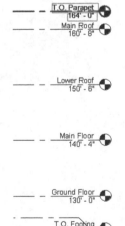

   You should be able to identify the names for each level.

   T.O. Parapet
   164' - 0"

   Main Roof
   160' - 8"

   Lower Roof
   150' - 6"

   Main Floor
   140' - 4"

   Ground Floor
   130' - 0"

   T.O. Footing
   126' - 0"

4. 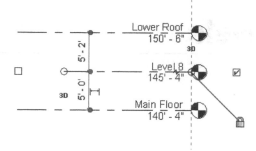 Select the **Level** tool from the Architecture ribbon.

5. Place a level **5'-0"** above the Main Floor.

6.  Note the elevation value for the new level.

7. Close without saving.

Exercise 4-25

Story vs. Non-Story Levels

Drawing Name: **story_levels.rvt**
Estimated Time to Completion: 15 Minutes

Scope
Understanding the difference between story and non-story levels
Converting a non-story level to a story level

Solution

1. 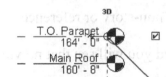   Activate the **South Elevation**.

2.   Select each level and place a check in the square that appears. This will turn on visibility of the level name.

3.   Study the Main Floor level. Notice that it is the color black while all the other levels are blue.

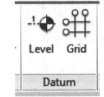

   The Main Floor level is a non-story or reference level. It does not have a view associated with it.

4.   Activate the Architecture ribbon.
   Select the **Level** tool on the Datum panel.

5.    On the Options bar: Uncheck **Make Plan View**. Set the Offset to **8' 0"**.

6. 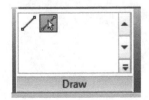   Select the **Pick** tool on the Draw panel.

7.

Select the Main Floor level.
Verify that the preview shows the level will be placed 8' 0" ABOVE the Main Floor level.

8.

The level is placed above the Main Floor.
Right click and select Cancel twice to exit the Level command.

9.

Select the Elbow control on the new level to add a jog.

10.

Note that the new level is also a non-story or reference level.
Check in the Project Browser and you will see that no views were created with the new level.

11.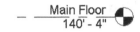

Activate the View ribbon.
Select the **Plan Views→Floor Plan** tool on the Create panel.

12.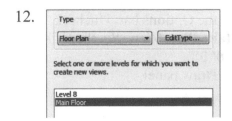

The reference levels are listed.
Select the **Main Floor** level and click **OK**.

13.  The Main Floor floor plan view will open.
Note that the Main Floor floor plan is now listed in the Project Browser; however, there is no ceiling plan for the Main Floor.

14. 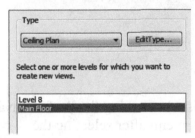 Activate the View ribbon.
Select the **Plan Views→Reflected Ceiling Plan** tool on the Create panel.

15. The reference levels are listed.
Select the **Main Floor** level and click **OK**.

16. The Main Floor ceiling plan view will open.

17. Activate the **South Elevation**.

18. 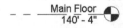 The Main Floor level is now the color Blue to indicate it has story views.

19. Close the file without saving.

Level and Grid Extents

All levels and grids (also known as datums) have 3D and 2D extents. Datums are Revit elements used as references while modeling. They are finite representations of infinite planes (vertical – grids, horizontal – levels) that are displayed as lines. They are not model elements. They are considered a special category, listed under annotation elements. Unlike other annotation elements, they are not view-specific.

If the 3D extents of the grids don't cross the entire elevation/section/3D view, they will not be displayed.

There are three functions or commands which allow you to control the extents of levels or grids:

- Maximize 3D Extents
- Reset 3D Extents
- Propagate Extents

You need to be familiar with all three of these commands when taking the Professional exam. All three functions are available on the right click menu after selecting the grid/level/section line.

The Maximize 3D Extents function expands the extents of the grid/level/section to the full boundaries of your model.

You cannot reset to 3D extents for a level if crop view is enabled and the level endpoints are not inside the crop area. To use the Reset function, toggle crop view off, use the Reset to 3D Extents and then enable crop view.

The Propagate Extents tool pushes any modifications you apply to a datum object from one view to other parallel views of your choosing. This tool does not work well on levels because the parallel views are essentially mirrored views of each other. For example, the orientation of the South elevation is the opposite of the North elevation; therefore, if you make a change to the extents at the right end of a level in the South elevation, those changes would be propagated to the left end in the North elevation.

The best way to apply the Propagate Extents tool is with the 2D extents of grids. Why only the 2D extents? Because changing the 3D extents affects the datum object throughout the project, independent of any specific view. The 2D extents controls the display of the datum line while 3D extents controls whether the line will appear in other referring views. When you need to adjust the level line in a specific view, but not in the entire model, the level should be set to 2D extents.

Exercise 4-26

Level and Grid Extents

Drawing Name: **datum_extents.rvt**
Estimated Time to Completion: 30 Minutes

Scope
Use Maximize 3D Extents
Use 2D Extents
Use Propagate Extents

Solution

1. Elevations (Building Elevation)
   East
   North
   South
   West

   Activate the **South Elevation**.

2.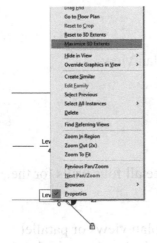

   Select Level 1.

   Right click and select **Maximize 3D Extents**.

3.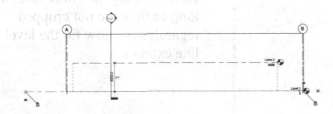

   Notice how the level extends past the building.

   Notice how the bubble at the end of Level 1 is unfilled and it indicates 3D.

4. Switch to a 3D view.

5. Change the display to Wireframe.

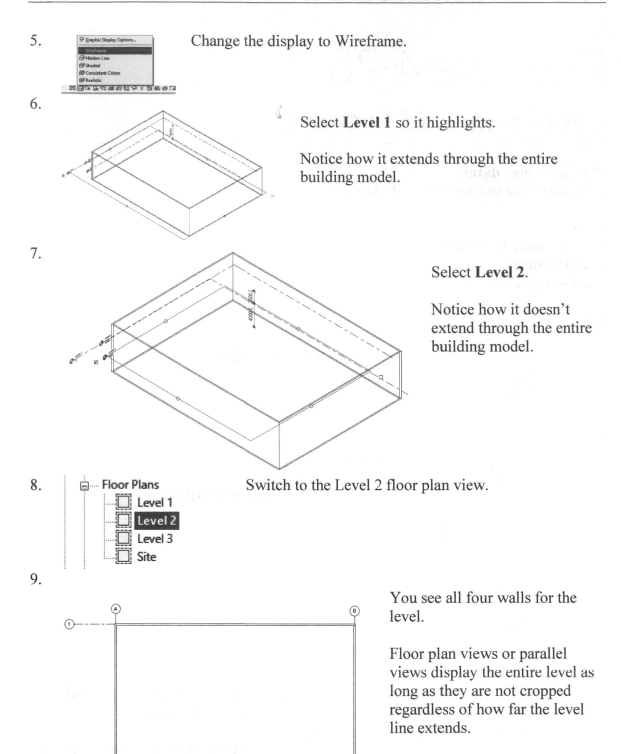

6. Select **Level 1** so it highlights.

Notice how it extends through the entire building model.

7. Select **Level 2**.

Notice how it doesn't extend through the entire building model.

8. Switch to the Level 2 floor plan view.

- Floor Plans
  - Level 1
  - Level 2
  - Level 3
  - Site

9. You see all four walls for the level.

Floor plan views or parallel views display the entire level as long as they are not cropped regardless of how far the level line extends.

10. 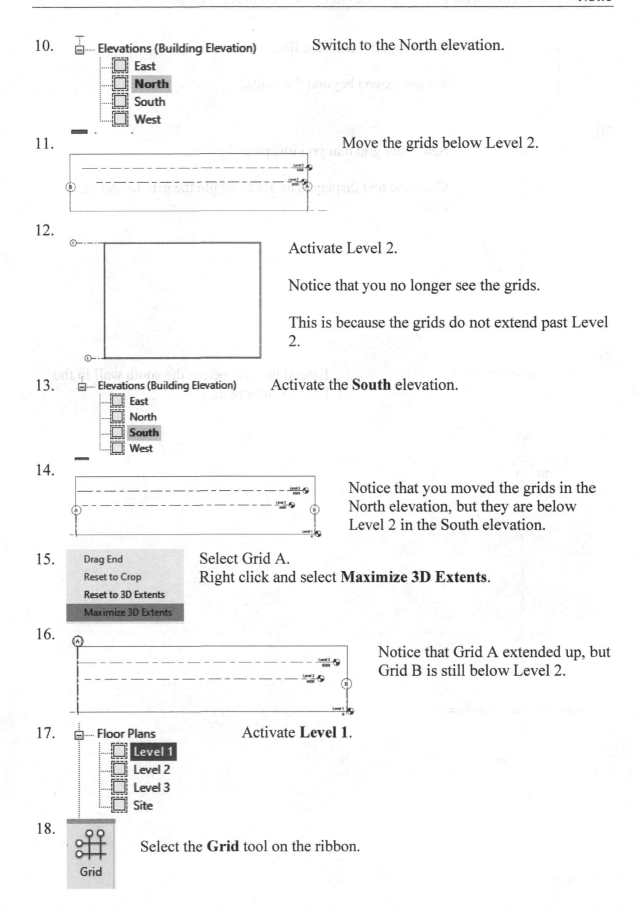 Switch to the North elevation.

     Elevations (Building Elevation)
        East
        North
        South
        West

11.   Move the grids below Level 2.

12. Activate Level 2.

Notice that you no longer see the grids.

This is because the grids do not extend past Level 2.

13. Activate the **South** elevation.

     Elevations (Building Elevation)
        East
        North
        South
        West

14. Notice that you moved the grids in the North elevation, but they are below Level 2 in the South elevation.

15. Select Grid A.
Right click and select **Maximize 3D Extents**.

    Drag End
    Reset to Crop
    Reset to 3D Extents
    Maximize 3D Extents

16. Notice that Grid A extended up, but Grid B is still below Level 2.

17. Activate **Level 1**.

     Floor Plans
        Level 1
        Level 2
        Level 3
        Site

18. Select the **Grid** tool on the ribbon.

    Grid

19.

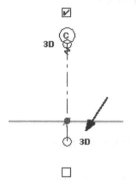

Draw a grid line inside the room.

Do not extend beyond the walls.

20.

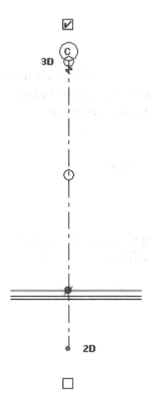

Select the grid that you just placed.

Click the text displayed as 3D to toggle the grid to use 2D extents.

21.

Extend the grid below the south wall in the Level 1 floor plan.

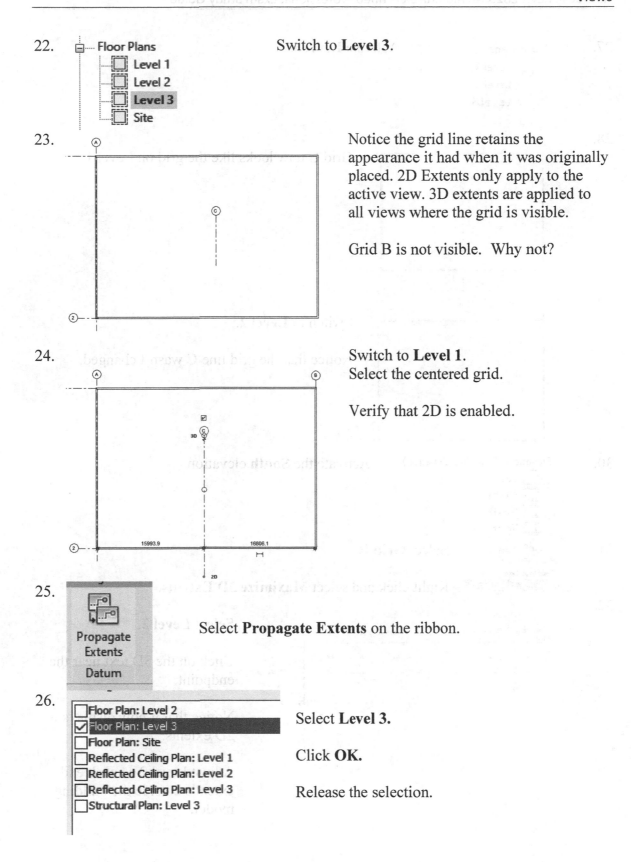

22. Floor Plans
    Level 1
    Level 2
    Level 3
    Site

Switch to **Level 3**.

23. Notice the grid line retains the appearance it had when it was originally placed. 2D Extents only apply to the active view. 3D extents are applied to all views where the grid is visible.

Grid B is not visible. Why not?

24. Switch to **Level 1**.
Select the centered grid.

Verify that 2D is enabled.

15993.9   16806.1

25. Propagate Extents Datum

Select **Propagate Extents** on the ribbon.

26. Floor Plan: Level 2
    ☑ Floor Plan: Level 3
    Floor Plan: Site
    Reflected Ceiling Plan: Level 1
    Reflected Ceiling Plan: Level 2
    Reflected Ceiling Plan: Level 3
    Structural Plan: Level 3

Select **Level 3.**

Click **OK.**

Release the selection.

27. 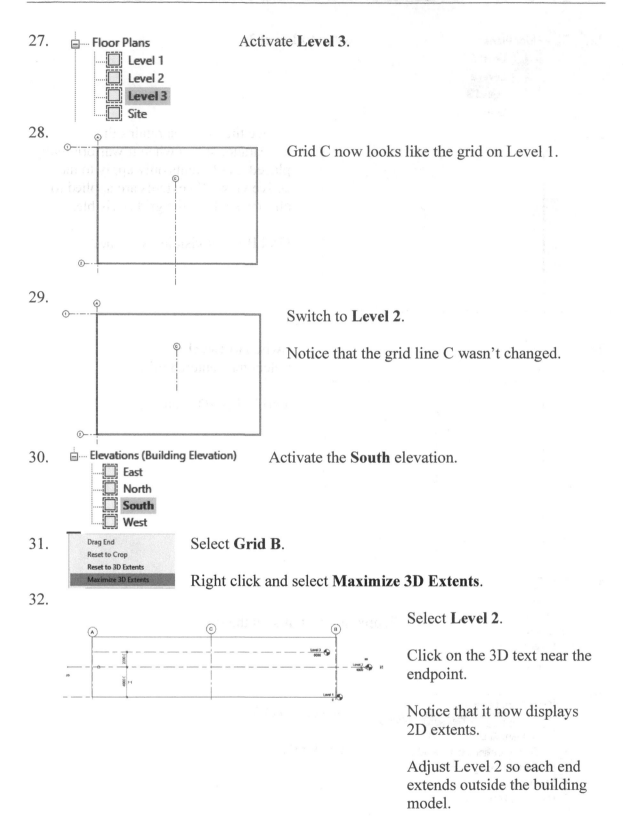 Activate **Level 3**.

28. Grid C now looks like the grid on Level 1.

29. Switch to **Level 2**.

   Notice that the grid line C wasn't changed.

30. Activate the **South** elevation.

31. Select **Grid B**.

   Right click and select **Maximize 3D Extents**.

32. Select **Level 2**.

   Click on the 3D text near the endpoint.

   Notice that it now displays 2D extents.

   Adjust Level 2 so each end extends outside the building model.

33.

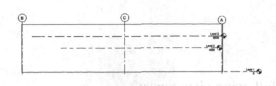

Switch to the **North** elevation.

Notice that Level 2 looks the same as it did before you extended it using 2D extents in the South elevation.

34. Save the file as *ex4-26.rvt*.

Practice Exam

1. Straight grid lines are visible in the following view types:

   A. ELEVATION
   B. PLAN
   C. 3D
   D. SECTION
   E. DETAIL

2. Which two shortcut keys launch the Visibility/Graphics dialog?

   A. VG
   B. VV
   C. VE
   D. VW
   E. F5

3. **True or False**: Objects hidden using the Temporary Hide/Isolate tool are not visible, but they are still printed.

4. Select the THREE options for Detail Level for a view:

   A. COARSE
   B. SHADED
   C. FINE
   D. HIDDEN
   E. WIREFRAME
   F. MEDIUM

5. **True or False**: If you delete a view, the annotations placed in the view are also deleted.

6. **True or False**: A camera cannot be placed in an elevation view.

7. To change the graphic appearance of your model from Hidden Line to Realistic, you:

   A. Modify the Rendering Settings in the View Control Bar
   B. Edit Visibility/Graphics Overrides
   C. Change graphic display options in View Properties
   D. Click Visual Styles on the View Control Bar

8. A story level is the color:

    A. Yellow
    B. Blue
    C. Black
    D. Green

9. If you create a level using the COPY or ARRAY tool, what type of level is created?

    A. Story
    B. Non-Story or Reference
    C. Elevation
    D. Plan

10. This type of level does not have a PLAN view associated to it:

    A. Story
    B. Non-Story or Reference
    C. Elevation
    D. Plan

11. In which view type can you place a level?

    A. PLAN
    B. LEGEND
    C. 3D
    D. CONSTRUCTION
    E. ELEVATION

12. To display or open a view: (Select all valid answers)

    A. Double click the view name in the project browser
    B. Double click the elevation arrowhead, the section head or the callout head
    C. Right click the elevation arrowhead, the section head or the callout head, and select Go to View
    D. Select the Surf tool from the ribbon

13. Select the icon used to lock a 3D view.

14. Scope boxes control the visibility of:

    A. Elements
    B. Plumbing Fixtures
    C. Object Styles
    D. Grid lines and levels

15. If you rotate a cropped view: (Select the answer which is TRUE)

   A. Any tags will automatically rotate with the view.
   B. The crop region will automatically resize.
   C. The Project North will change.
   D. Any hidden elements will become visible.

16. Depth Cueing applies to all the elements listed EXCEPT: (Select two answers)

   A. Model elements, like walls, doors, and floors.
   B. Shadows and Sketchy Lines
   C. Annotations, like dimensions and tags
   D. Linework and line weight

17. A _____ is a collection of view properties, such as view scale, discipline, detail level and visibility settings.

   A. View template
   B. Graphics display
   C. Crop Region
   D. Sheet

18. _____ provide(s) a way to override the graphic display and control of the visibility of elements that share common properties in a view.

   A. Hide/Isolate
   B. Hide Category
   C. Hide Element
   D. Filters

19. Select the Type properties for a Level. (Select all correct answers)

   A. Elevation Base
   B. Line Weight
   C. Elevation
   D. Name

20. Select the ONE item you cannot create a view type:

    A. Floor Plans
    B. Ceiling Plans
    C. Area Plans
    D. 3D Views

21. A _____ displays the view number and sheet number (if the view is included on a sheet) in the corresponding view.

    A. Title bar
    B. View Name
    C. Sheet
    D. View Reference

Floor Plan: Level 1	
Graphics	
View Scale	1/8" = 1'-0"
Scale Value 1:	96
Display Model	Normal
Detail Level	Coarse
Parts Visibility	Show Original
Detail Number	1
Rotation on Sheet	None

22. The View Scale, Detail Level, and Display is grayed out for the floor plan view. The reason is:

    A. The view is locked
    B. The view is pinned
    C. A view template has been assigned to the view
    D. The view range is applied.

23. T/F
Legends can be placed on more than one sheet.

24. Select the icon used to reveal hidden elements.

25. T/F
You place an annotation in a dependent view. It will not be visible in the parent view.

26. The tool to align views between sheets is called:

A. Align
B. Guide Grid
C. Align Grid
D. Pin

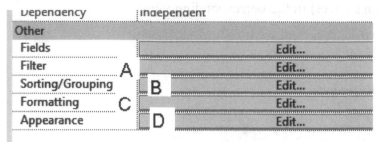

27.
To only see the doors placed on Level 1, which tab should be used?

28. A designer creates a ceiling in the Level 1 floor plan view and receives the following Warning. The Ceilings category is visible in the view. How can the designer view the ceiling?

Warning

None of the created elements are visible in Floor Plan: Level 1 View. You may want to check the active view, its Parameters, and Visibility settings, as well as any Plan Regions and their settings.

A. Select the Automatic Ceiling option when creating the ceiling.
B. Open the Level 1 reflected ceiling plan view.
C. Use the Sketch Ceiling option to model the ceiling.
D. Change the ceiling type that was created.

29. A designer needs to adjust the line weight for all walls in one floor plan view. Which dialog should the designer use?

A. View Specific Element Graphics
B. Graphic Display Options
C. Object Styles
D. Visibility/Graphic Overrides

30. What does the 3D next to the level mean?

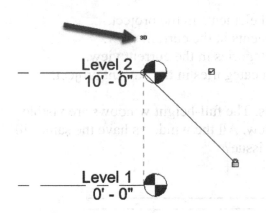

A. Level 2 is locked to Level 1.
B. A Floor Plan view in the model is associated to Level 2.
C. Any changes to the extents of Level 2 will propagate to other views.
D. A 3D view in the model is associated to Level 2.

31. A designer wants to control several view settings with a view template. After assigning a view template to a view, the View Scale can still be adjusted in the View Control Bar. Why is this happening?

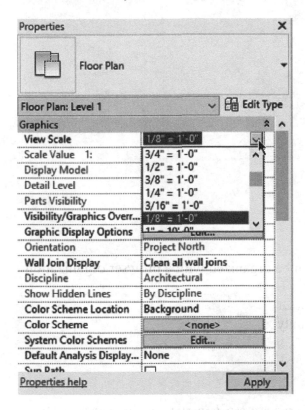

A. Include is not selected for the View Scale parameter in the view template.
B. A plan region is being used to control the View Scale.
C. Override is selected for the View Scale in the Visibility/Graphic Overrides dialog.
D. Override is selected next to View Scale in the Properties palette.

32. What can be accomplished in the Visibility/Graphic Overrides dialog? (Select all correct answers – there is more than one.)

    A. Change the line styles of individual elements in the project.
    B. Change the color of individual elements in the current view.
    C. Control the visibility of element categories in the current view.
    D. Control the line weights of element categories in the current project.

33. A designer is adding windows to exterior walls. The full-height windows are visible, but the smaller windows are not showing in the view. All the windows have the same sill height. What should the designer check to fix this issue?

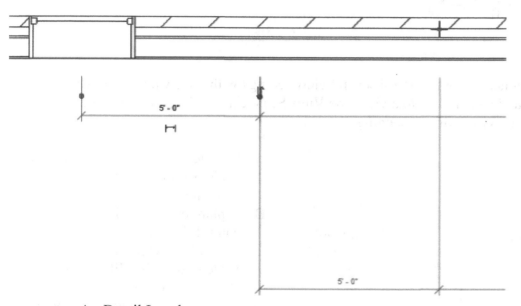

    A. Detail Level
    B. Visual Style
    C. Visibility/Graphic Overrides
    D. View Range

34. Which branches in the Project Browser can be organized with a browser organization scheme?
    A. Any branch in the Project Browser
    B. Views, Sheets, and Schedules only
    C. Views and Sheets only
    D. Views only

Answers:
    1) A, B & D; 2) A & B; 3) True; 4) A, C & F; 5) True; 6) False; 7) C; 8) B; 9) B; 10) B; 11) E; 12) A, B, and C; 13) D
    14) D; 15) A; 16) C & D; 17) A; 18) D; 19) A& B – C & D are instance properties; 20) C; 21) D; 22) C; 23) T; 24) B; 25) F; 26) B; 27) A; 28) B; 29) D; 30) C; 31) A; 32) C & D; 33) D; 34) B

Collaboration and Project Management

This lesson addresses the following certification exam questions:

- Demonstrate how to copy and monitor elements in a linked file
- Apply interference checking in Revit
- Using Shared Coordinates
- Import DWG, PDF, and Image files
- Assess review warnings
- Worksets
- Linked files

In most building projects, you need to collaborate with outside contractors and with other team members. A mechanical engineer uses an architect's building model to lay out the HVAC (heating and air conditioning) system. Proper coordination and monitoring ensure that the mechanical layout is synchronized with the changes that the architect makes as the building develops. Effective change monitoring reduces errors and keeps a project on schedule.

Worksets are used in a team environment when you have many people working on the same project file. The project file is located on a server (a central file). Each team member downloads a copy of the project to their local machine. The person is assigned a workset consisting of building elements that they can change. If you need to change an element that belongs to another team member, you issue an Editing Request which can be granted or denied. Workers check in and check out the project, updating both the local and server versions of the file upon each check in/out.

Project sharing is the process of linking projects across disciplines. You can share a Revit Structure model with an MEP engineer.

You can link different file formats in a Revit project, including other Revit files (Revit Architecture, Revit Structure, Revit MEP), CAD formats (DWG, DXF, DGN, SAT, SKP), and DWF markup files. Linked files act similarly as external references (XREFs) in AutoCAD. You can also use file linking if you have a project which involves multiple buildings.

It is recommended to use linked Revit models for

- Separate buildings on a site or campus.

- Parts of buildings which are being designed by different design teams or designed for different drawing sets.

- Coordination across different disciplines (for example, an architectural model and a structural model).

Linked models may also be appropriate for the following situations:

- Townhouse design when there is little geometric interactivity between the townhouses.

- Repeating floors of buildings at early stages in the design, where improved Revit model performance (for example, quick change propagation) is more important than full geometric interactivity or complete detailing.

You can select a linked project and bind it within the host project. Binding converts the linked file to a group in the host project. You can also convert a model group into a link which saves the group as an external file.

Exercise 5-1

Monitoring a Linked File

Drawing Name: **i_multiple_disciplines.rvt**
Estimated Time to Completion: 40 Minutes

Scope

Link a Revit Structure file.
Monitor the levels in the linked file.
Reload the modified Structure file.
Perform a coordination review.
Create a Coordination Review report.

Solution

1. If you have an Autodesk online Account, the sign in for the user account is used for your user name. If you don't want to use the name displayed by the Autodesk account, you need to change your profile to display a different name. This name is used in comments for revisions, etc.

    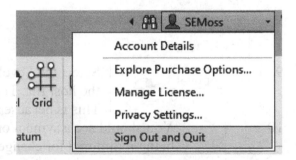

2. Open the *i_multiple_disciplines.rvt* file.

3. 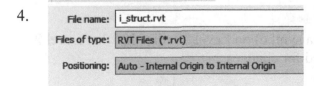 Activate the **Insert** ribbon.
    Select the **Link Revit** tool on the Link panel.

4. File name: i_struct.rvt
    Files of type: RVT Files (*.rvt)
    Positioning: Auto - Internal Origin to Internal Origin

    Locate the *i_struct* file.
    Set the Positioning to **Auto-Internal Origin to Internal Origin**.
    Click **Open**.

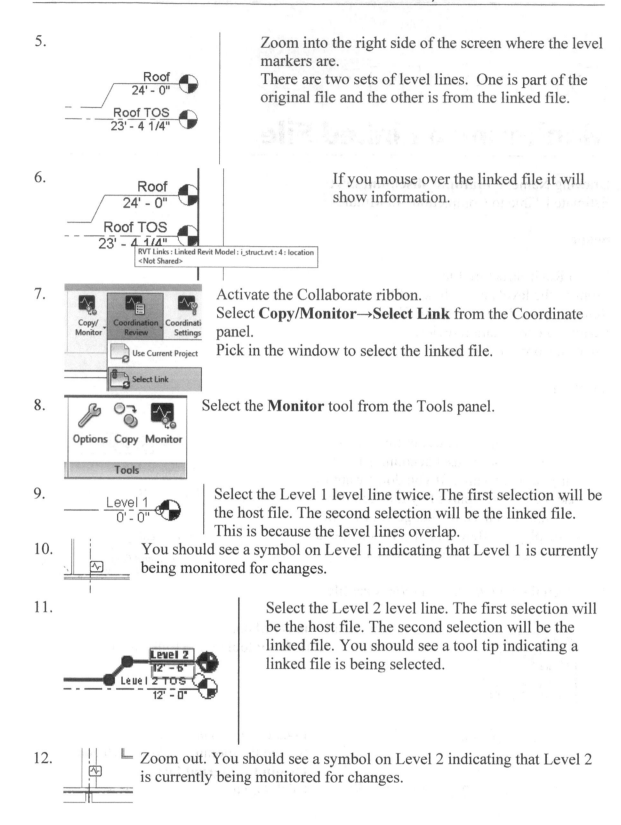

5. Zoom into the right side of the screen where the level markers are.
There are two sets of level lines. One is part of the original file and the other is from the linked file.

6. If you mouse over the linked file it will show information.

RVT Links : Linked Revit Model : i_struct.rvt : 4 : location <Not Shared>

7. Activate the Collaborate ribbon.
Select **Copy/Monitor**→**Select Link** from the Coordinate panel.
Pick in the window to select the linked file.

8. Select the **Monitor** tool from the Tools panel.

9. Select the Level 1 level line twice. The first selection will be the host file. The second selection will be the linked file. This is because the level lines overlap.

10. You should see a symbol on Level 1 indicating that Level 1 is currently being monitored for changes.

11. Select the Level 2 level line. The first selection will be the host file. The second selection will be the linked file. You should see a tool tip indicating a linked file is being selected.

12. Zoom out. You should see a symbol on Level 2 indicating that Level 2 is currently being monitored for changes.

13.

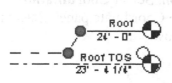

Select the Roof level line.
The first selection will be the host file.
The second selection will be the linked file.
You should see a tool tip indicating a linked file is being selected.

14.

Zoom out. You should see a symbol on the Roof level indicating that it is currently being monitored for changes.

15.

Warning

Elements already monitored

If you try to select elements which have already been set to be monitored, you will see a warning dialog.
Simply close the dialog and move on.

16.

✔ ✖

Finish Cancel

Copy/Monitor

Select **Finish** on the Copy/Monitor panel.

17.

Manage
Links

Activate the **Insert** ribbon.
Select **Manage Links** from the Link panel.

18.

| Revit | CAD Formats | DWF Markups | Point Clouds |

Linked File	Status	Reference Type
i_struct.rvt	Loaded	Overlay

Select the **Revit** tab.
Highlight the *i_struct.rvt* file.

19.

Reload From...

Select **Reload From**.

20.

File name: i_struct_revised

Files of type: RVT Files (*.rvt)

Locate the *i_struct_revised* file.
Click **Open**.

21.

Warning - can be ignored

Instance of link needs Coordination Review

Show   More Info   Expand >>

A dialog will appear indicating that the revised file requires Coordination Review.
Click **OK**.

22.

| Revit | CAD Formats | DWF Markups | Point Clouds |

Linked File	Status	Reference Type
i_struct_revised.rvt	Loaded	Overlay

Note that the revised file has replaced the previous link.
This is similar to when a sub-contractor or other consultant emails you an updated file for use in a project.
Click **OK**.

23. Activate the **Collaborate** ribbon. Select **Coordination Review→Select Link** from the Coordinate panel. Select the linked file in the drawing window.

24. A dialog appears. Expand the notations so you can see what was changed.

Message	Action
New/Unresolved	
Levels	
Maintain Position	
Level moved by 0' - 6"	Postpone
i_struct_revised.rvt : Levels : Level : Roof TOS : id 185763	
Levels : Level : Roof : id 378728	
Level moved by 0' - 6"	Postpone
i_struct_revised.rvt : Levels : Level : Level 2 TOS : id 32567	
Levels : Level : Level 2 : id 9946	

25. Highlight first change. Select **Move Level Roof** from the drop-down list.

Message	Action	Comment
New/Unresolved		
Levels		
Maintain Position		
Level moved by 0' - 6"	Postpone	Add comment
i_struct_revised.rvt : Levels : Level : Roof TOS : id 185763	Postpone / Reject / Accept difference / Move Level 'Roof'	
Levels : Level : Roof : id 378728		
Level moved by 0' - 6"		

26. Select the **Add Comment** button.

Levels		
Maintain Position		
Level moved by 0' - 6"	Move Level 'Roof'	Add comment
i_struct_revised.rvt : Levels : Level : Roof TOS : id 185763		

27. Type **Approved**. Click **OK**.

Edit Comment

Approved

Note that Revit automatically adds your user name and a time stamp.

Comment

Approved smoss@peralta.edu, 2/18/2021 9:36:44 AM

28. Highlight the second change. Select **Move Level 'Level 2'**.

Levels	
Maintain Position	
Level moved by 0' - 6"	Move Level 'Roof'
i_struct_revised.rvt : Levels : Level : Roof TOS : id 185763	
Levels : Level : Roof : id 378728	
Level moved by 0' - 6"	Postpone
i_struct_revised.rvt : Levels : Level : Level 2 TOS : id 32567	Postpone / Reject / Accept difference / Move Level 'Level 2'
Levels : Level : Level 2 : id 9946	

29. Select the **Add Comment** button.

Add comment

30. 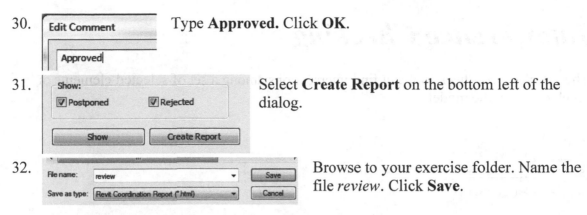 Type **Approved.** Click **OK**.

31. Select **Create Report** on the bottom left of the dialog.

32. 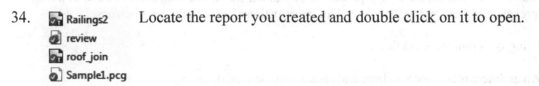 Browse to your exercise folder. Name the file *review*. Click **Save**.

33. Click **OK** to close the Coordination Review dialog box.

    If you close the Coordination Review dialog before you create the report and then re-open it to create the report, the report will be blank because all the issues will have been resolved.

34. Locate the report you created and double click on it to open.

    Railings2
    review
    roof_join
    Sample1.pcg

35. This report can be emailed or used as part of the submittal process.

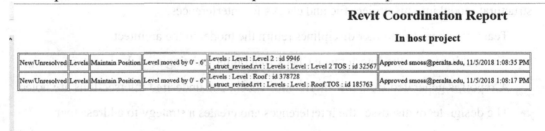

				Revit Coordination Report	
				In host project	
New/Unresolved	Levels	Maintain Position	Level moved by 0' - 6"	Levels : Level : Level 2 : id 9946 i_struct_revised.rvt : Levels : Level : Level 2 TOS : id 32567	Approved smoss@peralta.edu, 11/5/2018 1:08:35 PM
New/Unresolved	Levels	Maintain Position	Level moved by 0' - 6"	Levels : Level : Roof : id 378728 i_struct_revised.rvt : Levels : Level : Roof TOS : id 185763	Approved smoss@peralta.edu, 11/5/2018 1:08:17 PM

36. Close the file without saving.

Interference Checking

The Interference Check tool can find intersections among a set of selected elements or all elements in the model.

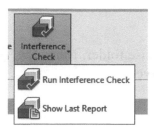

Typical Workflow for Interference Checking

This tool can be used during the design process to coordinate major building elements and systems. Use it to prevent conflicts and reduce the risk of construction changes and cost overruns.

The following is a common workflow:

- An architect meets with a client and creates a basic model.

- The building model is sent to a team that includes members from other disciplines, such as structural engineers. They work on their own version of the model. Then the architect links the structural model into his project file and checks for interferences.

- Team members from other disciplines return the model to the architect.

- The architect runs the Interference Check tool on the existing model.

- A report is generated from the interference check, and undesired intersections are noted.

- The design team discusses the interferences and creates a strategy to address them.

- One or more team members are assigned to fix any conflicts.

Elements Requiring Interference Checking

Some examples of elements that could be checked for interference include the following:

- Structural girders and purlins

- Structural columns and architectural columns

- Structural braces and walls

- Structural braces, doors, and windows

- Roofs and floors

- Specialty equipment and floors

- A linked Revit model and elements in the current model

Exercise 5-2
Interference Checking

Drawing Name: **c_interference_checking.rvt**
Estimated Time to Completion: 15 Minutes

Scope
Describe Interference Checks.
Check and fix interference conditions in a building model.
Generate an interference report.

Solution

1. If this dialog comes up when you open the file, click **OK**.

2. Activate the **Collaborate** ribbon.
    Select the **Interference Check→ Run Interference Check** tool on the Coordinate panel.

3. 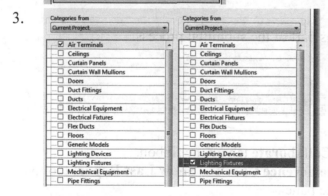 Enable **Air Terminals** in the left panel.
    Enable **Lighting Fixtures** in the right panel.

If you select all in both panels, the check can take a substantial amount of time depending on the project and the results you get may not be very meaningful.

4. Click **OK**.

5. Highlight the Lighting Fixture in the first error.

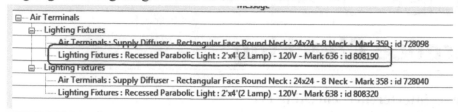

6. **Show**   Click the **Show** button.

7. 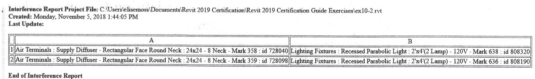 The display will update to show the interference between the two elements.

8. Show   Export...   Refresh   Select **Export**.

9. File name: interference-check
   Files of type: Revit Interference Report (*.html)
   Save

   Browse to your exercise folder.
   Name the file **interference check.**

   Click **Save**.

10. Click **Close** to close the dialog box.

11. in_place_mass
    interference-check
    Lincoln Memorial

    Locate the report you created and double click on it to open.

12. This report can be emailed or used as part of the submittal process.

    Interference Report Project File: C:\Users\elisemoss\Documents\Revit 2019 Certification\Revit 2019 Certification Guide Exercises\ex10-2.rvt
    Created: Monday, November 5, 2018 1:44:05 PM
    Last Update:

	A	B
1	Air Terminals : Supply Diffuser - Rectangular Face Round Neck : 24x24 - 8 Neck - Mark 358 : id 728040	Lighting Fixtures : Recessed Parabolic Light : 2'x4'(2 Lamp) - 120V - Mark 638 : id 808320
2	Air Terminals : Supply Diffuser - Rectangular Face Round Neck : 24x24 - 8 Neck - Mark 359 : id 728098	Lighting Fixtures : Recessed Parabolic Light : 2'x4'(2 Lamp) - 120V - Mark 636 : id 808190

    End of Interference Report

13. Interference Check
    Run Interference Check
    Show Last Report

    Activate the **Collaborate** tab on the ribbon.
    Select the **Interference Check→ Show Last Report** tool on the Coordinate panel.

14. Highlight the first lighting fixture. Verify that you see which fixture is indicated in the graphics window. Close the dialog.

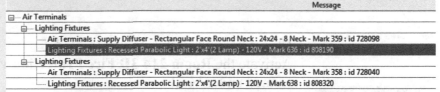

15. 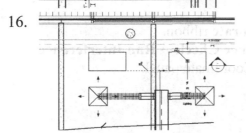 Move the lighting fixture to a position above the air terminal.
Arrange the lighting fixtures and the air diffuser so there should be no interference.

16. Move the second lighting fixture on the right so it is no longer on top of the air terminal.
Rearrange the fixtures in the room to eliminate any interference.

17. 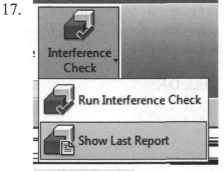 Activate the **Collaborate** ribbon.
Select the **Interference Check→ Show Last Report** tool on the Coordinate panel.

18. Select **Refresh**.

19. The message list is now empty. Close the dialog box.

20. 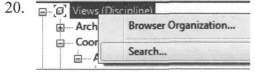 Highlight **Views** in the Project Browser.
Right click and select **Search**.

21. **Find:** Room 214

Next Previous

☐ Match case

Type **Room 214**.
Click **Next**.

22. ⊟ Mechanical
    ⊟ FP
        ⊟ 3D Views
            Room 214 3D Fire Protection

Click **Close** when the view is located.
Activate the **Room 214 3D Fire Protection view** located under Mechanical.

23.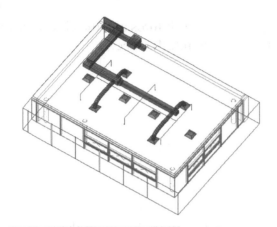

Zoom in so you can see the room.
Window around the entire room so everything is selected.

24. Interference Check

    Run Interference Check

Activate the **Collaborate** ribbon.
Select the **Interference Check→ Run Interference Check** tool on the Coordinate panel.

25.
Categories from
Current Selection

☑	Air Terminals
☑	Duct Fittings
☑	Ducts
☑	Flex Ducts
☑	Mechanical Equipment
☑	Pipe Fittings
☑	Pipes
☑	Sprinklers

Categories from
Current Selection

☑	Air Terminals
☑	Duct Fittings
☑	Ducts
☑	Flex Ducts
☑	Mechanical Equipment
☑	Pipe Fittings
☑	Pipes
☑	Sprinklers

Note that the interference check is being run on the current selection only and not on the entire project.

Click **OK**.

26.

Group by:	Category 1, Category : ▼	
		Message
⊟ Ducts		
⊟ Pipes		
	Ducts : Round Duct : Taps - Mark 1746 : id 613000	
	Pipes : Pipe Types : Standard - Mark 953 : id 736960	
⊟ Pipes		
	Ducts : Round Duct : Taps - Mark 1752 : id 613012	
	Pipes : Pipe Types : Standard - Mark 952 : id 736959	
⊟ Pipes		
	Ducts : Rectangular Duct : Mitered Elbows / Tees - Mark 1754 : id 613170	
	Pipes : Pipe Types : Standard - Mark 970 : id 736995	
⊟ Pipes		
	Ducts : Rectangular Duct : Mitered Elbows / Tees - Mark 1740 : id 612990	
	Pipes : Pipe Types : Standard - Mark 967 : id 736988	

Review the report.
Click **Close**.

27. Close all files without saving.

Shared Coordinates

Coordinates in a file will only be shared, or the files will have shared coordinates after a process of transferring the coordinate system used in a file into another. It does not matter if in two files the location is exactly the same. They will be not sharing coordinates unless the sharing coordinates process has been carried out.

Normally there is one file, and only one, that is the source for sharing coordinates. From this file the location of different models is transferred to them, and after that all files will be sharing coordinates, and be able to be linked with the "Shared Coordinates option".

Revit uses three coordinate systems. When a new project is opened, the three systems overlap.

They are:

- The Survey Point
- The Project Base Point
- The Internal Origin

The Survey Point is represented by a blue triangle with a small plus symbol in the center. This is the point that stores the universal coordinate system, or a defined global system of the project to which all the project structures will be referred. The Survey Point is a real-world relation to the Revit model. It represents a specific point on the Earth, such as a geodetic survey marker or a point of reference based on project property lines. The survey point is used to correctly orient the project with other coordinate systems, such as civil engineering software.

- It is the origin that Revit will use in case of share coordinates between models.
- The origin point used when inserting linked files with the Shared Coordinates option.
- The origin point used when exporting with Shared Coordinates option.
- The origin point to which spot coordinates and spot elevations are referenced, if the Survey Point is the coordinate origin in the type properties.
- The origin point to which level elevation is referred, if the SP is the Elevation Base in the level type properties.
- When the view shows the True North, it is oriented according to the Survey Point settings.

The Project Base Point is represented by a blue circle with an x. The position of this point is unique for each model, and this information is not shared between different models. The Project Base point could be placed in the same location as the Survey Point, but it is not usual to work in that way. This point is used to create a reference for positioning elements in relation to the model itself. By default, the Project Base Point is the origin (0,0,0) of the project. The Project Base Point should be used as a reference point for measurements and references across the site. The location of this point does not affect the shared site coordinates, so it should be located in the model where it makes sense to the model, usually at the intersection of two gridlines or at the corner of a building.

- The origin point to which spot coordinates and spot elevations are referenced, if the Project Base Point is the coordinate origin in the type properties.
- The origin point to which level elevation is referred, if the Project Base Point is the Elevation Base in the level type properties.
- When the view shows the Project North, it is oriented according to the Project Base Point.
- If it is not necessary, it is better not to move this point, so that it always is coincident with the third point: the Internal Origin.
- If we have moved the Project Base Point to a different location, it can be placed back to the initial position by right clicking on it when selected, and use **Move to Startup Location.**

The Internal Origin is now visible starting with the 2022 release of Revit. The Internal Origin is coincident with the Survey Point and the Project Base Point when a new project is started. It is displayed with an XY icon. Prior to this release, users sometimes resorted to placing symbols at the internal origin location to keep track of the origin.

- The project should be modeled in a restricted area around the internal origin point. The model should be inside a 20 miles radius circle around the internal origin, so that Revit can compute accurately.

- This is the origin point that is used when inserting external files using the "Internal Origin to Internal Origin" option.

- This is the origin point that is used when copy/pasting model objects from one file into another using the "aligned" option.

- The origin point to which spot coordinates and spot elevations are referenced, if Relative is the coordinate origin in the type properties.

- The origin point used when exporting with Internal Coordinates option.

- Revit API and Dynamo use this point as the coordinates origin point for internal computational calculations.

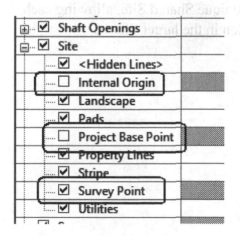

The visibility of these three systems is controlled under the Site category on the Model tab of the Visibility Graphics dialog.

Moving the project base point is like changing the global coordinates of the entire project.

Another way of looking at it is—if you modify the location of the project base point, you are moving the model around the earth; if you modify the location of the survey point, you are moving the earth around the model.

The benefits of correctly adjusting shared coordinates are not limited to aligning models for coordination but can also position the project in the real world if using proper geodetic data from a surveyor landmark or a provided file with the real-world information present.

To fully describe where an object (a building) sits in 3D space (its location on the planet), we need four dimensions:

- East/West position (the X coordinate)
- North/South position (the Y coordinate)
- Elevation (the Z coordinate)
- Rotation angle (East/West/True North)

These four dimensions uniquely position the building on the site and orients the building relative to a known benchmark as well as other landmarks. Revit allows you to assign a unique name to these four coordinates. Revit designates this collection as a **Shared Site**. You can designate as many Shared Sites as you like for a project. This is useful if you are planning a collection of buildings on a campus. By default, every project starts out with a single Shared Site which is named Internal. The name indicates that the Shared Site is using the Internal Coordinate System. Most projects will only require a single Shared Site, but if you are managing a project with more than one building or construction on the same site, it is useful to define a Shared Site for each instance of a linked file.

For example, in a resort, there may be different instances of the same cabin located across the site. You can use the same Revit file with the cabin model and copy the link multiple times. Each cabin location will be assigned a unique Shared Site, allowing each instance of the linked file to have its own unique location in the larger Shared Coordinate system.

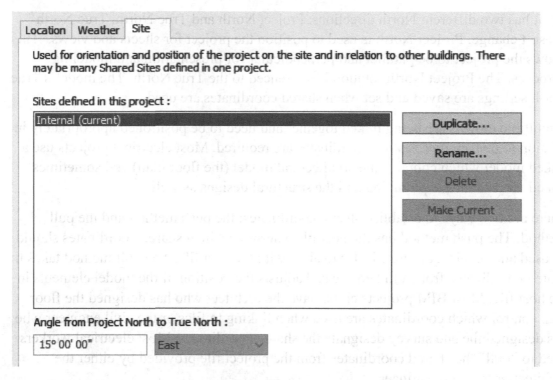

Shared Sites are named using the Location Weather and Site dialog.

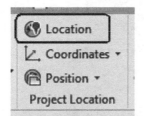

The dialog is opened using the Location tool on the Manage tab on the ribbon.

Why setting up shared coordinates is important:

- To properly coordinate the project models to work across platforms, including BIM Track.
- To display the project in its real-world location (including BIM and GIS data overlay).
- To link files together that have different base points (arbitrary coordinate systems).

Shared Coordinates allow you to adjust the reference point from project origin to a shared site point, to make it appear in a different position or with a different angle based on the internal base point. It is important to note that nothing **in the Revit model moves when using this system,** even if it looks like it moves and/or rotates on the screen. It is just applying an adjustment to the point of origin, so in fact, all elements keep their relation to the internal base point when using this option.

Revit has two different North directions: Project North and True North. True North doesn't change. Project North is used to position the project for sheets and views. This allows the project to be parallel and perpendicular for presentation and modeling purposes. The Project North rotation is referenced to the True North. The model's True North settings are saved and set when shared coordinates are used.

If multiple models are being linked together and need to be positioned appropriately in relation to one another, shared coordinates are required. Most electrical projects use a linked model which contains the architectural model (the floor plan) and sometimes linked models for the plumbing and the structural designs as well.

There are two ways to establish shared coordinates: the push method and the pull method. The push method has the host file determine which shared coordinates should be used and requires all linked files to align with the host file. The pull method takes the shared coordinates from a linked file and adjusts the position of the model elements in the host file. Most BIM projects either have the architect who has designed the floor plan control which coordinates are used when linking to files or the civil engineer who has designed the site survey designate the shared coordinates. Most electrical workers need to "pull" the shared coordinates from the project file provided by either the architect or the civil engineer.

The steps to pull shared coordinates are as follows:

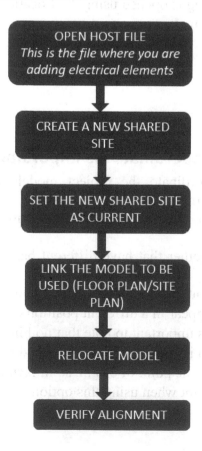

Exercise 5-3

Understanding Shared Coordinates

Drawing Name: new
Estimated Time: 5 minutes

Scope:
- Shared Coordinate System
- Site Point
- Base Point
- Internal Origin
- Views
- Visibility/Graphics

Solution

1.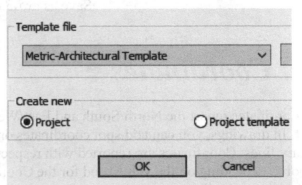

   Start a new project using the *Metric-Architectural* template.

2.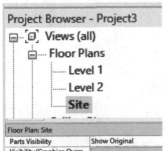

   Open the Site floor plan.

3.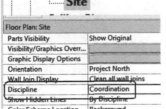

   Set the Discipline to **Coordination**.

4.

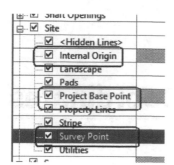

Type **VV** to open the Visibility/Graphics dialog.

On the Model Categories tab:

Enable:
- Internal Origin
- Project Base Point
- Survey Point

These are located under the Site category.

Click **OK**.

The Internal Origin, Project Base Point and Survey Point are all visible.

They are overlaid on top of each other.

5.　　　　　　　　　　Save as *ex5-3.rvt*.

Spot Coordinates

Spot coordinates report the North/South and East/West coordinates of points in a project. In drawings, you can add spot coordinates on floors, walls, toposurfaces, and boundary lines. Coordinates are reported with respect to the survey point or the project base point, depending on the value used for the Coordinate Origin type parameter of the spot coordinate family.

Exercise 5-4

Placing a Spot Coordinate

Drawing Name: spot_coordinate.rvt.
Estimated Time: 20 minutes

Scope:
- ❑ Spot Coordinate
- ❑ Specify Coordinates at a Point
- ❑ Report Shared Coordinates

Solution:

We can add a spot coordinate symbol to the project base point to keep track of its location.

1. 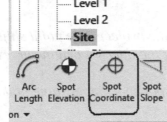 Open the **Site** floor plan.

2. 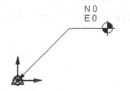 Switch to the Annotate ribbon.

   Select the **Spot Coordinate** tool on the Dimension panel.

3. Select the Project Base Point.

   Left click to place the spot coordinate.
   That is the circular symbol.

   Cancel out of the command.

4.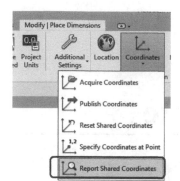

Switch to the Manage ribbon.

Select **Report Shared Coordinates** on the Project Location panel.

Select the Project Base Point.

5.

The coordinates are displayed on the Options bar.

Cancel out of the command.

6.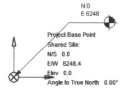

Select the Project Base Point.

Notice that the coordinates that are displayed are blue. This means they are temporary dimensions which can be edited.

7.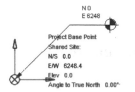

Click on the **E/W** dimension.
Change it to **6248.4.**
Click **ENTER**.

8.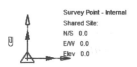

The location of the project base point shifted to be 6248.4 m east of the internal origin/survey point.

Notice that the spot coordinate value updated.

9.

Left click on the Survey Point.
That's the triangle symbol.
Notice that dimensions are black, which means they cannot be edited.

Collaboration and Project Management

10. 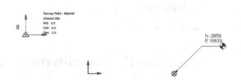 Drag the survey point away from the internal origin (the XY icon).

Notice that the Project Base Point value updates.

Remember the Survey Point represents a real and known benchmark location in the project (usually provided by the project survey). The Project Base Point is simply a known point on the building (usually chosen by the project team).

Think of the Survey Point as the coordinates in the *World* and the Project Base Point as the local building coordinates.

The Survey Point is always located at 0,0, while the Project Base Point is relative to the Survey Point.

It is called a shared coordinate system because it represents the coordinate system of the world around us and is shared by all the buildings on the site.

11. 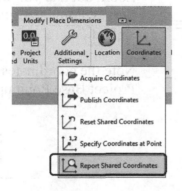 Select **Report Shared Coordinates** on the Project Location panel.

Select the Internal Origin (the XY icon).

12.  The coordinates are displayed on the Options bar.

The values will be different from mine.

Notice that the Internal Origin is located relative to the Survey Point. The Internal Coordinate System represents the building's "local" coordinates.

Cancel out of the command.

5-23

13.

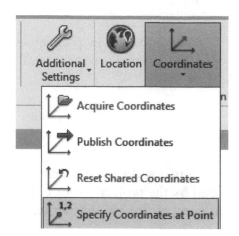

Usually, you will be provided the coordinate information for a project from the civil engineer.

Select **Specify Coordinates at Point**.

Select the Survey Point (the triangle symbol.).

14.

Fill in the Shared Coordinates:

For North/South: **11988.8.**
For East/West: **4876.8.**
For Elevation: **6096.**
For Angle from Project North to True North: **15° East**.

Click **OK**.

15.

Zoom out and you will see that the Survey Point's position has shifted.

Note the coordinates for the Project Base Point updated.

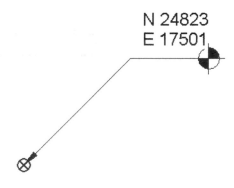

N 24823
E 17501

16.   Save as *ex5-4.rvt*.

Exercise 5-5

Understanding Location

Drawing Name: Simple Building.rvt
Estimated Time: 20 minutes

Scope:
- Linking Files
- Shared Coordinate System
- Survey Point
- Base Point
- Internal Origin
- Visibility/Graphics
- Spot Elevation

Solution:

1. 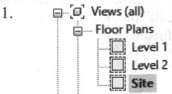 Open the **Site** floor plan.

2. 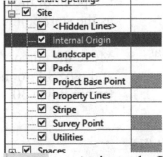 *The Survey Point and the Project Base Point are visible, but not the Internal Origin.*

    Open the Visibility/Graphics dialog by typing **VV**.

3.  On the Model Categories tab:

    Enable: **Internal Origin**

    This is located under the Site category.

    Click **OK** and close the dialog.

4. Activate the **Insert** ribbon.

    Select **Link Revit**.

5.

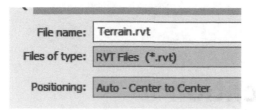

Locate the *Terrain.rvt* file.

Set the Positioning to **Auto – Center to Center**.

Click **Open**.

Many users will be tempted to align the files using Internal Origin to Internal Origin. This is not the best option when we plan to reposition the building on the site plan. Origin to Origin works well when linking MEP files to the Architectural file.

Notice that the linked file's coordinate system is slightly offset from the host file's coordinate system.

Next, we will re-position the linked file into the correct relative position. If we do it this way, we are not moving the building, we are moving the terrain. We will still use the terrain to establish the Shared Coordinate system, which is the "pull" method. The "pull" method pulls the information from the linked file.

6.

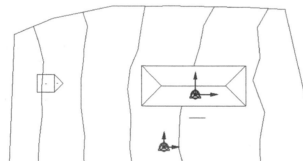

Select the Move tool on the tab on the ribbon.
Select the bottom midpoint of the building as the start point.
Move the cursor to the left and type 15' to move the terrain left 15'.

Select the linked file.

Select the **Move** tool on the tab on the ribbon.

Select the bottom midpoint of the building as the start point.

Move the cursor down and type 30' to move the terrain down 30'.

7.

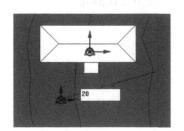

Select the linked file.

Select the Rotate tool on the tab on the ribbon.

Start the angle at the horizontal 0° level.
Rotate up and type **20°**.

8.

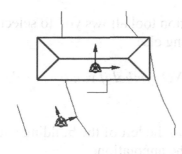

We have repositioned the terrain on the XY plane, but we have not changed the elevation.

Click on the Project Base Point on the building.

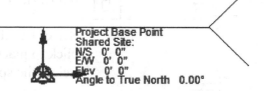

Notice that it has not changed.

9.

Click on the Project Base Point for the terrain file.

It shows as 0,0 still as well.

10.

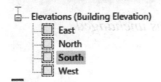

Open the **South** elevation.

11.

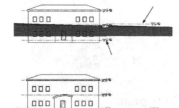

Notice that the levels are not aligned between the files.

12.

*Use the **ALIGN** tool to move the Level 1 in the terrain file to align with Level 1 in the simple building file.*

Select the **ALIGN** tool on the MODIFY ribbon.

Select the Level 1 that is on the simple building file as the target.
Select the Level 1 on the terrain file to be moved.

Cancel out of the ALIGN command.

13.

Spot Elevation

Switch to the Annotate ribbon.

Select **Spot Elevation**.

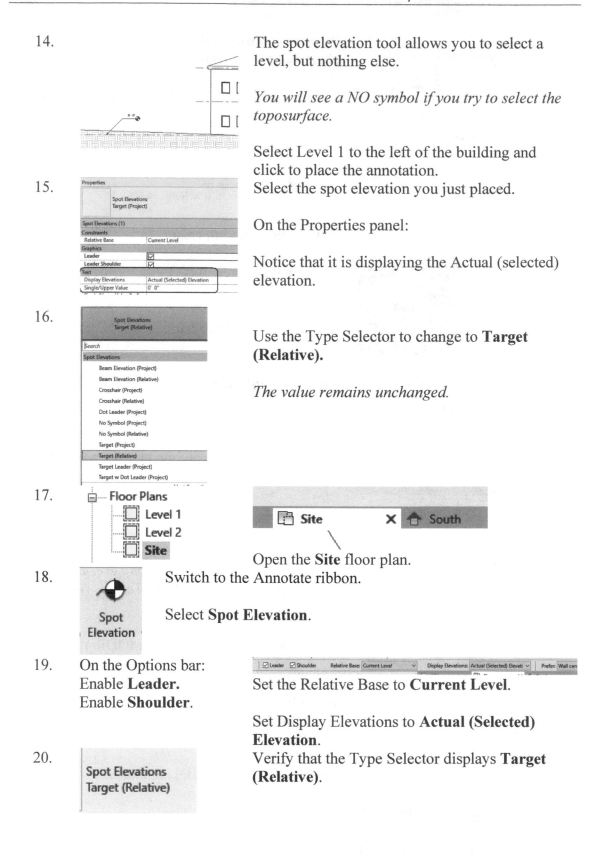

14. The spot elevation tool allows you to select a level, but nothing else.

You will see a NO symbol if you try to select the toposurface.

Select Level 1 to the left of the building and click to place the annotation.

15. Select the spot elevation you just placed.

On the Properties panel:

Notice that it is displaying the Actual (selected) elevation.

16. Use the Type Selector to change to **Target (Relative).**

The value remains unchanged.

17. Open the **Site** floor plan.

18. Switch to the Annotate ribbon.

Select **Spot Elevation**.

19. On the Options bar:
Enable **Leader.**
Enable **Shoulder**.

Set the Relative Base to **Current Level**.

Set Display Elevations to **Actual (Selected) Elevation**.

20. Verify that the Type Selector displays **Target (Relative).**

21.

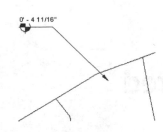

Click on the toposurface.

The elevation is displayed in black.
This means it cannot be modified. The value is
generated by the actual location of the
toposurface.

The value you see may be different.

22.

Verify that both site points display 0,0 as their coordinates.

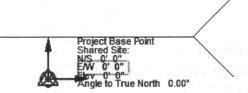

23.

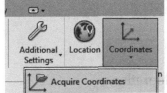

Switch to the Manage ribbon.

Select **Acquire Coordinates** and then select the linked terrain file.

24.

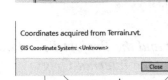

Click **Close**.

25.

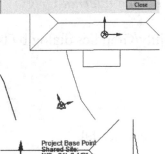

The Survey Point in the host file (simple building) moved. It is now aligned with the Survey Point on the linked file (terrain).

26.

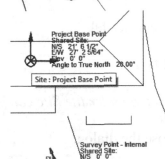

Select the Project Base Point for the simple building.

Notice the change to the coordinates.

27.

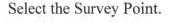

Select the Survey Point.

Notice it is set to 0,0.
The spot elevation value is unchanged because
the coordinates were acquired from the linked
file and the spot elevation is reporting the
elevation of the linked file.

28.   Save the file as *ex5-5.rvt*.

Exercise 5-6
Linking Files using Shared Coordinates

Drawing Name: terrain.rvt
Estimated Time: 10 minutes

Scope:
- Linking Files
- Shared Coordinate System
- Survey Point
- Base Point
- Internal Origin
- Visibility/Graphics

Solution

1.   Open the **Site** floor plan.

        Views (all)
            Floor Plans
                Level 1
                Site

2.

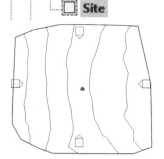

The Survey Point and the Project Base Point are visible, but not the Internal Origin.

Open the Visibility/Graphics dialog by typing **VV**.

3.

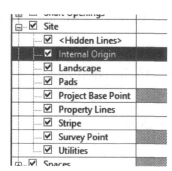

On the Model Categories tab:

Enable: **Internal Origin**

      This is located under the Site category.

Click **OK** and close the dialog.

4.   Use **FILTER** to select just the Survey Point.

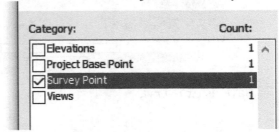

Category:	Count:
☐ Elevations	1
☐ Project Base Point	1
☑ Survey Point	1
☐ Views	1

Notice that it is still located at 0,0.

Click **ESC** to release the selection.

5.   Select the **Project Base Point**.

It is also located at 0,0.

The file wasn't changed because we acquired the coordinates from the linked file and applied them to the host file.

6.   Switch to the **Insert** tab on the tab on the ribbon.

Select **Link Revit**.

7.   Select *Simple Building 2.rvt*. Set the Positioning to **Auto -By Shared Coordinates.**

File name:	Simple Building 2.rvt
Files of type:	RVT Files (*.rvt)
Positioning:	Auto - By Shared Coordinates

Simple Building 2 has the coordinates assigned that we defined in the previous exercise.

Click **Open**.

8.    The linked file is positioned in the correct location.

9.   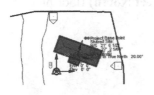 Click on the building.
Notice that Project Base Point is using the Shared Site coordinates.

10.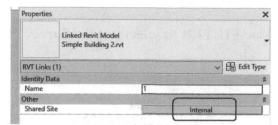

With the linked file selected, notice in the Properties palette, it is using the Internal Shared Site definition.

11. Save as *ex5-6.rvt*.

Exercise 5-7

Defining a Shared Site

Drawing Name: terrain 2.rvt
Estimated Time: 30 minutes

Scope
- Linking Files
- Shared Coordinate System
- Survey Point
- Base Point
- Shared Site

Solution

1. Views (all)
   Floor Plans
      Level 1
      Site

   Open *Building A.rvt*.
   Open the **Site** floor plan.

2.

   Verify that the Survey Point and Base Point are located at 0,0.

   Close the file.

3. Floor Plans
      Platform 1
      Platform 2
      Platform 3
      Site (Dimensioned)
      Site (Project North)
      Site (True North)
      Site Datum

   Open the *terrain 2. rvt* file.

   Open the **Site (Project North)** floor plan.

4.   The site plan has building pads for three buildings at three different locations on the site.

5.   Switch to the **Insert** tab on the tab on the ribbon.

     Select **Link Revit**.

6.

File name:	Building A.rvt
Files of type:	RVT Files (*.rvt)
Positioning:	Auto - Center to Center

Select Building A.*rvt*.
Set the Positioning to **Auto -Center to Center.**

Click **Open**.

7. *Note the position of Building A on the site plan.*

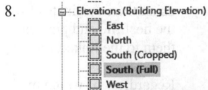

Use the **ALIGN** tool to move Building A onto the A platform. Align Grids 1 to each other and Grids A to each other.

Remember to select the Platform grid first as the target and then the building grid.

8.
- ⊟ **Elevations (Building Elevation)**
  - ▢ East
  - ▢ North
  - ▢ South (Cropped)
  - ▢ **South (Full)**
  - ▢ West

Open the **South (Full)** elevation view.

9.

Notice that the Building A elevation needs to be adjusted.

10.

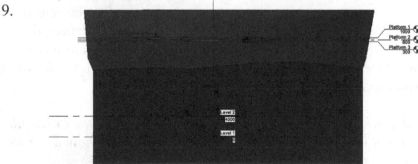

Use the **ALIGN** tool to align Level 1 for Building A to Platform 1.

11. 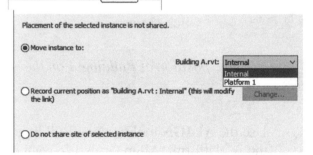 Open the **Platform 1** view.

12. Verify that Building A is situated properly on the platform.

13. Select the linked file (Building A).

Click **<Not Shared>** on the Properties palette.

14. We can either push coordinates or pull coordinates.

If we select **Move instance to Internal**, we are using the coordinate system from the linked files and pulling them into the host file.

If we select **Move instance to Platform 1**…, we are using the coordinates from the host file and pushing them to the linked file.

Regardless of which method we use, after the selection, they will be sharing the same coordinate system.

Since we started with the terrain file, it is standard to let the topo file control which coordinate system is used in a project.

Remember that whichever option you select does not change the linked file or the host file local internal coordinates. We are only defining how we want the files to interact relative to each other.

15.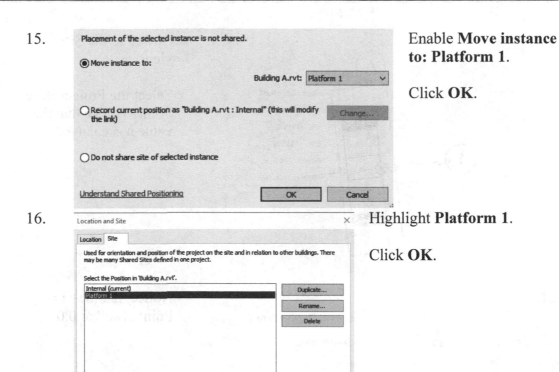

Enable **Move instance to: Platform 1**.

Click **OK**.

16.

Highlight **Platform 1**.

Click **OK**.

17.

The coordinates on the linked file updates.

Save the file as *ex5-7.rvt*.

18.

Select the linked file. In the Properties palette, it shows the Shared Site as Platform 1.

19. Close all open files.

Open *Building A.rvt*.

20.

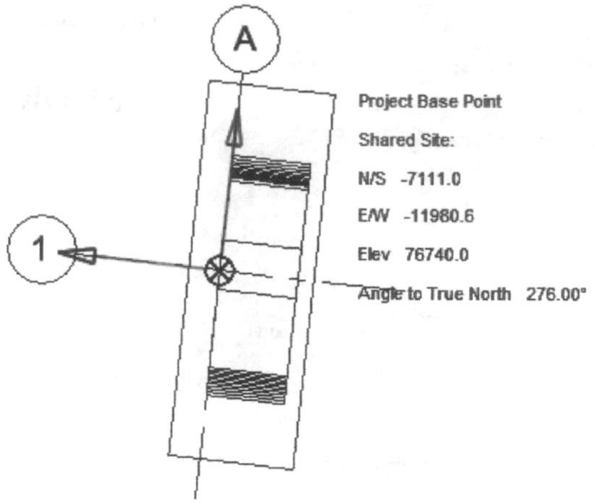

Project Base Point

Shared Site:

N/S -7111.0

E/W -11980.6

Elev 76740.0

Angle to True North 276.00°

Switch to the **Site** plan view.

Select the Project Base Point. Notice that the value has changed.

21.

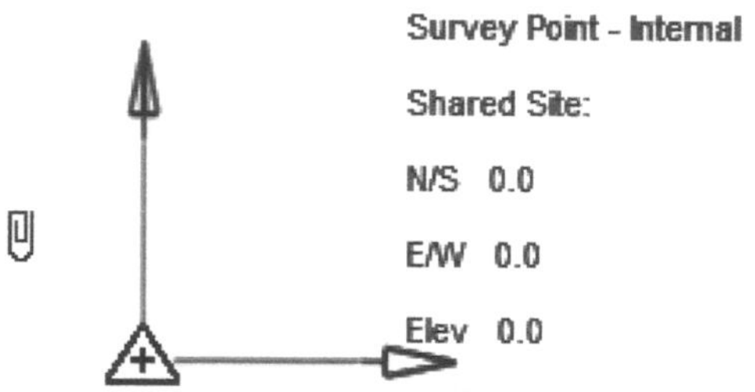

Survey Point - Internal

Shared Site:

N/S 0.0

E/W 0.0

Elev 0.0

Notice that the Survey Point is still at 0,0.

22.

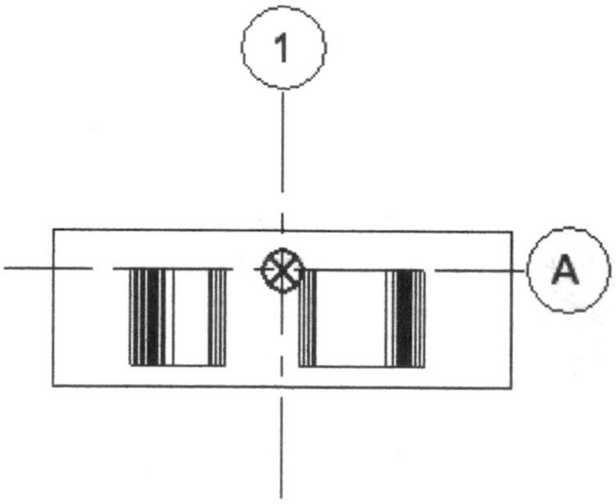

Open the Level 1 view.

The building is oriented horizontally and vertically on the screen to make it easy to work on.

23.

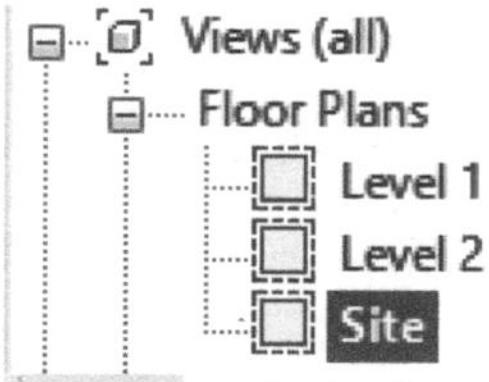

- Views (all)
  - Floor Plans
    - Level 1
    - Level 2
    - Site

Open the **Site** plan.

24.

Link Revit

Switch to the **Insert** tab on the tab on the ribbon.

Select **Link Revit**.

25.

File name:	Terrain 2.rvt
Files of type:	RVT Files (*.rvt)
Positioning:	Auto - By Shared Coordinates

Select *terrain 2.rvt.*
Set the Positioning to **Auto - By Shared Coordinates.**

Click **Open**.

26. Notice that the terrain is oriented perfectly to the building.

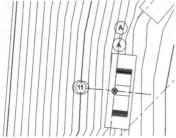

This is because we are using the same shared coordinates between the two files.

Close all files.

Exercise 5-8
Using Project North

Drawing Name: **Import Site.Dwg**
Estimated Time to Completion: 40 Minutes

Scope

Link an AutoCAD file.
Set Shared Coordinates.
Set Project North.

Every project has a project base point ⊗ and a survey point △, although they might not be visible in all views because of visibility settings and view clippings. They cannot be deleted.

The project base point defines the origin (0,0,0) of the project coordinate system. It also can be used to position the building on the site and for locating the design elements of a building during construction. Spot coordinates and spot elevations that reference the project coordinate system are displayed relative to this point.

The survey point represents a known point in the physical world, such as a geodetic survey marker. The survey point is used to correctly orient the building geometry in another coordinate system, such as the coordinate system used in a civil engineering application.

Solution

1. Start a new project file using the **Metric-Architectural** template.

2. Activate the **Site** floor plan.

3.  Activate the **Insert** ribbon.
    Select the **Link CAD** tool on the Link panel.

4. Select the *Import Site* drawing. Set Colors to **Preserve**. Set Import Units to **Auto-Detect**. Set Positioning to **Auto - Center to Center**. Click **Open**.

File name:	Import Site.dwg	∨
Files of type:	DWG Files (*.dwg)	∨

Colors:	Preserve ∨	Positioning:	Auto - Center to Center ∨
Layers/Levels:	All ∨	Place at:	Level : Level 1 ∨
Import units:	Auto-Detect ∨ 1.000000		☑ Orient to View
	☑ Correct lines that are slightly off axis		Open Cancel

5.
Other	⌃
Shared Site	\<Not Shared\>
Enable Cutting in Views	☐

Select the imported site plan. In the Properties pane, select the Shared Site button.

6.
○ Publish the current shared coordinate system to "Import Site.dwg." This will modify all Named Positions of the linked model.

◉ Acquire the shared coordinate system from "Import Site.dwg." This will modify the current model and all Named Positions of other linked models.

Record selected instance as being at Position:

Import Site.dwg : DefaultLocation   [Change...]

Enable **Acquire the shared coordinate system...** Select **Change**.

7.
Location	Site

Define Location by:

Internet Mapping Service ∨

800 Fallon St Oakland, CA 94607

Activate the Location tab. Enter the Project Address. Click **Search** and then click **OK**.

You must be connected to the internet in order to input an address.

8. [Reconcile]   Click **Reconcile**.

9.

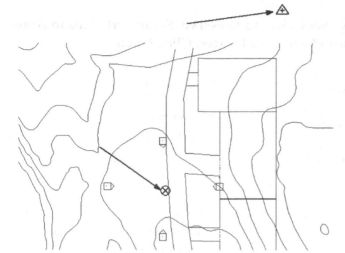

Open the **Site** plan view.

Note that the survey point and project base point shift position.

10.

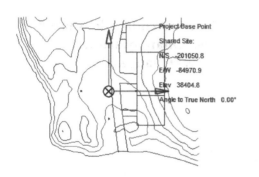

Select the Project Base point.

This is the symbol that is a circle with an X inside.

11.

Project Base Point (1)	
Identity Data	
N/S	-141666.8
E/W	-68891.1
Elev	41404.8
Angle to True North	90.00°

Change the Angle to Truth North to **90.00**.
Click **Apply**.
The site plan will rotate 90 degrees.

12.

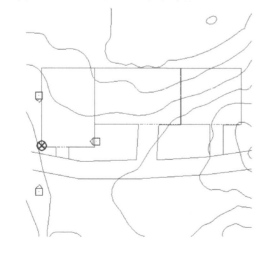

Move the import site plan so that the corner of the rectangle is coincident with the project base point.
Use the **MOVE** tool.
Select the corner of the building and then the Project Base Point.

13. Autodesk Revit 2017

Warning - can be ignored

Shared Sites in the link "Import Site.dwg" have been modified, but not saved back to the link. Upon reopening, instances of the link will return to their last Saved Positions. You can Save the link later via the Manage Links dialog.

Show   More Info   Expand >>

Save Now   OK   Cancel

Select **OK**.

If you select Save Now, this will modify the linked file.

14. Use the PIN tool to prevent the site drawing from shifting around.

Select the linked file first, then select the PIN.

15. Link Revit

Activate the Insert tab on the ribbon.
Select **Link Revit** from the Link panel.

16. File name: i_shared_coords

Files of type: RVT Files (*.rvt)

Positioning: Manual - Base point

Select the *i_shared_coords* file.
Set the Positioning to **Manual - Base point**.
Click **Open**.

17. Place the Revit file to the left of the site plan.

18. Select the **Align** tool on the Modify panel on the Modify tab on the ribbon.

19.   Select the vertical line above the project base point.
       Then select Grid line 1 on the i_shared_coords imported file.

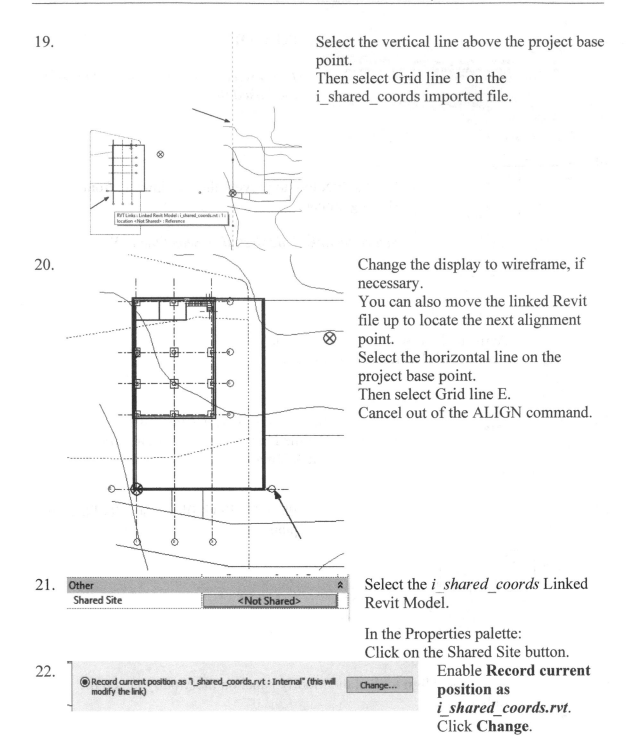

RVT Links : Linked Revit Model : i_shared_coords.rvt : 1 :
location <Not Shared> : Reference

20.   Change the display to wireframe, if necessary.
       You can also move the linked Revit file up to locate the next alignment point.
       Select the horizontal line on the project base point.
       Then select Grid line E.
       Cancel out of the ALIGN command.

21. | Other | ⌃ |
       |---|---|
       | Shared Site | <Not Shared> |

       Select the *i_shared_coords* Linked Revit Model.

       In the Properties palette:
       Click on the Shared Site button.

22.   ⦿ Record current position as "i_shared_coords.rvt : Internal" (this will modify the link)   [Change...]

       Enable **Record current position as i_shared_coords.rvt**.
       Click **Change**.

23.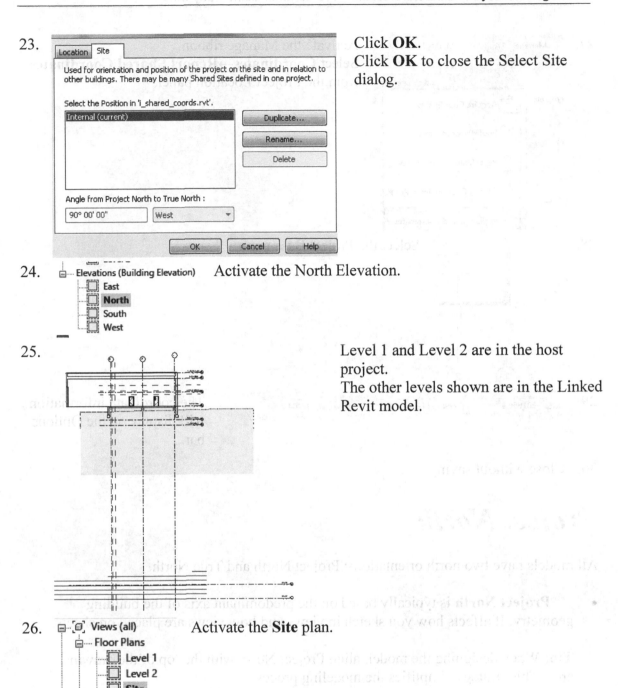

Click **OK**.
Click **OK** to close the Select Site dialog.

24. Activate the North Elevation.

25. Level 1 and Level 2 are in the host project.
The other levels shown are in the Linked Revit model.

26. Activate the **Site** plan.

27. 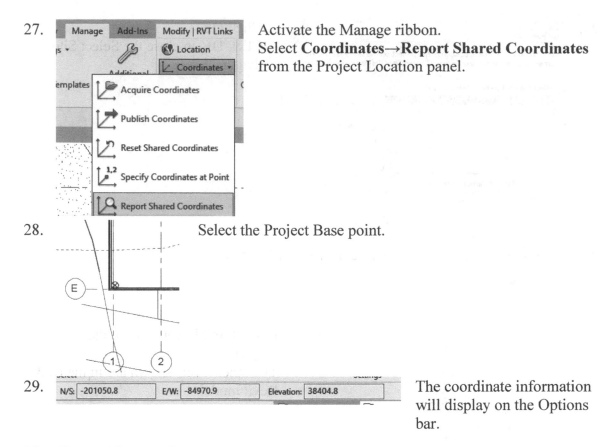 Activate the Manage ribbon.
Select **Coordinates→Report Shared Coordinates** from the Project Location panel.

28. Select the Project Base point.

29. N/S: -201050.8 E/W: -84970.9 Elevation: 38404.8 The coordinate information will display on the Options bar.

30. Close without saving.

Project North

All models have two north orientations: Project North and True North.

- **Project North** is typically based on the predominant axis of the building geometry. It affects how you sketch in views and how views are placed on sheets.

Tip: When designing the model, align Project North with the top of the drawing area. This strategy simplifies the modeling process.

- **True North** is the real-world north direction based on site conditions.

Tip: To avoid confusion, define True North only after you begin modeling with Project North aligned to the top of the drawing area and after you receive reliable survey coordinates.

All models start with Project North and True North aligned with the top of the drawing area, as indicated by the survey point △ and the project base point ⊗ in the site plan view.

You may want to rotate True North for the following reasons:

- to represent site conditions
- for solar studies and rendering to ensure that natural light falls on the correct sides of the building model
- for energy analysis
- for heating and cooling loads analysis

Exercise 5-9

Using True North vs Project North

Drawing Name: parcel map.rvt
Estimated Time: 20 minutes

Scope

- Project North
- True North

Solution

1.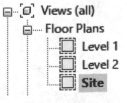

   Open the **Site** floor plan.

2.

   On the Properties panel:

   Verify that the Orientation is set to **Project North**.

3. 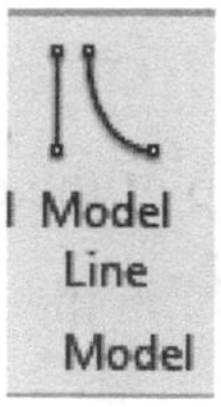 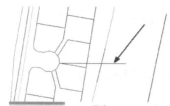 Select **Model Line** from the Architecture ribbon.

4.  Draw a horizontal line from the east quadrant of the cul de sac to the right.

5.  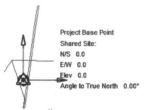 Place an Angular dimension to determine the orientation of the parcel map.

6. 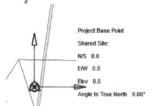 The angle value is 9°.

Delete the model line and the angular dimension.

7. Select the Project Base Point.

Project Base Point
Shared Site:
N/S 0.0
E/W 0.0
Elev 0.0
Angle to True North 0.00°

8. Change the Angle to True North to **9°**.

Project Base Point
Shared Site:
N/S 0.0
E/W 0.0
Elev 0.0
Angle to True North 9.00°

9.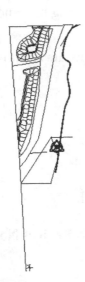

Window around all the elements in the view.

10.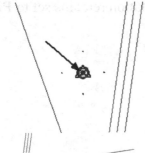

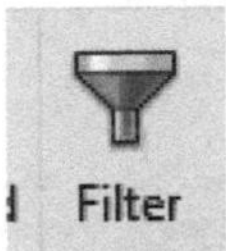

Select **Filter** on the ribbon.

11.

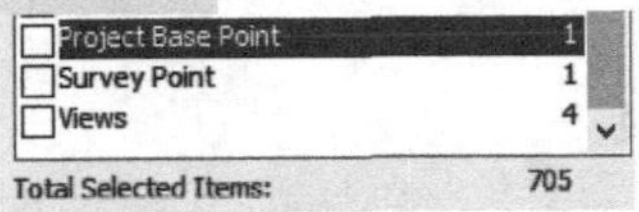

Scroll down to the bottom and uncheck:

- Project Base Point
- Survey Point
- Views

Click **OK**.

12.  Select **Rotate** on the Modify tab on the ribbon.

13.

Click **Place** next to Center of Rotation on the Options bar.

Disable **Copy**.

14. Select the Project Base Point.

15. Select a horizontal point to the right of the Project Base Point as the starting point of the angle.

Drag the mouse up to set the end angle at 9°.

16.   You can ignore the warning box and close it.

Warning: 1 out of 63

Line is slightly off axis and may cause inaccuracies.

17.   The parcel map now appears to show the parcels oriented orthogonally.

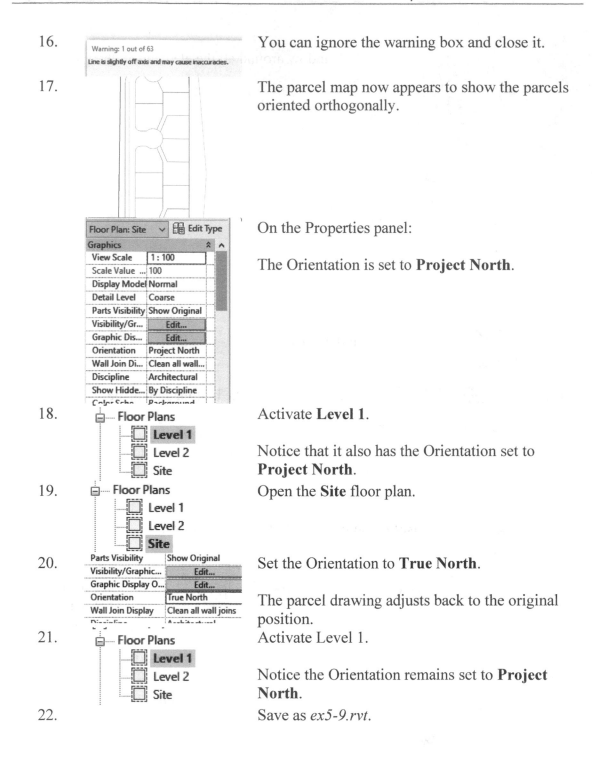

On the Properties panel:

The Orientation is set to **Project North**.

18.   Activate **Level 1**.

Notice that it also has the Orientation set to **Project North**.

19.   Open the **Site** floor plan.

20.   Set the Orientation to **True North**.

The parcel drawing adjusts back to the original position.

21.   Activate Level 1.

Notice the Orientation remains sct to **Project North**.

22.   Save as *ex5-9.rvt*.

Importing Files

Revit allows you to add raster images or PDF files into a building model. These can be inserted into 2D views – elevations, floor plans, ceiling plans, and sections.

AutoCAD DWG files can also be imported or linked into a Revit building model.

Linking AutoCAD files allows them to be used as underlays for views. If the AutoCAD DWG is modified, it will automatically update in the Revit file if it is reloaded.

If you import an AutoCAD file, it is inserted into the building model. It will not update if the external file is changed.

Exercise 5-10

Import DWG

Drawing Name: property survey.dwg
Estimated Time: 10 minutes

Scope

- ❑ Import CAD
- ❑ Property Line
- ❑ Toposurface

Solution

1.  Start a New project using the Metric-Architectural template.

   Click **OK**.

2. Switch to the Insert ribbon.

   Select **Import CAD**.

3.

Select the *property survey.dwg*.

Set Colors to **Preserve**.
Set Positioning to **Auto- Origin to Internal Origin**.
Set Layers/Levels to **All**.
Set Import Units to **feet**.
Click **Open**.

4.

Open the **Site** plan view.

5.

Switch to the Massing & Site ribbon.

Select **Property Line**.

6.

Select **Create by sketching**.

How would you like to create the property lines?

→ Create by entering distances and bearings

→ Create by sketching

7.

Select the Pick Line tool.

Select the elements in the property survey drawing.

8.

You should see a completed outline.

Select the Green Check on the ribbon to complete.

If you get a warning, zoom in and see if you missed any portions of the outline.

Use **Edit Sketch** to add the missing sections.

9.

Select **Toposurface**.

10.

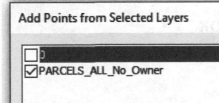

Select **Create from Import →Select Import Instance**.

Select the imported drawing.

11.

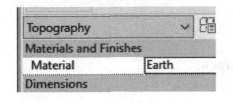

Enable the **PARCELS_All_No_Own**er layer.

Click **OK**.

12.

Set the Material to **Earth** in the Properties Panel.

13.  Select **Green Check** to complete the toposurface.

14.

Switch to a 3D view.

Change the display to **Shaded**.

Save as *ex5-10.rvt*.

Exercise 5-11

Import PDF

Drawing Name: bldg._8_2.pdf
Estimated Time: 5 minutes

Scope

❑ Import PDF

Solution

1. 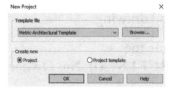 Start a New project using the Metric-Architectural template.

Click **OK**.

2.  Switch to the Insert ribbon.

Select **Import PDF**.

3.  Select *bldg._8_2.pdf*.

Click **Open.**

4.

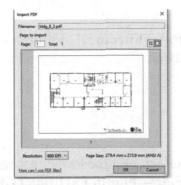

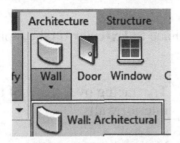

Set the Resolution to **600 DPI**.

Click **OK**.

Left click to place the PDF in the view.

Left click anywhere in the window to release the selection.

5.

Select the **Wall:Architectural** tool from the Architecture ribbon.

6.

Use the Type Selector on the Properties panel to select the **Interior – 79 mm Partition (1-hr)** wall.

7.

Set the Location Line to: **Finish Face: Exterior** on the Options bar.

8.

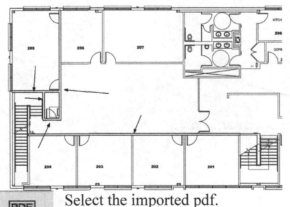

Try tracing over a few of the interior walls.

Notice that you cannot snap to any of the PDF elements.

Exit out of the command.

9.

Select the imported pdf.

Toggle **Enable Snaps** on the ribbon so it is ON.

10.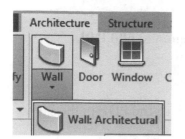

Select the **Wall:Architectural** tool from the Architecture ribbon.

11.

Use the Type Selector on the Properties panel to select the **Interior – 79 mm Partition (1-hr)** wall.

12. Set the Finish Face to **Exterior** on the Options bar.

13.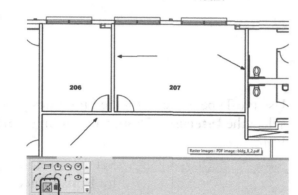

Try tracing over a few of the interior walls.

Notice that now you can snap to the PDF elements to place walls.

14.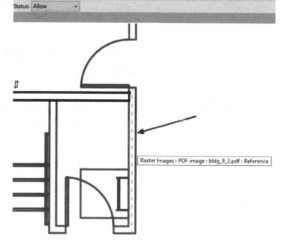

You can also use the **Pick Lines** tool to place walls.

15. Save as *ex5-11.rvt*.

Exercise 5-12

Import Image

Drawing Name: wrexham.jpg
Estimated Time: 5 minutes

Scope

❑ Import Image

Solution

1. 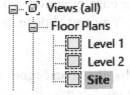 Start a New project using the Metric-Architectural template.

   Click **OK**.

2.

   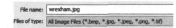 Open the **Site** plan view.

3.  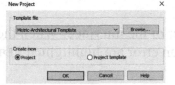 Switch to the Insert ribbon.

   Select **Import Image**.

4. Select *wrexham.jpg*.

   Click **Open**.

   Left click to place in the view.

5.  Use the grips at the corners to enlarge the image.

6. 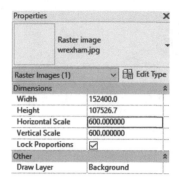 On the Properties panel, change the Horizontal Scale to **600**.

Notice that if Lock Proportions is enabled, the Vertical Scale will be the same as the Horizontal Scale.

7. Save as *ex5-12.rvt*.

Review Warnings

At any time when working on a project, you can review a list of warning messages to find issues that might require review and resolution.

Warnings display in a dialog in the lower-right corner of the interface. When the warning displays, the element or elements that have an error are highlighted in a user-definable color.

Unlike error messages, warning messages do not prohibit the current action. They just inform you of a situation that may not be your design intent. You can choose to correct the situation or ignore it.

The software maintains a list of warning messages that are displayed and ignored while you are working. The Warnings tool lets you view the list at your convenience to determine if the conditions described in the warnings still exist.

If you receive a file from an outside source, it may be worth your time to review any warnings to determine if there are any issues that may require some resolution.

Exercise 5-13
Review Warnings

Drawing Name: review warnings.rvt
Estimated Time: 30 minutes

Scope

- ❑ Review Warnings
- ❑ Export Warnings
- ❑ Areas
- ❑ Stairs
- ❑ Rooms
- ❑ Floors
- ❑ Element ID

Solution

1.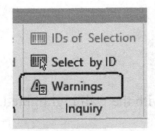
   Switch to the Manage ribbon.

   Select **Warnings** from the Inquiry panel.

2.
   A list of warnings is displayed.

   Expand **Warning 1**.

3.

Place a check next to the first wall under Warning 1.

Click **Show**.

4.

Click **Close**.

Click **Show** again.

5.

Click **OK**.

6.

A view will open with the wall highlighted.

Click **Close** to close the dialog.

7.

Switch to a 3D view.

8.

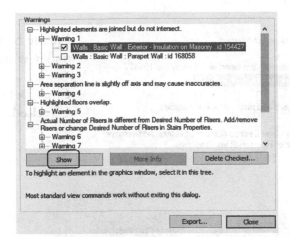

Open the Warnings dialog again.

Place a check next to the first wall under Warning 1.

Click **Show**.

9.

The wall highlights.

10.

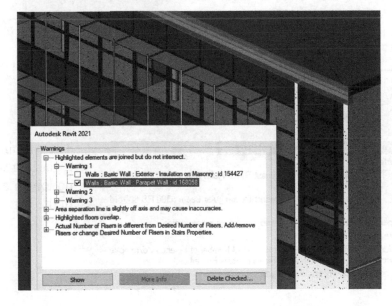

Place a check on the second wall for Warning 1.

The parapet wall highlights.

You can ignore this warning.

11.

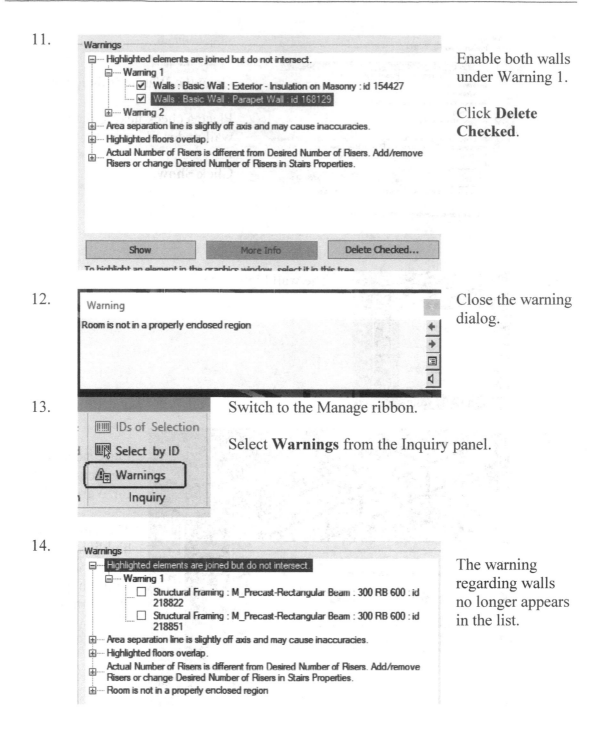

Enable both walls under Warning 1.

Click **Delete Checked**.

12.

Close the warning dialog.

13.

Switch to the Manage ribbon.

Select **Warnings** from the Inquiry panel.

14.

The warning regarding walls no longer appears in the list.

15.

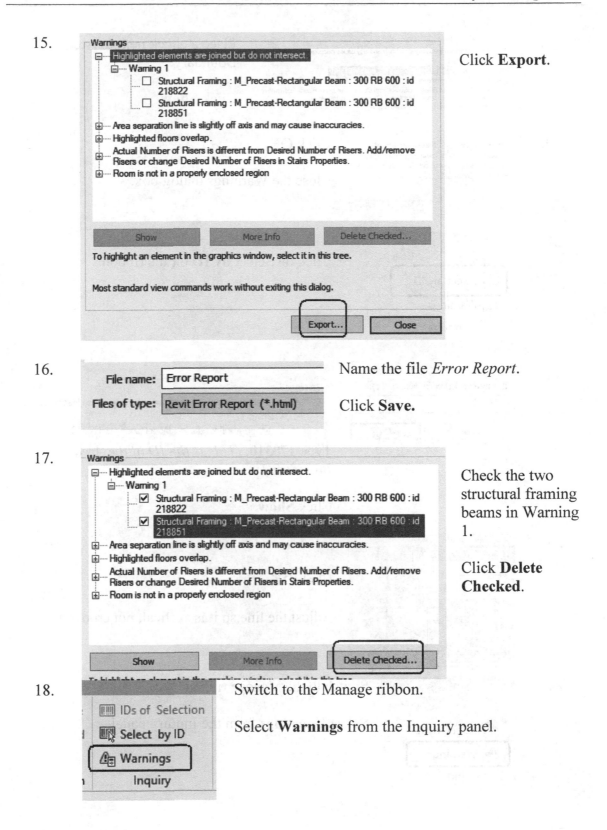

Click **Export**.

16.

File name:	Error Report
Files of type:	Revit Error Report (*.html)

Name the file *Error Report*.

Click **Save.**

17.

Check the two structural framing beams in Warning 1.

Click **Delete Checked**.

18.

IIIII IDs of Selection

Select by ID

Warnings

Inquiry

Switch to the Manage ribbon.

Select **Warnings** from the Inquiry panel.

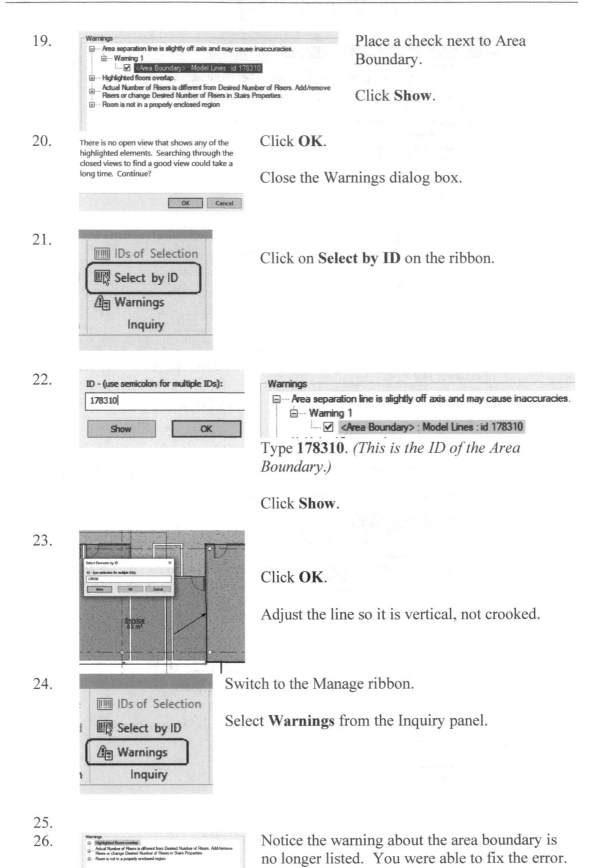

19. Place a check next to Area Boundary.

    Click **Show**.

20. Click **OK**.

    Close the Warnings dialog box.

21. Click on **Select by ID** on the ribbon.

22. Type **178310**. *(This is the ID of the Area Boundary.)*

    Click **Show**.

23. Click **OK**.

    Adjust the line so it is vertical, not crooked.

24. Switch to the Manage ribbon.

    Select **Warnings** from the Inquiry panel.

25.

26. Notice the warning about the area boundary is no longer listed. You were able to fix the error.

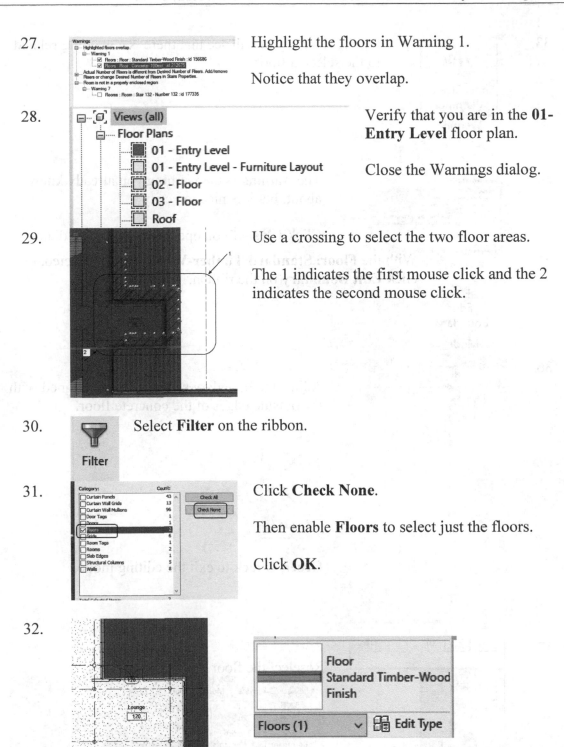

27. Highlight the floors in Warning 1.

Notice that they overlap.

28. Verify that you are in the **01-Entry Level** floor plan.

Close the Warnings dialog.

29. Use a crossing to select the two floor areas.

The 1 indicates the first mouse click and the 2 indicates the second mouse click.

30. Select **Filter** on the ribbon.

31. Click **Check None**.

Then enable **Floors** to select just the floors.

Click **OK**.

32. Select the **Floor: Standard Timber-Wood Finish**.

33.

On the ribbon, you will see that there is a warning related to the selected floor.

34.

The warning is the warning you already knew about, but it is nice to verify it.

Click **Close** if you opened the Warning dialog.

35.

With the **Floor: Standard Timber-Wood Finish** selected, click **Edit Boundary** on the ribbon.

36.

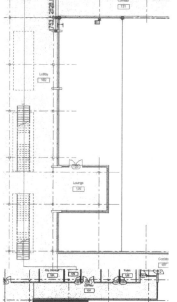

Adjust the boundary so the lines are aligned with the outside edges of the concrete floor.

Green Check to exit the editing mode.

37.

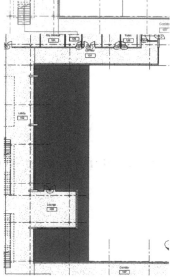

Reselect the floor.

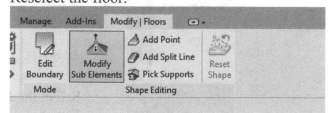

Check to see if the Warning icon appears on the ribbon.

If it does, edit the boundary again and adjust so it no longer overlaps the concrete floor.

38.

Switch to the Manage ribbon.

Select **Warnings** from the Inquiry panel.

39.

Expand Warnings 1 through 5.

Can you tell if these warnings apply to only one element or how many elements need to be modified?

40.

Place a check next to the Stairs under Warning 1.

Click **Show**.

Close the dialog.

41.

Select the stairs.

Scroll down in the Properties panel.

Note that the Desired Number of Risers is 26, but the Actual Number of Risers is 25.

42.

Desired Number of Risers is too small.
Computed Actual Riser Height is greater than
Maximum Riser Height allowed by type.

Change the Desired Number of Risers to 25.

This error will appear.

Click **Close**.

43.

Click **Edit Type** on the Properties panel.

44.

Select **Duplicate**.

45.

Change the Name to **200mm max riser 300 mm tread**.

Click **OK**.

46.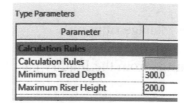

Change the Maximum Riser Height to **200mm**.

Click **OK**.

47.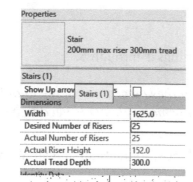

The Stair now shows the Desired Number of Risers equal to the Actual Number of Risers.

The Actual Riser Height is 152 mm which is about 6 inches high, well within code.

48.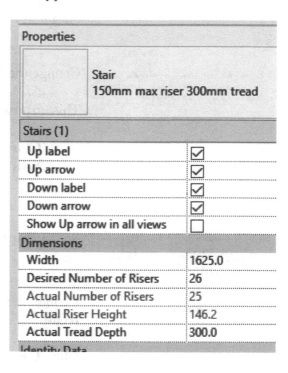

Select the stairs above the stairs you just modified.

This stair has the same error – the desired and actual number of risers do not match.

Use the Type Selector to change the Stair to the new type.

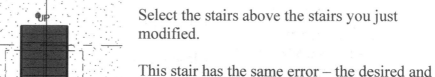

Properties	
Stair 150mm max riser 300mm tread	
Stairs (1)	
Up label	☑
Up arrow	☑
Down label	☑
Down arrow	☑
Show Up arrow in all views	☐
Dimensions	
Width	1625.0
Desired Number of Risers	26
Actual Number of Risers	25
Actual Riser Height	146.2
Actual Tread Depth	300.0
Identity Data	

49.

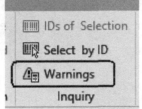

Locate the stair in Room 114.

Right click and select **Select All Instances→In Entire Project**.

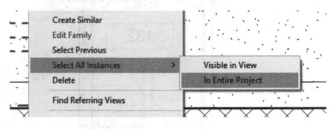

Change the stairs to use the **200mm max riser 300 mm tread**.

Release the selection.

50. Switch to the Manage ribbon.

Select **Warnings** from the Inquiry panel.

51. Expand Warning 1.

Click **Show**.

Close the dialog.

52.

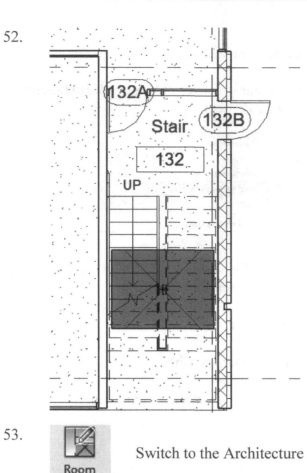

The room is located in Stair 132, but there is no wall on the south side, so the room is not fully enclosed.

53.

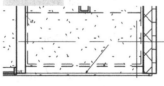

Switch to the Architecture ribbon.

Select **Room Separator** on the Room & Area panel.

54.

Draw a line between the two vertical lines to enclose the room.

55.

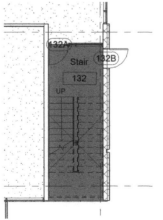

Use a crossing and a filter to select the room again.

The room size has adjusted using the walls and the room separator.

56. 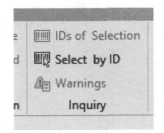 Switch to the Manage ribbon.

Notice that **Warnings** is now grayed out.

This is because you have resolved all the warnings.

57. Save as *ex5-13.rvt*.

Worksets

Worksets are used in a team environment when you have many people working on the same project file. The project file is located on a server (a central file). Each team member downloads a copy of the project to their local machine. The person is assigned a workset consisting of building elements that they can change. If you need to change an element that belongs to another team member, you issue an Editing Request which can be granted or denied. Workers check in and check out the project, updating both the local and server versions of the file upon each check in/out.

In the certification exam, you may be asked a question regarding how to control the display of worksets or how to identify elements in a workset.

Other Hints:

- Name any sheets you create. That way you can distinguish between your sheet and other sheets.
- Create a view that is specific to your changes or workset, so that you have an area where you can keep track of your work.
- Create one view for the existing phase and one view for the new construction phase. That way you can see what has changed. Name each view appropriately.
- Use View Properties to control the phase applied to each view.
- Check Editing Requests often.
- Duplicate any families you need to modify, rename, and redefine. If you modify an existing family, it may cause problems with someone else's workset.

Exercise 5-14

Worksets

Drawing Name: **worksets.rvt**
Estimated Time to Completion: 15 Minutes

Scope

Use of Worksets
Workset Visibility

Solution

1.

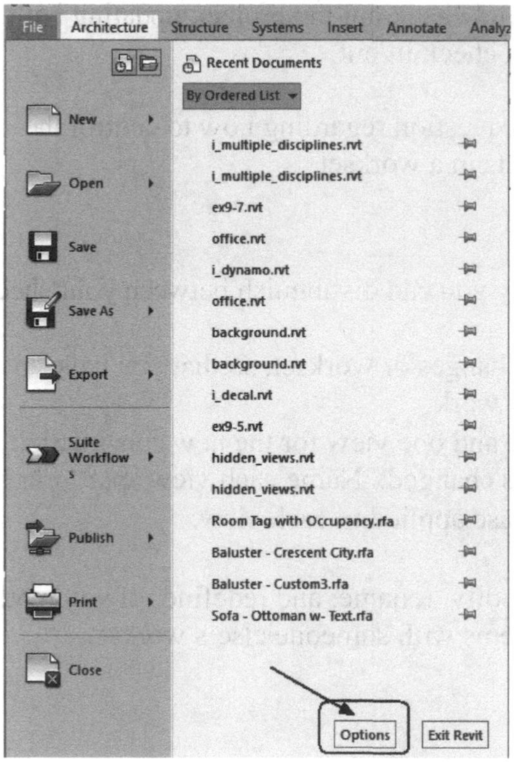

Before you can use Worksets, you need to set Revit to use your name.

Close any open projects.

Go to **Options**.

2.

Username

EliseMoss

You are currently signed in. Your Autodesk ID is used as the username. If you need to change your username, you will need to sign out.

If you are using an Autodesk account, you will see the user name assigned to that account. To use a different name, you need to modify your Autodesk profile to use the desired name. If you sign out, you may lose access to your Revit license.

3. Collaborate | Select the **Collaborate** ribbon.

4. Collaborate | Select the **Collaborate** tool.
There will be a slight pause while Revit checks to see if you are connected to the Internet.

5.

You are enabling collaboration. This will allow multiple people to work on the same Revit model simultaneously.

How would you like to collaborate?

⦿ **Within your network**
Collaborate on a local or wide area network (LAN or WAN). The model will be converted to a workshared central model.

○ **In the cloud**
Collaborate with controlled permissions among project members. The model will be cloud workshared in the project you select.

How can I collaborate using the cloud?

OK   Cancel

Enable **Within your network**.
Click **OK**.

Collaborate in Cloud

Collaborate in Cloud is only available for those users who purchase a separate subscription for Collaboration for Revit.

6. Worksets | Select **Worksets** on the Worksets panel.

7.

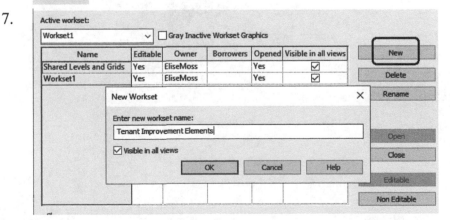

Active workset:

Workset1 ☐ Gray Inactive Workset Graphics

Name	Editable	Owner	Borrowers	Opened	Visible in all views
Shared Levels and Grids	Yes	EliseMoss		Yes	☑
Workset1	Yes	EliseMoss		Yes	☑

New Workset ✕

Enter new workset name:

Tenant Improvement Elements

☑ Visible in all views

OK   Cancel   Help

New
Delete
Rename
Open
Close
Editable
Non Editable

Create a new worksets by clicking **New**.

Type **Tenant Improvement Elements** for the name.

Click **OK**.

8.

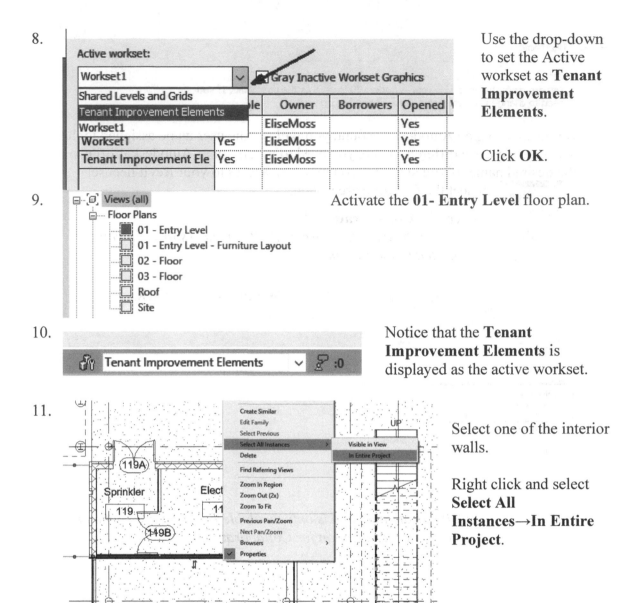

Use the drop-down to set the Active workset as **Tenant Improvement Elements**.

Click **OK**.

9. Activate the **01- Entry Level** floor plan.

10. Notice that the **Tenant Improvement Elements** is displayed as the active workset.

11. Select one of the interior walls.

Right click and select **Select All Instances→In Entire Project**.

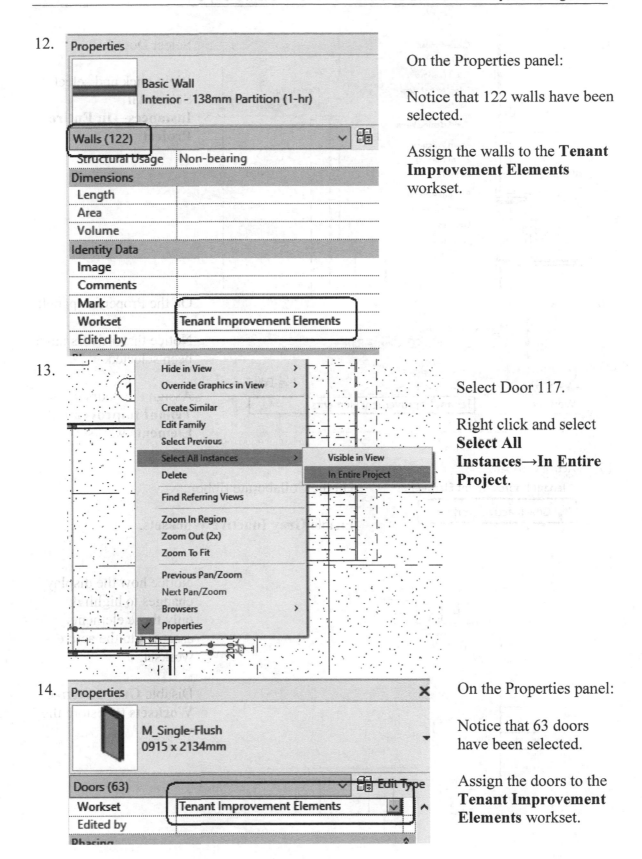

12. On the Properties panel:

Notice that 122 walls have been selected.

Assign the walls to the **Tenant Improvement Elements** workset.

13. Select Door 117.

Right click and select **Select All Instances→In Entire Project**.

14. On the Properties panel:

Notice that 63 doors have been selected.

Assign the doors to the **Tenant Improvement Elements** workset.

15.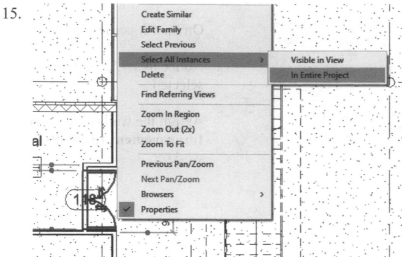

Select Door 118.

Right click and select **Select All Instances→In Entire Project**.

16.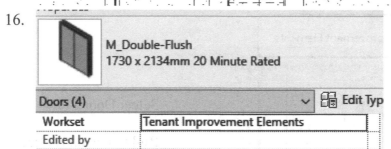

On the Properties panel:

Notice that 4 doors have been selected.

Assign the doors to the **Tenant Improvement Elements** workset.

17.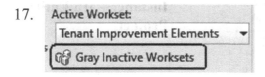

On the Collaborate ribbon:

Enable **Gray Inactive Worksets**.

18.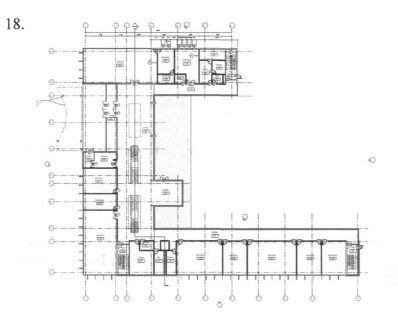

Notice how the display changes to highlight only those elements assigned to the active workset.

Disable **Gray Inactive Worksets** to restore the view.

19. Select **Worksets** on the Worksets panel.

Worksets   If a dialog appears asking to save to Central, click **OK**.

20.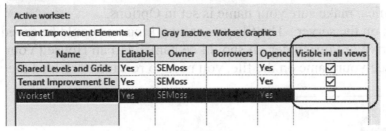

Uncheck Visible in all views for Workset 1.

Click **OK**.

21.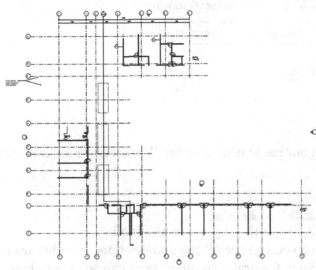

Notice how the view changes.

22. Save the file.

23. This is the first time that the project has been saved since Worksharing was enabled. This project will therefore become the central model. Do you want to save this project as the central model?

If you want to save the file as the central model with a different name and/or different file location, click No and use the Save As command.

Click **No**.

Save as *ex5-14.rvt*.

Workset Checklist

1. Before you open any files, make sure your name is set in Options.
2. The main file is stored on the server. The file you are working on should be saved locally to your flash drive or your folder. It should be named Urban House [Your Name]. If you don't see your name on the file, you need to re-save or re-load the file.

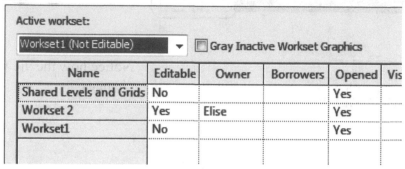

3. Go to Worksets and verify that your name is next to the Workset you are assigned.

4. Verify that the Active work set is the work set you are assigned.
5. Re-load the latest from Central so you can see all the updates done by other users.
6. When you are done working, save to Central – so other users can see your changes.
7. When you close, relinquish your work sets, so others can keep working.

The Worksharing Display Options allow you to color code elements in a view to distinguish between which elements different team members own as well as identifying which worksets different elements have been assigned.

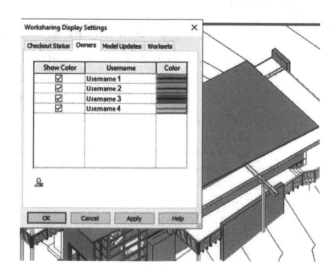

Workset Visibility

In a workshared project, you can control the visibility of worksets in project views.

Note: To improve performance, hide worksets that are not required for current work in the local model.

When you create a workset, use the Visible in all views option of the New Workset dialog to indicate whether the workset displays in all views of the model. This setting is reflected in the Visible in all views column of the Worksets dialog.

This global setting defines the default behavior for each workset in project views. You can override the visibility of each workset for individual views.

All elements that are not in the active workset can display as gray in the drawing area.

Temporary elements, such as temporary dimensions and controls, do not display in gray. This option has no effect on printing, but it helps to prevent adding elements to an undesired workset.

Exercise 5-15
Controlling Workset Visibility

Drawing Name: **Simple House_Central.rvt**
Estimated Time to Completion: 25 Minutes

Scope

Use of Worksets
Worksharing Display Options

Solution

1.   Open the **Simple House_Central** file.

2.   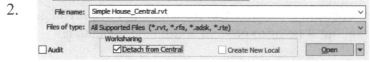   Enable Detach from
                                     Central.

                                     Click **Open**.

                                     This turns off worksharing
                                     in the file and converts the
                                     file to a local file.

3.   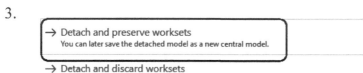   Click **Detach and
                                     preserve worksets**.

4.   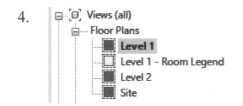   Activate **Level 1**.

Notice that the file name is updated.

5. Select **Worksharing Display Settings** on the View Control bar.

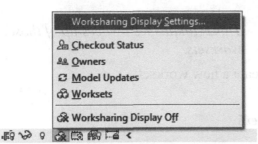

6. 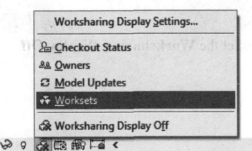 On the Checkout Status tab:

    Enable the **Show Color** display for all the options.

    Click **OK**.

7. Enable display of **Worksets**.

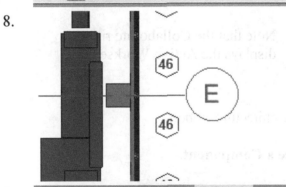

8. What color is Column Grid E?

    It displays as Green which indicates it is owned by the author.

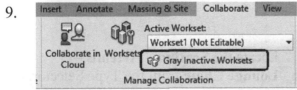

9. Switch to the Collaborate ribbon.

    Enable **Gray Inactive Worksets**.

10. Select the **Worksets** tool from the Collaborate ribbon.

11. 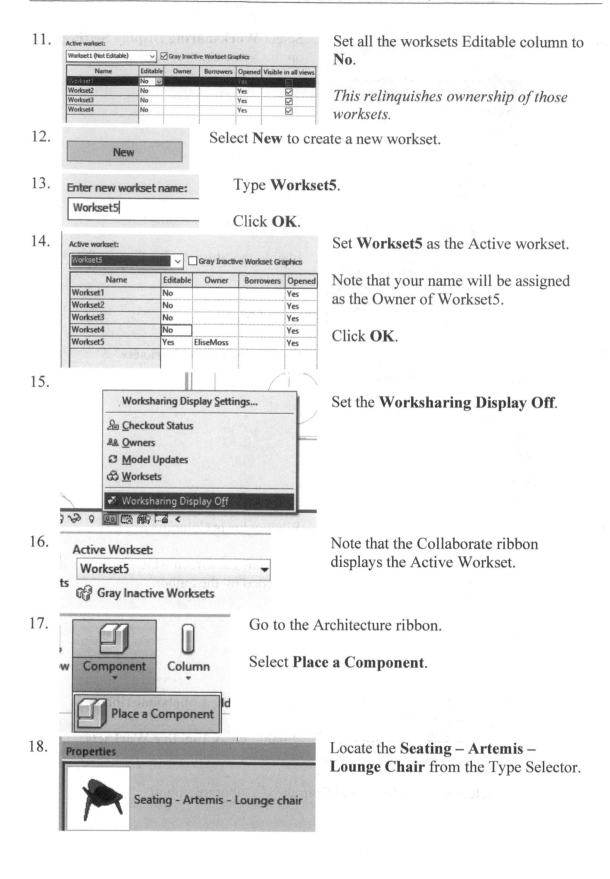 Set all the worksets Editable column to **No**.

This relinquishes ownership of those worksets.

12. Select **New** to create a new workset.

13. Type **Workset5**.

Click **OK**.

14. Set **Workset5** as the Active workset.

Note that your name will be assigned as the Owner of Workset5.

Click **OK**.

15. Set the **Worksharing Display Off**.

16. Note that the Collaborate ribbon displays the Active Workset.

17. Go to the Architecture ribbon.

Select **Place a Component**.

18. Locate the **Seating – Artemis – Lounge Chair** from the Type Selector.

19.

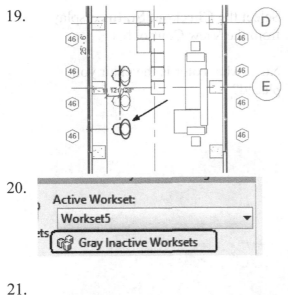

Place a chair below the other two chairs.

Use the SPACE bar to rotate the chair prior to placing.

Note that the chair you placed is not grayed out.

20.

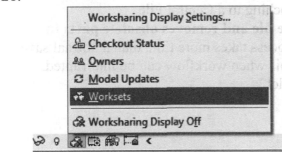

Select the **Gray Inactive Worksets** toggle on the Collaborate ribbon.

Note that the display changes so nothing is grayed out.

21.

Select the **Worksets** display from the View Control bar.

22.

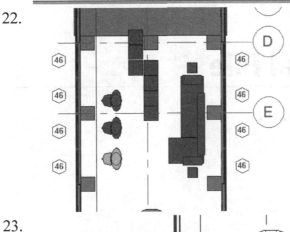

What color is the chair you placed?

23.

Select the **Owners** display from the View Control bar.

Now, what color is the chair you placed?

24.

Select the **Checkout Status** display from the View Control bar.

Now, what color is the chair you placed?

25. Close the file without saving.

Compacting a Central File

When worksets are active, all the users are synching to a Central File.
The process of compacting **rewrites the entire file and removes obsolete parts in order to save space**. Because the Compact process takes more time than a normal save, it is strongly recommended that you only do this when workflow can be interrupted. This option reduces file size of the central model.

Exercise 5-16
Compacting a Central File

Drawing Name: **Simple House_Central.rvt**
Estimated Time to Completion: 35 Minutes

Scope

Use of Worksets
Re-linking Central File
Insert a Component
Synchonizing a Central File

Solution

1. Open the **Simple House_Central** file.
2.

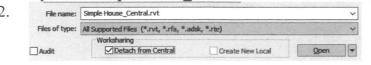

Enable Detach from Central.

Click **Open**.

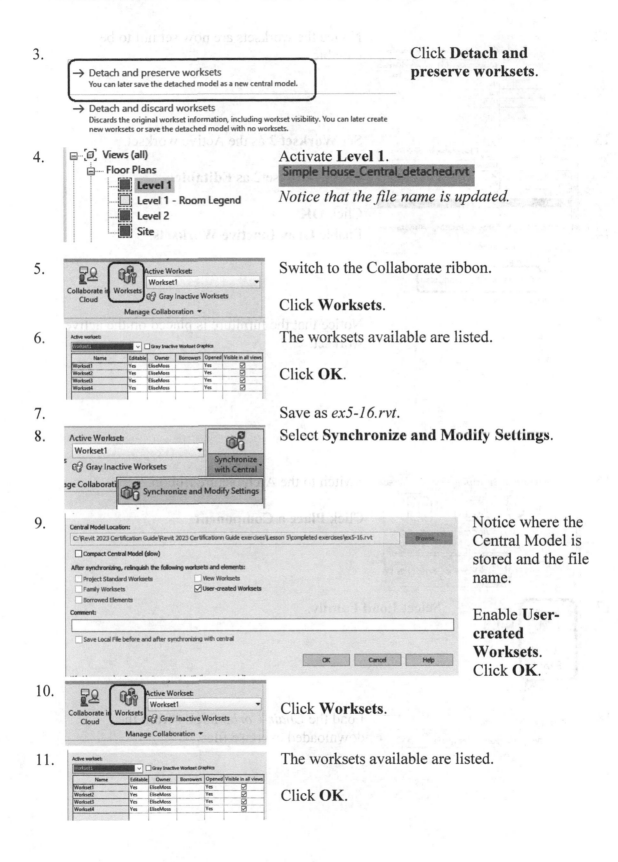

3.
→ Detach and preserve worksets
You can later save the detached model as a new central model.

→ Detach and discard worksets
Discards the original workset information, including workset visibility. You can later create new worksets or save the detached model with no worksets.

Click **Detach and preserve worksets**.

4.
⊟ Views (all)
  ⊟ Floor Plans
    Level 1
    Level 1 - Room Legend
    Level 2
    Site

Activate **Level 1**.
Simple House_Central_detached.rvt

Notice that the file name is updated.

5.
Active Workset:
Workset1
Collaborate in Cloud Worksets Gray Inactive Worksets
Manage Collaboration ▾

Switch to the Collaborate ribbon.

Click **Worksets**.

6.
Active workset:
Workset1 ☐ Gray Inactive Workset Graphics

Name	Editable	Owner	Borrowers	Opened	Visible in all views
Workset1	Yes	EliseMoss		Yes	☑
Workset2	Yes	EliseMoss		Yes	☑
Workset3	Yes	EliseMoss		Yes	☑
Workset4	Yes	EliseMoss		Yes	☑

The worksets available are listed.

Click **OK**.

7. Save as *ex5-16.rvt*.

8.
Active Workset:
Workset1
Gray Inactive Worksets
Synchronize with Central
age Collaborati
Synchronize and Modify Settings

Select **Synchronize and Modify Settings**.

9.
Central Model Location:
C:\Revit 2023 Certification Guide\Revit 2023 Certificationn Guide exercises\Lesson 5\completed exercises\ex5-16.rvt   Browse...
☐ Compact Central Model (slow)
After synchronizing, relinquish the following worksets and elements:
☐ Project Standard Worksets   ☐ View Worksets
☐ Family Worksets   ☑ User-created Worksets
☐ Borrowed Elements
Comment:

☐ Save Local File before and after synchronizing with central
OK   Cancel   Help

Notice where the Central Model is stored and the file name.

Enable **User-created Worksets**.
Click **OK**.

10.
Active Workset:
Workset1
Collaborate in Cloud Worksets Gray Inactive Worksets
Manage Collaboration ▾

Click **Worksets**.

11.
Active workset:
Workset1 ☐ Gray Inactive Workset Graphics

Name	Editable	Owner	Borrowers	Opened	Visible in all views
Workset1	Yes	EliseMoss		Yes	☑
Workset2	Yes	EliseMoss		Yes	☑
Workset3	Yes	EliseMoss		Yes	☑
Workset4	Yes	EliseMoss		Yes	☑

The worksets available are listed.

Click **OK**.

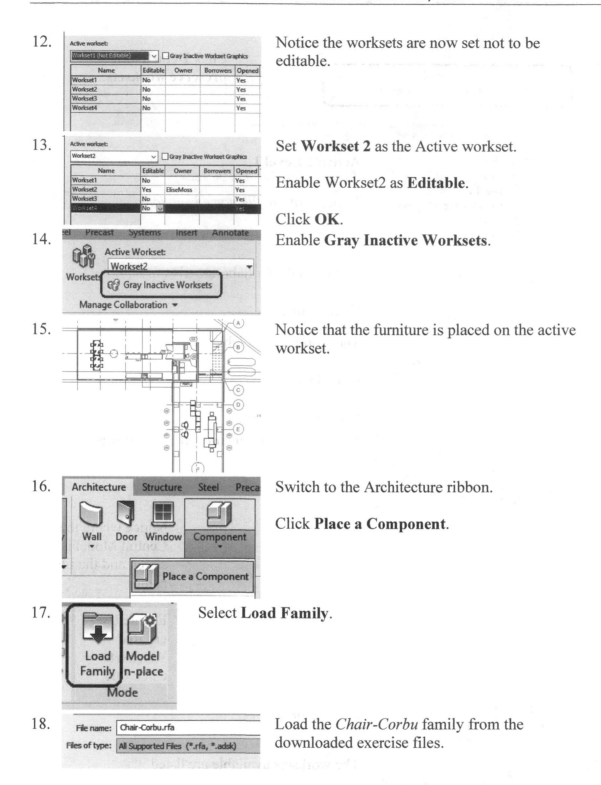

12. Notice the worksets are now set not to be editable.

13. Set **Workset 2** as the Active workset.

Enable Workset2 as **Editable**.

Click **OK**.

14. Enable **Gray Inactive Worksets**.

15. Notice that the furniture is placed on the active workset.

16. Switch to the Architecture ribbon.

Click **Place a Component**.

17. Select **Load Family**.

18. Load the *Chair-Corbu* family from the downloaded exercise files.

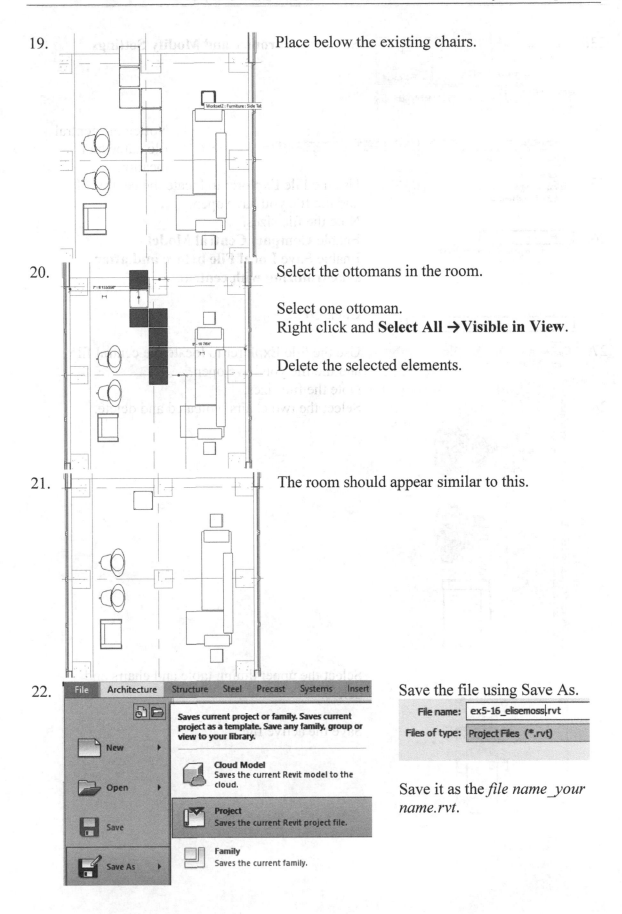

19.   Place below the existing chairs.

20.   Select the ottomans in the room.

Select one ottoman.
Right click and **Select All →Visible in View**.

Delete the selected elements.

21.   The room should appear similar to this.

22.   Save the file using Save As.

File name: ex5-16_elisemoss.rvt

Files of type: Project Files (*.rvt)

Save it as the *file name_your name.rvt*.

23. 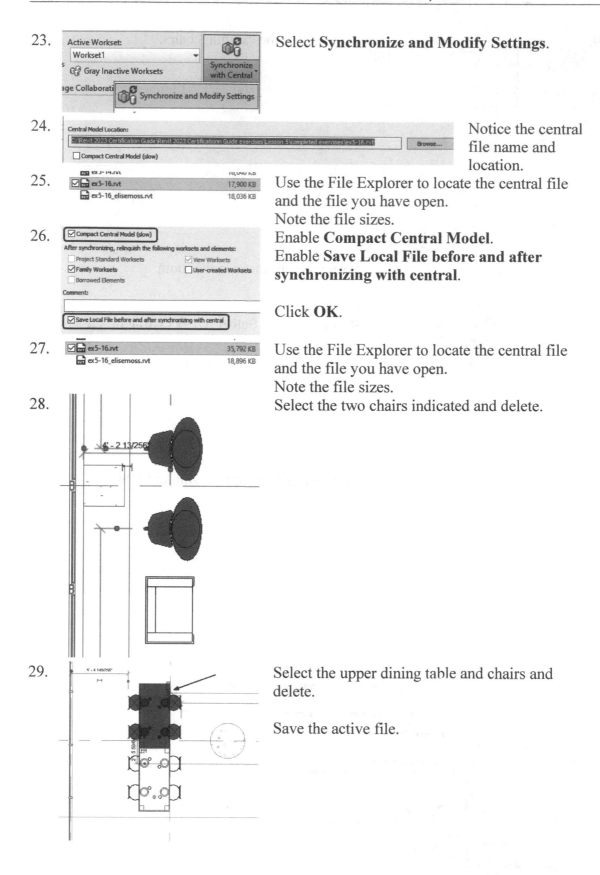 Select **Synchronize and Modify Settings**.

24. Notice the central file name and location.

25. Use the File Explorer to locate the central file and the file you have open.
Note the file sizes.

26. Enable **Compact Central Model**.
Enable **Save Local File before and after synchronizing with central**.

    Click **OK**.

27. Use the File Explorer to locate the central file and the file you have open.
Note the file sizes.

28. Select the two chairs indicated and delete.

29. Select the upper dining table and chairs and delete.

    Save the active file.

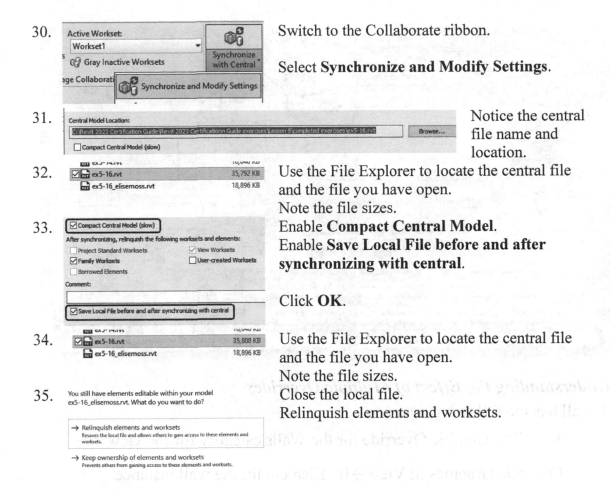

30. Switch to the Collaborate ribbon.

    Select **Synchronize and Modify Settings**.

31. Notice the central file name and location.

32. Use the File Explorer to locate the central file and the file you have open.
    Note the file sizes.

33. Enable **Compact Central Model**.
    Enable **Save Local File before and after synchronizing with central**.

    Click **OK**.

34. Use the File Explorer to locate the central file and the file you have open.
    Note the file sizes.

35. Close the local file.
    Relinquish elements and worksets.

Element Visibility Hierarchy

Most users are unaware that Revit applies a hierarchy to how elements are displayed and their visibility.

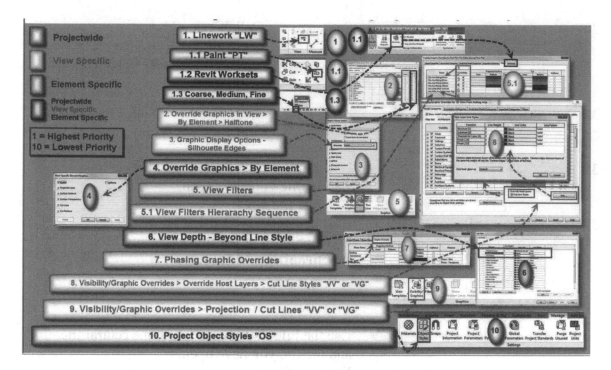

Understanding the Effect of Multiple Overrides

A wall has the following overrides:

- Visibility/Graphic Override for the Walls category for the view

- Override Graphics in View→By Element for the wall instance

- Phasing graphic override

- View filter overriding the graphics for walls over a specific thickness

With the four overrides defined, the element override on the wall instance is visible because it is the highest in the hierarchy:

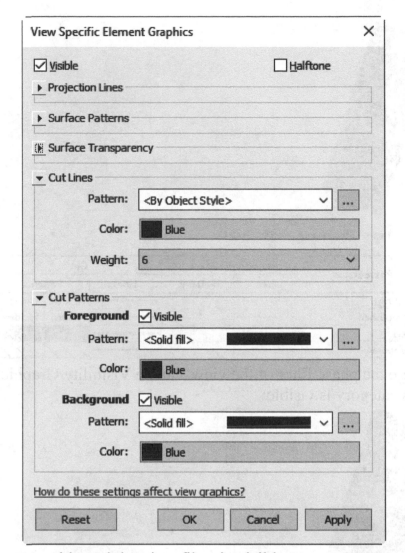

Remove the element override and the view filter is visible:

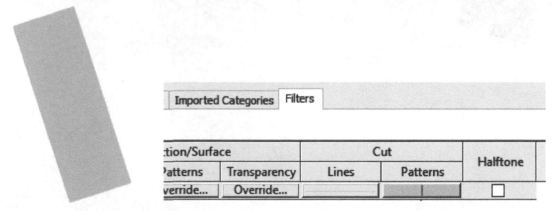

Remove the view filter and the phasing graphic override is visible:

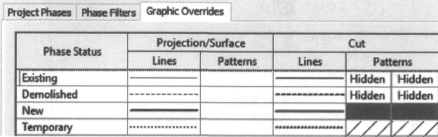

Phase Status	Projection/Surface		Cut		
	Lines	Patterns	Lines	Patterns	
Existing	———————		———————	Hidden	Hidden
Demolished	--------------		--------------	Hidden	Hidden
New	———————		———————		
Temporary	················		················	/////	/////

Change the phase filter of the view and the Visibility/Graphic override for the Walls category is visible:

Exercise 5-17
Element Visibility Hierarchy

Drawing Name: **graphic overrides.rvt**
Estimated Time to Completion: 90 Minutes

Scope

Apply several different display overrides to the same wall.
Remove the display overrides to see which overrides have priority.

Solution

1.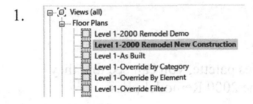
    In the Project Browser, there are several views that show different overrides.

    Open the **Level 1- 2000 Remodel New Construction** floor plan.

2.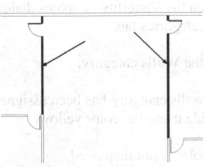
    Note the two walls that are magenta.

    Select one of the vertical magenta walls.

3. | Phasing | |
    | --- | --- |
    | Phase Created | 2000 Remodel |
    | Phase Demolished | None |

    The walls are assigned the 2000 Remodel phase.

4.

    Switch to the Manage ribbon.

    Click **Phases**.

5.

Phase Status	Projection/Surface		Cut	
	Lines	Patterns	Lines	Patterns
Existing				Hidden
Demolished	- - - - - - - -		- - - - - - - -	Hidden
New				
Temporary				

Click the **Graphic Overrides** tab.

Note that New construction has been assigned the color Magenta.

Click **OK**.

6.
Floor Plans
    Level 1-2000 Remodel Demo
    Level 1-2000 Remodel New Construction
    Level 1-As Built
    Level 1-Override by Category
    Level 1-Override By Element
    Level 1-Override Filter

Open the **Level 1 – Override by Category** floor plan view.

7.

The two walls are displayed as magenta. This is the Phases graphic override.

Select the two walls.

8.

Phasing	
Phase Created	2000 Remodel
Phase Demolished	None

In the Properties palette, you can see that they are assigned the 2000 Remodel phase.

9.

Vertical Circulation					
Walls					
<Hidden Lines>					
Common Edges					
Non-Core Layers					

Type **VG** to open the Visibility Graphics dialog. On the Model Categories tab:

Scroll down to the **Walls** category.

Notice that the walls category has been assigned a graphic override using the color yellow.

However, this color is not displayed.
That is because the phase override is higher on the hierarchy than the category override.

Click **OK** to close the dialog.

10.

Phasing	
Phase Created	As-Built
Phase Demolished	None

Select the two magenta walls and change the Phase Created to **As-Built**.

Click to release the selection.

The walls color changes to the new phase assignment.

11. 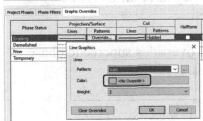   Switch to the Manage ribbon.

Click **Phases**.

12. 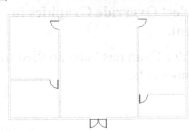   Click the **Graphic Overrides** tab.

Set the Color for the Existing phase to **No Override** in the Lines columns.

Click **OK**.

13.   The walls now appear as yellow.

Yellow is the color assigned to the walls category.

14.   Select the two walls indicated and assign them to the 2000 Remodel phase.

15. 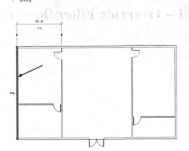   Open the **Level 1 – Override by Element** floor plan view.

16.   Select one of the outer walls.

Notice it was created in the As-Built phase.

However, the color is not yellow. The Visibility Graphics override is view-specific.

Release the selection.

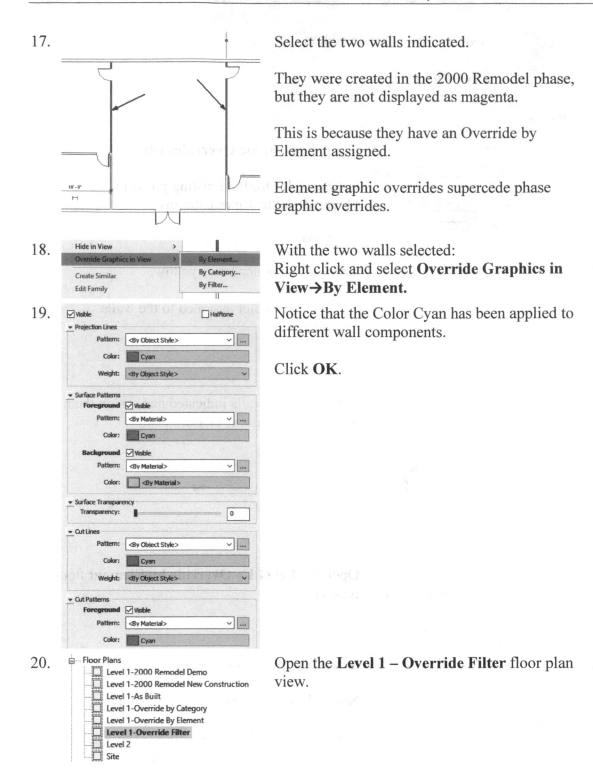

17.   Select the two walls indicated.

They were created in the 2000 Remodel phase, but they are not displayed as magenta.

This is because they have an Override by Element assigned.

Element graphic overrides supercede phase graphic overrides.

18.   With the two walls selected:
Right click and select **Override Graphics in View→By Element.**

19.   Notice that the Color Cyan has been applied to different wall components.

Click **OK**.

20.   Open the **Level 1 – Override Filter** floor plan view.

21.

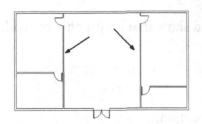

The outer walls are assigned the As-Built Phase. The indicated walls are assigned the 2000 Remodel Phase but have a graphic override by element applied.

Type **VG** to bring up the dialog box.

22.

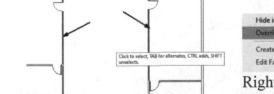

Select the **Filters** tab.

There is a filter called walls that has assigned the color red to some walls.

23.

All document filters are defined and modified here | Edit/New...

Click **Edit/New**.

24.

Filter Rules

AND (All rules must be true) | Add Rule | Add Set

Walls | Phase Created

equals | 2000 Remodel

The filter changes the color for walls assigned to the **2000 Remodel** phase.

Click **OK**.

25.

Vertical Circulation
Walls
  <Hidden Lines>
  Common Edges
  Non-Core Layers

Go to the Model Categories tab.

Scroll down to Walls.
No Graphic overrides have been applied.

Click **OK** to close the dialog.
Select the two walls.

26.

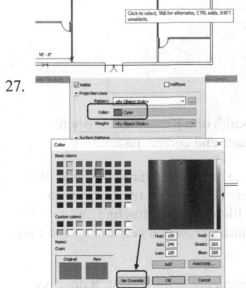

Hide in View	>	
Override Graphics in View	>	By Element...
Create Similar		By Category...
Edit Family		By Filter...

Right click and select **Override Graphics in View→By Element.**

27.

☑ Visible   ☐ Halftone
▼ Projection Lines
   Pattern: <By Object Style>
   Color: Cyan
   Weight: <By Object Style>
▼ Surface Patterns

Color

Basic colors:

Custom colors:

Name: Cyan
Hue: 120   Red: 0
Sat: 240   Green: 255
Lum: 120   Blue: 255

Original   New

Add   PANTONE...

No Override   OK   Cancel

Click in each color box.

Click **No Override**.

Click **OK**.

28.

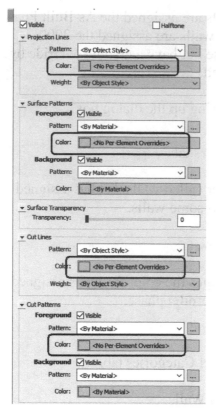

The box should show that no graphic overrides are active.

Click **OK**.

Release the selection.

29.

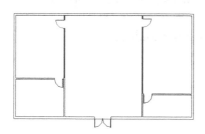

The filter override is now applied.

The filter graphic override has a higher place on the hierarchy than the phase graphic override.

Save as *ex5-17.rvt*.

Audit

Use the Audit function periodically to maintain the health of a Revit model, when preparing to upgrade the software, or as needed to locate and correct issues.

Auditing an Autodesk Revit file helps you maintain the integrity of large files and files that more than one user works on. The auditing process checks your file integrity to determine that no corruption has occurred within the file's element structure and automatically fixes any minor file errors that it encounters. You can perform an audit on your Revit file when you open it in the application.

The Audit function scans, detects, and fixes corrupt elements in the model. It does not provide feedback on which elements are fixed.

Auditing a model can be time-consuming, so be prepared to wait while the process completes.

As a best practice, audit the model weekly. If the model is changing rapidly, audit it more often.

In addition to auditing models, you can also use this function to audit families and templates.

Exercise 5-18

Audit a File

Drawing Name: **audit.rvt**
Estimated Time to Completion: 5 Minutes

Scope
Audit a Revit building project

Solution

1.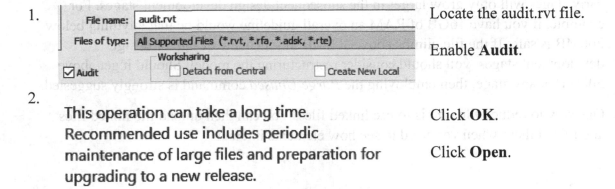
Locate the audit.rvt file.

Enable **Audit**.

2. This operation can take a long time. Recommended use includes periodic maintenance of large files and preparation for upgrading to a new release.

Click **OK**.

Click **Open**.

3. Save the file and close.

Purge

Purging a Revit model assists in removing unused families, views, and objects from a project. It is recommended that the model should be purged after every submittal and milestone to remove any remaining elements that have accumulated in the project.

Revit's automated purging method only removes certain elements within your project; therefore, you still need to go through your models and manually remove any unwanted area schemes, views, groups, and design options. You should also replace in-place families with regular component families to further reduce the file size.

When you consider using **Purge Unused** it is important to understand how this works, mainly because in most cases should you require to remove any unwanted families, objects and views, you would need to employ the 3 step Purge process. The reason for this is when first using *Purge Unused* you can only see the items that are not being currently used in the project. This means that any families that are listed may be using materials that are in the project. So when you purge out a family, its materials now might be unused in the project, and when you next launch the Purge Unused dialog you may see additional materials that can be purged from the project. As a recommendation it is also necessary to select the *Check None* button as your first step when using *Purge Unused*, then look for and select what you would like to purge. This is important because by default everything that can be purged is pre-selected. This means if you have an element in your project template, but you haven't used it yet in the project, it will be removed and you'll have to re-create it or re-load it.

I try to discourage my students from using the Purge tool because they invariably find themselves needing families they have purged and then they have to spend a great deal of time trying to locate them and reload them.

Always remember to leave space for file growth because files that become large in the early stages will only grow larger in the subsequent design development stages. For example, if you have 16GB of RAM an overall guideline would be that anything below 200MB is safe. If the file climbs above 300MB in schematic design or design development stages, you should consider restructuring the model. Should it get above 500MB at any stage, then employing the *Purge Unused* command is strongly suggested.

One way to reduce file size is to use linked files. You can then unload the linked files and reload them when you need to see how elements interact.

Exercise 5-19

Purge a File

Drawing Name: **purge.rvt**
Estimated Time to Completion: 5 Minutes

Scope
Purge a Revit building project

Solution

1.

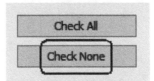

Switch to the Manage ribbon.

Select **Purge Unused** from the Settings panel.

2.

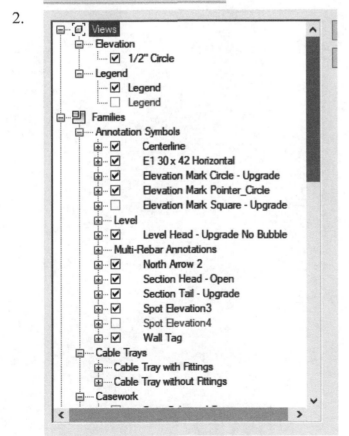

Expand the list.

Notice there is an unused Legend.

This means the legend has not been placed on a sheet, but you may still want to use it.

Browse through the list.

Click **Check None**.

That way you don't accidentally delete an element you may need later.

3.

⊞···· Arrowhead
⊞···· Elevation Tag
···· Extensible Storage Schemas
⊟···· Linear Dimension Style
····☑ Diagonal - 3/32" Arial
····☑ Linear w Center - 3/32" Arial
⊞···· Material Assets
⊟···· Materials

Locate the **Linear Dimension Style**.

Select the two dimension styles listed.

Click **OK**.

4. Save as *ex5-19.rvt*.

Practice Exam

1. Worksharing allows team members to:
    A. Share families from different projects
    B. Share views from different projects
    C. Work on the same parts of a project simultaneously
    D. Work on different parts of the same project

2. Set Phase Filters for a view in the:
    A. Properties pane
    B. Design Options
    C. Manage ribbon
    D. Project Browser

3. The Coordination Review tool is used when you use:
    A. Worksets
    B. Linked Files
    C. Interference Checking
    D. Phase

4. The Worksets option under the worksharing display menu will show:

    A. The color that corresponds with each workset
    B. The color that corresponds with the active workset
    C. The color that corresponds with the owner of the active workset
    D. The color of all worksets that can be edited (checked out).

5. True/False
    You can ignore review warnings, but you cannot ignore an error.

6. The Coordination Review is useful for:

    A. Checking what monitored elements were changed in the host file
    B. Aligning the origin of one project with a linked project
    C. Performing multiple copy operations from a linked project
    D. Copying a large number of elements from one project to another

7. When you link to a CAD drawing:

    A. You bring in an image of the CAD file which cannot be updated
    B. You automatically import any external references used by that CAD file.
    C. If the CAD file is modified, you will see any changes when you reload the file.
    D. The CAD file is added to the Revit project as decal.

8. You are using worksets. When you make changes to the local file, these changes:

    A. Cannot be saved to the Central Model.
    B. Can only be saved to the Central Model if the changes are minor.
    C. Are reflected in the Central Model only after synchronizing.
    D. Are automatically reflected in the Central Model.

9. When reviewing errors, why isn't using the Remove Constraints button the best method to resolve the error?

    A. It might not actually remove the constraints
    B. It may add more constraints
    C. It's too easy a solution
    D. It may remove elements that are required in the design

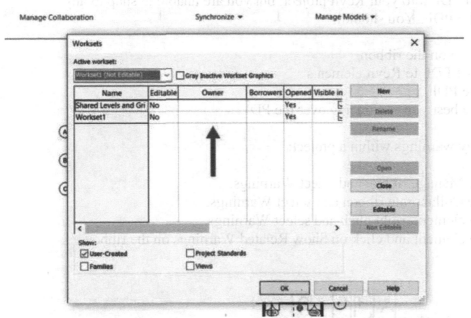

10. What will appear in the Owner field for Workset1 if you change the Editable field to Yes?

   A. The User name you set in the Options will appear in the Owner field
   B. The original creator of the workset will be assigned as the Owner
   C. Workset1 will not be assigned an Owner, but it will still be editable
   D. The User you select from the drop-down list will be assigned the owner

11. I choose Import CAD to bring in an AutoCAD drawing file with the floor plan. My vendor just emailed me that he made some changes to the floor plan. What now?

   A. An asterisk will appear next to the drawing file name indicating there are unsaved changes
   B. The drawing will automatically update
   C. A warning will pop up the next time I reopen the Revit project advising me to reload the drawing
   D. The changes are not reflected in my Revit project, and I do not see any warnings or indications that the file may have changed.

12. When creating a local version of a centralized model, what happens if you select the Detach from Central option when opening the file?

   A. You will have a read-only version of the model and can't make changes but can review.
   B. Changes you make will automatically synchronize with the Central Model.
   C. Changes you make to the local model won't be synchronized and the local file will now be independent of the Central model.
   D. Your username will not be assigned to any worksets.

13. You import a PDF into your Revit project, but you are unable to snap to any elements in the PDF. You should:

   A. Enable Snap on the ribbon
   B. Convert the PDF to Revit elements
   C. Reload the PDF
   D. Just do the best you can tracing over the PDF

14. To review any warnings within a project:

   A. Go to the Manage ribbon and select Warnings.
   B. Go to the Collaborate ribbon and select Warnings.
   C. Select an element, right click and select Warnings.
   D. Select an element and click on Show Related Warnings on the ribbon.

1 : 100

15. Which icon should be used to manage the display settings for worksets?

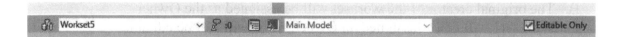

16. Editable Only is enabled on the status bar. This means:

   A. Only elements in the Main Model design option can be edited.
   B. Only elements in the host file can be edited.
   C. Only elements in the active workset can be edited.
   D. Only elements in the worksets with editing enabled can be selected or edited.

17. The Phase of a view is set to New Construction and the Phase Filter is set to Show Previous + New. What will control the graphic display of Existing elements?

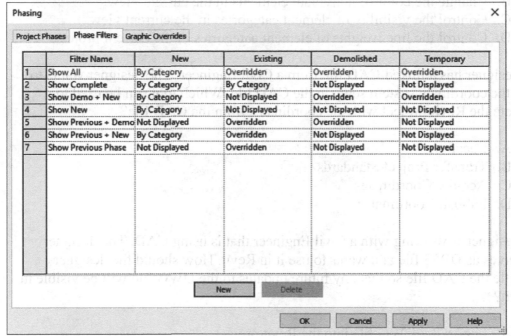

	Filter Name	New	Existing	Demolished	Temporary
1	Show All	By Category	Overridden	Overridden	Overridden
2	Show Complete	By Category	By Category	Not Displayed	Not Displayed
3	Show Demo + New	By Category	Not Displayed	Overridden	Overridden
4	Show New	By Category	Not Displayed	Not Displayed	Not Displayed
5	Show Previous + Demo	Not Displayed	Overridden	Overridden	Not Displayed
6	Show Previous + New	By Category	Overridden	Not Displayed	Not Displayed
7	Show Previous Phase	Not Displayed	Overridden	Not Displayed	Not Displayed

A. The settings for that category in the Visibility/Graphic Overrides dialog.
B. Existing elements will not be visible in the view.
C. The Existing settings on the Graphic Overrides tab.
D. Object Styles for the New Construction phase.

18. A designer wants to check for clashes in a project. Which tool should be used?
A. Run Interference Check
B. Copy/Monitor
C. Review Clashes
D. Coordination Review

19. A designer needs to adjust the line weight for all walls in one floor plan view. Which dialog should the designer use?

A. View Specific Element Graphics
B. Graphic Display Options
C. Object Styles
D. Visibility/Graphic Overrides

20. What do design options help accomplish?

A. Explore various design alternatives in the same model.
B. Combine elements together to easily place multiple times.
C. Create layers or subcomponents from individual elements.
D. Change the display of elements based on the design.

21. What can be accomplished in the Visibility/Graphic Overrides dialog?

    A. Change the line styles of individual elements in the project.
    B. Change the color of individual elements in the current view.
    C. Control the visibility of element categories in the current view.
    D. Control the line weights of element categories in the current project.

22. A designer has a linked CAD file from a Civil Engineer. The designer wants to use the coordinate system setup in the CAD file. Which tool can be used to update the Revit project with the coordinates from the CAD file?

    A. Location
    B. Transfer Project Standards
    C. Acquire Coordinates
    D. Publish Coordinates

23. A designer is working with a Civil Engineer that is using CAD. The designer receives as DWG file and wants to use it in Revit. How should the designer handle the CAD file so that any future updates to the DWG file will be visible in Revit?

    A. Import the DWG file into the Revit project.
    B. Link the RVT file to the CAD file.
    C. Link the DWG file to the Revit project.
    D. Import the RVT file into the CAD file.

24. What is the phase filter setting for a typical demolition plan?

    A. Show Demo + New
    B. Show Demo
    C. Show Previous + Demo
    D. Show Previous + New

25. An interior designer and architect are working on the same project. The interior designer wants to link the architectural model and use the levels that were already created. Which tool should the interior designer use to be able to duplicate the levels and be notified when a move occurs?

    A. Copy/Monitor
    B. Coordination Review
    C. Duplicate as Dependent
    D. Link Levels

26. What does it mean to compact a central model?

    A. The process of compacting rewrites the entire file and removes obsolete parts in order to save space. This reduces the file size.

    B. During the compacting process, unused families and family types are removed from the project. This reduces the file size.

    C. It is the same thing as zipping a file. It allows users to compact all elements of a model into one file so you can send it to other members of the design team.

    D. Compact is another word for archive. It allows users to archive models at milestone submittals.

27. A designer is adding windows to exterior walls. The full-height windows are visible, but the smaller windows are not showing in the view. What should the designer check to fix this issue?

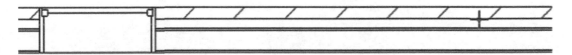

    A. Detail Level
    B. Visual Style
    C. Visibility/Graphic Overrides
    D. View Range

28. T F Object Styles control the visibility of each view independently.

29. Once worksharing has been enabled, what is the next step when using this feature and sharing models with others?

    A. Set up permissions
    B. Share a local copy
    C. Create a central model
    D. Purge elements

30. When is a good time to use the purge tool?

    A. Right after you create the project to remove the unnecessary elements that are located in the project template.

    B. At the middle and end of your projects.

    C. After you have been working in a model for a little while and can safely purge out elements that you most likely will not use.

    D. You should never use the purge unused option; it will only cause errors to occur.

Answers

    1) D; 2) A; 3) B; 4) A; 5) T; 6) A; 7) C; 8) C; 19) D; 10) A; 11) D; 12) C; 13) A; 14) A; 15) C; 16) D; 17) C;18) A; 19) D; 20) A; 21) C; 22) C; 23) C; 24) C; 26) A; 27) D; 28) F; 29) C; 30) C